**Physics of Separated
Flows –
Numerical, Experimental,
and Theoretical Aspects**

Edited by
Klaus Gersten

W0043817

Notes on Numerical Fluid Mechanics (NNFM)   Volume 40

Physics of
Separated Flows –
Numerical, Experimental,
and Theoretical Aspects

DFG Priority Research Programme
1984–1990

Edited by
Klaus Gersten

Mathematics Subject Classification: 76-XX

Produced by W. Langelüddecke, Braunschweig
Printed on acid-free paper

ISSN 0179-9614
ISBN 978-3-531-07640-9   ISBN 978-3-663-13986-7 (eBook)
DOI 10.1007/978-3-663-13986-7

FOREWORD

This volume contains 37 contributions in which the research work is summarized which has been carried out between 1984 and 1990 in the Priority Research Program "Physik abgelöster Strömungen" of the Deutsche Forschungsgemeinschaft (DFG, German Research Society). The aim of the Priority Research Program was the intensive research of the whole range of phenomena associated with separated flows. Physical models as well as prediction methods had to be developed based on detailed experimental investigations. It was in accordance with the main concept of the research program that scientists working on problems of separated flows in different technical areas of application participated in this program. The following fields have been represented in the program: aerodynamics of wings and bodies, aerodynamics of automobiles, turbomachinery, ship hydrodynamics, hydraulics, internal flows, heat exchangers, bio-fluid-dynamics, aerodynamics of buildings and structures.

In order to concentrate on problems common in all those areas the emphasis of the program was on basic research dealing with generic geometric configurations showing the fundamental physical phenomena of separated flows.

The engagement and enthusiasm of all participating scientists are highly appreciated. The program was organized such that all researchers met once a year to report on the progress of their work. Special thanks ought to go to Prof. E.A. Müller (Göttingen), Prof. H. Oertel jun. (Braunschweig), Dr. W. Schmidt (Dornier), Dr. H.-W. Stock (Dornier) and Dr. B. Wagner (Dornier), who had the functions of referees on those annual meetings.
In addition to the annual meetings workshops on special topics have been held regularly for smaller working groups. The following scientists deserve thanks for having been chairmen of those working groups: Prof. D. Hummel (Three-dimensional separated flows), Prof. W. Nitsche (Experimental methods for separated flows), Prof. I. Teipel (Unsteady separated flows), Prof. K. Gersten (Turbulence modelling of separated flows).

The final scientific symposium of this DFG Program has been held on 17th and 18th February, 1992 at the Ruhr-Universität Bochum.

It is worth mentioning that altogether 195 scientific publications have come out of this Priority Research Program.

On behalf of all participating scientists I thank very much the Forschungsgemeinschaft for making available the financial support. Dr. Lachmeier, representing the DFG, deserves very special thanks, because without his assistance, his scientific and administrative knowledge and his enthusiasm this research program would not have been so successful.

K. Gersten

CONTENTS

TWO-DIMENSIONAL FLOWS WITH PRESSURE INDUCED SEPARATION

UNSTEADY FLOWS WITH SEPARATION

THREE-DIMENSIONAL FLOWS PAST BLUFF BODIES AND CIRCULAR CYLINDERS

THREE-DIMENSIONAL FLOWS WITH VORTEX DOMINATED SEPARATION

OVERVIEW

K. Gersten

Institut für Thermo- und Fluiddynamik, Ruhr-Universität Bochum,
Universitätsstr. 150, D-4630 Bochum 1

STRUCTURE OF CONTENT

Although the topic "Physics of Separated Flows" was meant to cover laminar as well as turbulent flows, almost all contributions are dealing with turbulent separated flows. The investigations cover numerical, experimental, and theoretical aspects of the physical phenomena of separated flows.

The content of this book has been put in order mainly by the geometry of the configurations and hence of the structure of the separated flows. The volume is divided into three main topics: two-dimensional, unsteady and three-dimensional separated flows, although in many contributions more than one of these topics are covered. In order to get a better structure of the content, different sections have been provided for two-dimensional flows with fixed separation points and for two-dimensional flows with pressure induced separation. Furthermore, as far as three-dimensional flows are concerned, it has been distinguished between separated flows past bluff bodies and circular cylinders and vortex-dominated separation.

It is obvious that the state of the art of theoretical and numerical methods is more advanced for two-dimensional separated flows than for three-dimensional separated flows. Hence, the more complicated the flow becomes, the more experimental work will dominate. Therefore, each section starts with theoretical and numerical contributions and will then continue to more complicated cases where exprimental investigations may play the dominating part.

As far as the experiments are concerned, it is worth mentioning that in many cases laser-Doppler anemometry has been put into action. It is mainly the merit of this priority research program, that now several institutes at German universities have at their disposal those very sophisticated and expensive equipments for experimental investigations of separated flows.

TWO-DIMENSIONAL FLOWS WITH FIXED SEPARATION POINT

Configurations under consideration were backward-facing steps or surface-mounted obstacles like square ribs or vertical flat plates. The separation points were fixed due to the sharp edges of the geometry. The following models have been used for prediction methods: large-eddy simulation (*Friedrich* and *Arnal*), second-order Reynolds stress turbulence model, k-ε turbulence model (*Perić* and *Tropea*; *Cordes, Rodi* and *Cho*) as well as discrete vortex model (*Fernholz* and *Jaroch*). The large-eddy simulation technique has reached a mature stage and can be considered as a valuable tool for the prediction of complex flows. The use of the second-moment turbulence closure leads to fairly good simulation of the mean velocity and turbulence field around the obstacles. It is far better than the simulation by the k-ε model whose poor representation of the flow field is attributed to inappropriate modelling of the production rate in the equation of the turbulent kinetic energy. The investigation by

Cordes et al. shows that the standard k-ε model can be improved considerably by resolving the viscosity-affected regions near the wall with a one-equation model based on empirical length-scale prescriptions instead of using wall functions. In this context it is worth mentioning that the wall function changes from the logarithmic law of the wall to a square-root law when approaching the separation point (*Gersten, Klauer* and *Vieth*). The discrete vortex model is able to produce consistent results for the pressure and the velocity field, but unable to predict the turbulence intensities and correlations.

Experimental investigations of separated flows are far more difficult than those of attached flows. That is particularly true for skin-friction measurements. *Fernholz, Dengel* and *Hess* performed comparative skin-friction measurements in flows with separation and instantaneous reverse flow by using various methods (Preston tube, surface fence, single film, McCroskey hot-film probe, wall pulsed wire). The differences in the results of the different measuring techniques are explained and the failure of devices under these severe conditions is displayed.

Skin-friction measurements as well as flow-visualization experiments in the near-wall region of a separated region indicate the existence of large coherent structures, whose structures differ considerably from those observed in attached turbulent boundary layers (*Wagner* and *Fernholz*).

Very comprehensive flow field investigations have been carried out in a backward-facing step flow with variation of the wall inclination angle by using laser-Doppler anemometry (*Ruck* and *Makiola*). The experiments can both, contribute to the understanding of the phenomenology of separated flows and establish a comprehensive data base for the validation of numerical codes. The data are also relevant to two-dimensional wide-angle diffusers (see also *Nitsche* and *Haberland*).

Two experimental investigations are dealing with the effects of particles in the separated flow past a backward-facing step. Particles in the diameter range from 1 to 70 μm were suspended in the air flow, and the changes of the particle velocity were shown in comparison to the velocity field of the continuous phase deduced from the 1 μm particles (*Ruck* and *Makiola*). This is of fundamental importance for the concept of laser-Doppler anemometry and shows the strong influence which different particle sizes have on flow data measurements. These effects may especially be pronounced in separated flows with strong vortices. The other study (*Dohmann*) refers to the effects of polymer additives to the structure of turbulent motion. The position of the re-attachment point as well as the Reynolds stresses in the turbulent free shear layer are changed considerably by polymer additives.

TWO-DIMENSIONAL FLOWS WITH PRESSURE INDUCED SEPARATION

Transition from laminar to turbulent boundary-layer flow often occurs via a laminar separation bubble. The laminar boundary layer separates due to adverse pressure gradient, the separating laminar free shear layer becomes unstable and eventually turbulent which often leads to reattachment of the turbulent free shear layer and hence to a closure of the separation region. Such separation bubbles have been investigated by *Bestek, Gruber*, and *Fasel* for flows over smooth backward-facing steps. The direct numerical simulation, based on finite-difference solutions of the unsteady Navier-Stokes equations, could clarify the mechanism of natural unsteadiness of the separation bubbles caused by hydrodynamic instability of the separated boundary layer.

2

Another approach is the global stability analysis by *Morzyński* and *Thiele* where the non-parallel stability has been investigated numerically for the flow fields past the circular cylinder, the ellipse, and an airfoil. It has been concluded from this study that both, the onset of the Karman vortex street and the Tollmien-Schlichting waves are two different aspects of the same phenomenon represented by non-parallel flow stability theory.

Althaus and *Würz* carried out detailed experimental studies on laminar separation bubbles. Static pressure, hot wire anemometry and flow visualization data were aquired. The data were used to evaluate the applicability of existing separation bubble models.

Three investigations were dealing with the flow past airfoils at moderate and high angles of attack. Numerical calculations performed for wall boundary layers and wakes show that the inverse procedure is able to predict flow fields past airfoils with quite large separation region without any difficulties, when the viscous-inviscid interaction is taken into account. *Schalau* and *Thiele* carried out such calculations by applying four different turbulence models (Cebeci-Smith, Balwin-Lomax, Johnson-King and k-ϵ model's low-Reynolds number version by Lam and Bremhorst). Although the numerical method leads to convergence even for an airfoil NACA 4412 at an angle of attack of $\alpha = 12°$, the agreement between numerical and experimental results is not fully satisfying and could be increased by improving the turbulence models. The flow past the same airfoil at even $\alpha = 13,9°$ could be predicted quite well by *Cordes* et al.. They combined the k-ϵ model with a one-equation model for the near-wall region instead of using wall functions. In the prediction procedure, developed by *Isay* and *Marzi* for flows past airfoils at high angles of attack, an integral method has been used for the boundary-layer calculation. This procedure takes into account laminar separation bubbles, wall curvature effects and also very long separation bubbles. In one of the examples (airfoil NACA 631012, $\alpha = 13,8°$) the flow is separated practically over the entire upper surface of the airfoil. Further examples show that the wall curvature effects can play an important role. This result was also found by *Lieser* and *Ewald*, who tested various turbulence models (algebraic and two-equation models with wall functions) for transonic flows with separation. They found that turbulence models developed for attached flows have to be modified for separated flows and for high pressure gradients.

A very detailed and comprehensive experimental investigation on conical diffusers has been carried out by *Nitsche* and *Haberland*. By changing the diffusor angle cases with pressure induced separation (small angles) as well as cases with fixed separation position (large angles) have been covered. Mean flow results (velocity field, pressure and wall shear distributions) and turbulence quantities (Reynolds stress, turbulent energy and also turbulent diffusion by a new especially developed probe) are documented. The experiments have been compared with calculations using the standard k-ϵ turbulence model. The comparison shows that in separated flows the effect of the pressure on turbulent diffusion becomes increasingly important.

Considering the state of the art of turbulence modelling for separated flows leads to the conclusion that the situation is not fully satisfying, even for two-dimensional flows, and that improvements are still necessary. One main reason for discrepancies may be that the near-wall region changes dramatically during the transition from attached to separated flow. This is particularly valid for higher Reynolds numbers. *Gersten*, *Klauer*, and *Vieth* have shown that the wall function changes from the logarithmic law to the square-root law when the separation point is approached. Since all existing turbulence models do not have enough free model constants to satisfy both types of wall functions, increasing discrepancies might be expected near separation for growing Reynolds numbers. *Gersten*, *Klauer*, and *Vieth* determined the characteristic values of equilibrium boundary layers including the special case of zero wall-shear stress by using generalized wall functions. Based on these results an integral method was developed, which gives asymptotically correct results at separation.

Turbulent separating flows over humps in an open channel have been investigated by *Larsen* and *Kertzscher*. Position and extent of separation were the main results of the study, which was also extended to density stratified flows.

UNSTEADY FLOWS WITH SEPARATION

By definition unsteadiness of turbulent flows refers to the properly averaged mean flow. This mean flow can be periodic or transient. In the case of periodic oscillation it has to be distinguished between the organized and the random fluctuations. Three categories of unsteady turbulent flows will be considered:

1. Flows past bluff bodies with steady free stream and self-induced oscillations

2. Flows past oscillating bluff bodies

3. Internal flows with given unsteady inflow conditions.

The physical structure of separated flows and wakes behind plane and axisymmetric bodies have been investigated by *Geropp* and *Leder*. In order to detect the unsteady behaviour of the flow fields the velocities were measured by the so-called phase-locked LDA, a specially developed laser-Doppler anemometer. The detailed data of the fields of velocities and Reynolds stresses lead to a deeper insight into the physical mechanisms of vortex shedding past plane and axisymetric bluff bodies. The data can also be used for validation and improvement of prediction methods and turbulence models. Such a new turbulence model for separated two-dimensional flows past bluff bodies has been developed by *Geropp* and *Schumann*. The organized large scale motion is characterised by a new integral relation for the pressure velocity correlation. The random part of the fluctuation is described by an existing second-order closure model. It is shown for the example of the separated flow behind a vertical flat plate that this new extended model describes correctly the physics of the flow field.

The flow past a circular cylinder has been investigated experimentally and theoretically by *Teipel*. Lift and drag could be determined by the theory, a combination of the discrete vortex method and the boundary layer theory. *Fago* and *Mahrenholtz* studied the relation between the motion of vibrating circular or square cylinders and the oscillating flow structure with the aim to develop a mathematical model for such a fluidelastic system.

Two investigations concern internal flows where unsteady inflow conditions are prescibed. *Dibelius* and *Ahlers* studied the effect of periodic perturbations on the flow of a turbine blade. For the simulation of wakes being shed from upstream blade rows in turbomachines the oncoming flow in an especially built test rig was periodically disturbed by a wake generator. The position of separation is shifted downstream by the periodic wake flow as compared to the steady flow case. The experimental results led to better theoretical understanding of the flow phenomena and may lead to improvements of turbulence modelling.

According to *Affeld* et al. flow separations are most unwelcome in the flow through artificial heart valves, because separation increases the danger of generating a thrombus. The recirculating area behind separation allows a concentration of platelets, which may accumulate densely enough and get attached to the foreign material such that the formation of a thrombus becomes likely. Here flow separation plays a vital role for patients with artificial heart valves.

THREE-DIMENSIONAL FLOWS PAST BLUFF BODIES AND CIRCULAR CYLINDERS

The numerical prediction of a three-dimensional turbulent flow field with separation is one of the most difficult problems in fluid mechanics. The most promising concept of sufficient generality is the large-eddy simulation (LES). *Wengle* and *Werner* applied this method to the flow past a cube and past a square rib, both mounted on a wall of a two-dimensional channel. Large amounts of computing time are necessary, e.g. at least 500 CPU hours on a CRAY/Y-MP machine for one case. The numerical results show good agreement with experiments (see *Perić* and *Tropea*). The existence of a three-dimensional mean flow structure close to the square rib, as observed in the experiments, could be confirmed by LES with the consequence that surprisingly enough the two-dimensional square rib case turned out to be numerically more complicated than the three-dimensional case of the cube.

Three-dimensional flow effects have also been found in the separated region downstream of a swept backward-facing step by *Fernholz* et al.. An improved method for determining skin-friction distributions by means of oil-film interferometry has been applied in addition to other existing methods, see also *Fernholz, Dengel*, and *Hess*.

Three experimental investigations are dealing with three-dimensional effects on the flow past a circular cylinder of finite length. According to *Eisenlohr* and *Eckelmann* the appearance of discontinuities in the dependence of the Strouhal number on the Reynolds number is an end effect. The flow field around a circular cylinder of finite span, mounted on a flat plate, and its dependence on the approaching turbulent shear flow has been investigated by *Hölscher* and *Niemann*. Based on these results the authors have in an additional study developed a method to determine whether the pressure fluctuations on the cylinder wall are caused by self-induced alternating separations or by velocity fluctuations in the oncoming turbulent shear flow. These results are very important for the aerodynamics of buildings and structures.

In two studies a method has been developed to predict lift and drag forces on bluff bodies. The procedure is equivalent to existing methods for predicting the aerodynamic forces on airfoils or slender bodies, where the flow field is devided into various zones which can be treated separately: The inviscid outer flow can be calculated by means of potential theory leading to lift and pressure drag, the boundary-layer calculation leads to the friction, and by viscous-inviscid interaction also displacement effects of the boundary layer and the wake can be taken into account. In order to determine the displacement effect of the dead water region of bluff bodies it is assumed that the displacement surface of the wake has a certain universal form if it is properly scaled. The experimental work by *Gersten, Becker*, and *Demmer* on an axisymmetric body and a three-dimensional half body demonstrates this universal behaviour. By applying this concept to configurations of automobiles *Papenfuß* and *Dilgen* were able to predict drag and lift forces which agree very well with experiments.

In two investigations numerical solutions of the full Navier-Stokes equations for three-dimensional laminar flows with separation have been found. *Dallmann* and *Kordulla* considered the transonic flow around a hemisphere-cylinder configuration, where the various topological structures of three-dimensional separation could be studied in detail. The flow field near the body turned out to be almost stationary despite the unsteadiness of the wake flow. *Fiebig* and *Mitra* solved numerically the Navier-Stokes and energy equations for the laminar flow in a rectangular channel with wing-type longitudinal vortex generators. The results show that the spiraling motion induced by longitudinal vortices can drastically increase the heat transfer.

The four remaining investigations on vortex-dominated three-dimensional separation deal with turbulent flows and are mainly experimental. The work by *Gallus*, *Schulz*, and *Poensgen* refers to the flow separation in an annular compressor cascade. These data can be used to improve turbulence models and hence computational codes for turbomachines. The validation of codes for solving the Euler equation was the purpose of the experiments by *Ganzer* and *Kelm*. The transonic lee-side flow of a delta wing was measured by laser-2-focus velocimetry and by flow visualization (Schlieren, oil-film and laser-light-sheet techniques). The agreement between experimental results and numerical solutions of the Euler equations were good in areas where viscous effects are not important.

The last two investigations deal with vortex formations on wings. *Kommallein* and *Hummel* analyzed in great detail the flow field over an inclined slender delta wing using a 3D laser-Doppler anemometer. The results brought also new insight into the phenomenon of vortex breakdown. *Oelker* and *Hummel* investigated the aerodynamics of canard configurations. The interference of the various vortex systems generated by the canard, the wing and the fuselage has been studied by means of six-component force measurements, pressure distribution and flow field measurements as well as by flow visualizations. Since the formation of vortices is taken into account in the aerodynamic design of such configurations, it is essential that this particular type of three-dimensional flow separation is always under control in these practical applications.

CONCLUSIONS

In practice flow separation should be avoided or could be desirable depending on the application. Therefore, there is no doubt that one has to live with flow separation and one should try to understand this physical phenomenon in all its varieties.

As far as experiments are concerned, nowadays the expensive but very powerful tool of laserDoppler anemometry is available to investigate the complex velocity fields of separated flows.

Numerical simulation of separated flows is still a difficult problem in particular when the flow is turbulent. In the long run the large-eddy simulation will become the appropriate turbulence model which is flexible and general enough for describing the complex features of turbulent separated flows. For this an improvement of simulating the flow in the near-wall region in particular in the neighbourhood of separation lines is still needed. Simpler turbulence models lead at present already to acceptable results, when two-dimensional steady flows with not too large separated regions are considered, as for instance flows past airfoils of moderate angle of incidence. In those cases the entire flow field consists of various zones which can be connected with each other by the concept of viscous-inviscid interaction. This concept can be extended even to bluff bodies with good results, if the aerodynamic forces are of interest and the flow field in the wake is of minor importance.

Turbulent Backward-Facing Step Flow

Rainer Friedrich, Michel Arnal

Lehrstuhl für Fluidmechanik, Technische Universität München
8000 München 2, Fed. Rep. of Germany

Summary

Large-eddy simulations (LES) of the unsteady three-dimensional flow over a one-sided backstep (bfs) in a channel have been carried out to study the physics of separation and reattachment. The discussion of the influence of two different subgrid-scale (SGS) models, of grid refinement and grid anisotropy shows that the LES technique has reached a mature stage in which it can be considered as a valuable tool for the prediction of complex flows. The balance of Reynolds shear and normal stresses reveals the varying importance of turbulence production and redistribution in the free shear layer and the fact that the dissipation rate tensor is strongly anisotropic within the free shear layer and the reattachment zone.

Outline of the numerical technique

The LES starts from the low-pass filtered Navier-Stokes equations for an incompressible, constant-viscosity fluid. Filtering is achieved by integrating the basic equations over the mesh volume of an equidistant cartesian grid which leads to a central finite difference scheme in a natural way [1]. Staggering of the discrete flow variables is used. The explicit leapfrog scheme which is of second order accuracy with respect to convection terms serves for time integration. Each time step is split into two substeps where the first provides an approximate velocity field and the second corrects this field with the help of the exact pressure gradient obtained from the direct solution of a Poisson equation. A FFT in the spanwise direction converts the 3D Poisson problem for the pressure into a set of 2D Helmholtz problems which are solved with a cyclic reduction technique. Since the flow field is not box-like, the capacitance matrix technique has to be utilized which essentially means that the 3D Poisson problem must be solved twice at each time step [2].

Artificial *wall boundary conditions* are used in order to reduce the necessary number of grid points in the wall normal direction. *Periodicity conditions* hold in the spanwise direction in which there is no mean flow. There is a need to provide a sufficient width of the computational domain if large scale flow structures in the reattachment region are not to be distorted. In [3] a six step heights wide domain was found to avoid undesirable numerical effects which would otherwise have an impact on global flow quantities such as the reattachment length. The *outflow boundary condition* for the instantaneous velocity vector is obtained from a simplified convection equation thus realizing the concept of "frozen" turbulence [4]. Great efforts have to be made to specify time-dependent *inflow boundary conditions*. In the present case, an extra LES of a fully developed channel flow is employed for each step flow simulation with the same grid, time-step, subgrid-scale model and Reynolds number. A sufficiently long time-series of the velocity vector in each cell of a plane which is

normal to the main channel flow forms the proper inflow conditions for bfs flow.
Initial conditions must be specified in the entire computational domain. Their statistical part is taken from a separate 2D steady state calculation with a standard k-ε model of turbulence. Their fluctuating part is obtained from random numbers suitably weighted and made divergence free.

Two different eddy-viscosity-type SGS models have been implemented and tested following suggestions of [1] and [5] with slight modifications thereof, see [2,3,6]. The *geometry* of the computational domain (fig.1) consisted of a 1:2 sudden expansion with the inlet plane located 4 step heights (H) upstream of the backstep, the outlet plane 20 H downstream and a width of 8 H. The Reynolds number of the simulated flow, based on H and the inlet bulk velocity, was $Re_b = 1.55 \cdot 10^5$.

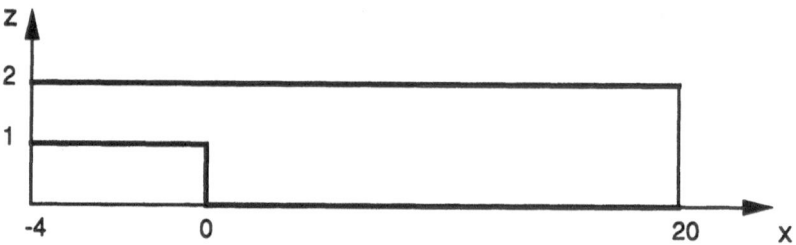

Fig.1: Geometry of bfs flow.

SGS model dependency

In this section the results of calculations with Schumann's two-part eddy-viscosity model (1,Schu) and Smagorinsky's model (2,Sma) are discussed. They were obtained on a $192 \times 64 \times 32$ grid resolving the (x, y, z)-computational domain. The mean wall pressure coefficient in fig.2 seems to be relatively insensitive to model changes, whereas the mean shear stress coefficient c_f shows up some differences. It indicates fully developed flow in the case of model 1 and recovering flow in the case of model 2.

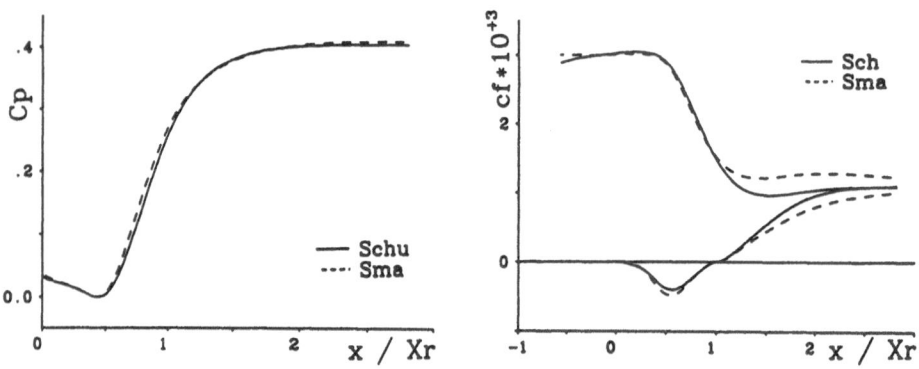

Fig.2: Wall pressure and shear stress coefficients for model 1 and model 2.

A definition of these coefficients is appropriate:

$$c_p = \left(\langle p \rangle_w - p_{ref}\right) / \left(0.5 \rho u_{ref}^2\right) \quad , \quad c_f = \langle \tau \rangle_w / \left(0.5 \rho u_{ref}^2\right) \quad . \tag{1}$$

The reference pressure and velocity are statistical quantities taken at $x = 0$, $z = 1.5$. This definition of c_p is in agreement with that of [7] which takes into account that the reattachment pressure rise depends on the conditions of the flow approaching separation.

The following series of figures 3,4 and 5 shows the model influence on profiles of the mean velocity and three of the Reynolds stress components (all have been non-dimensionalized with the mean centerline velocity of the channel flow). The latter are computed from the resolved scales and a model contribution. Three positions have been selected: the first, at $x = 0.24x_r$ lies well within the recirculation zone, the second, at $x = 0.94x_r$ is close to the mean reattachment point x_r and the third, at $x = 2.35x_r$, represents the relaxation region in which the fluid approaches the stage of fully developed channel flow. The model influence is noticeable in all the statistical quantities and at all positions. However, the differences in the results of the two models are not big and they will become unimportant on a refined grid. For the present grid it is hard to say which model gives a better prediction since we observe considerable scatter in the experimental data of [8]. Moreover, the inflow conditions in the experiment differed from those of the simulation and corresponded to a still developing channel flow with two boundary layers on the top and bottom walls of the channel. Inspection of the mean streamwise velocity profile in fig.5 gives an explanation for the differences in the c_f-distributions in fig.1 between the two models. While the Smagorinsky model seems to overpredict the shear stress on the upper wall it apparently underpredicts it on the reattachment side. The truth most probably lies in between the model results. Two more remarks can be made with respect to the presented data. While the contribution of Schumann's SGS model to the Reynolds stresses is about 20% that of Smagorinsky's model is 10 % in the mean. Due to its greater algebraic simplicity the latter needs roughly 5% less computer time and its implementation is much easier.

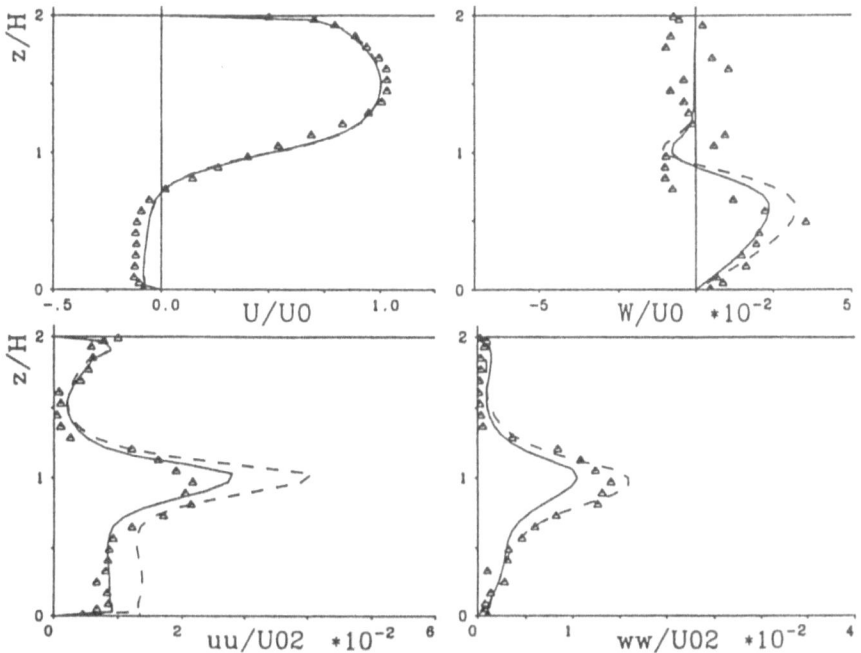

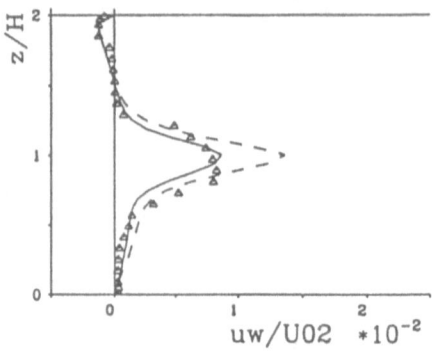

Fig.3
Influence of SGS model on mean velocity and Reynolds stress profiles at $x=0.24x_r$. (— Schumann's model, - - - Smagorinsky's model, Δ exp. of [8]).

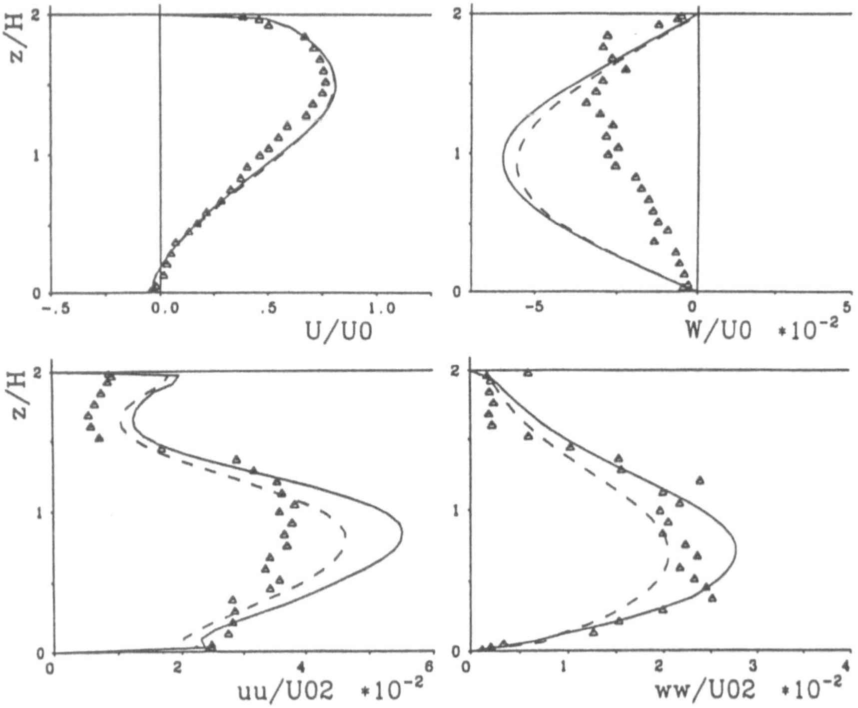

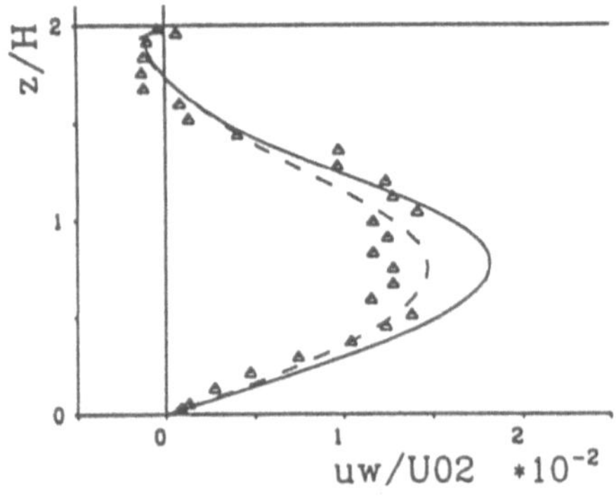

Fig.4
Influence of SGS model continued. $x=0.94x_r$.

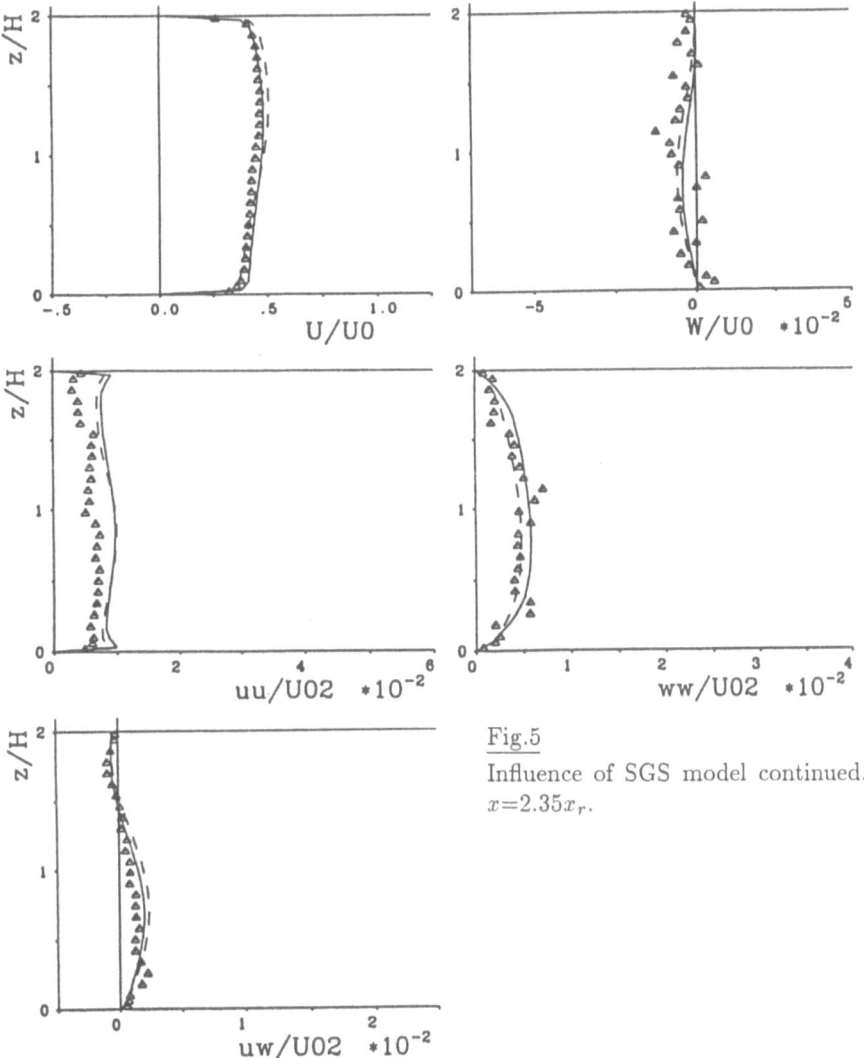

Fig.5

Influence of SGS model continued. $x=2.35x_r$.

Grid refinement

Using Schumann's model we have performed large-eddy simulations with 3 different grids, namely a coarse 80×16^2 grid, a 160×32^2 grid and a fine $320 \times 64 \times 48$ grid resolving the domain downstream of the step. In these cases we have chosen a reduced domain width of $4H$ in order to allow for the finest grid. The effect of grid refinement is demonstrated in fig.6 . The curves for the Reynolds stresses do not contain the model contribution. It is obvious that grid refining improves the results and that we are not too far from a grid independent solution.

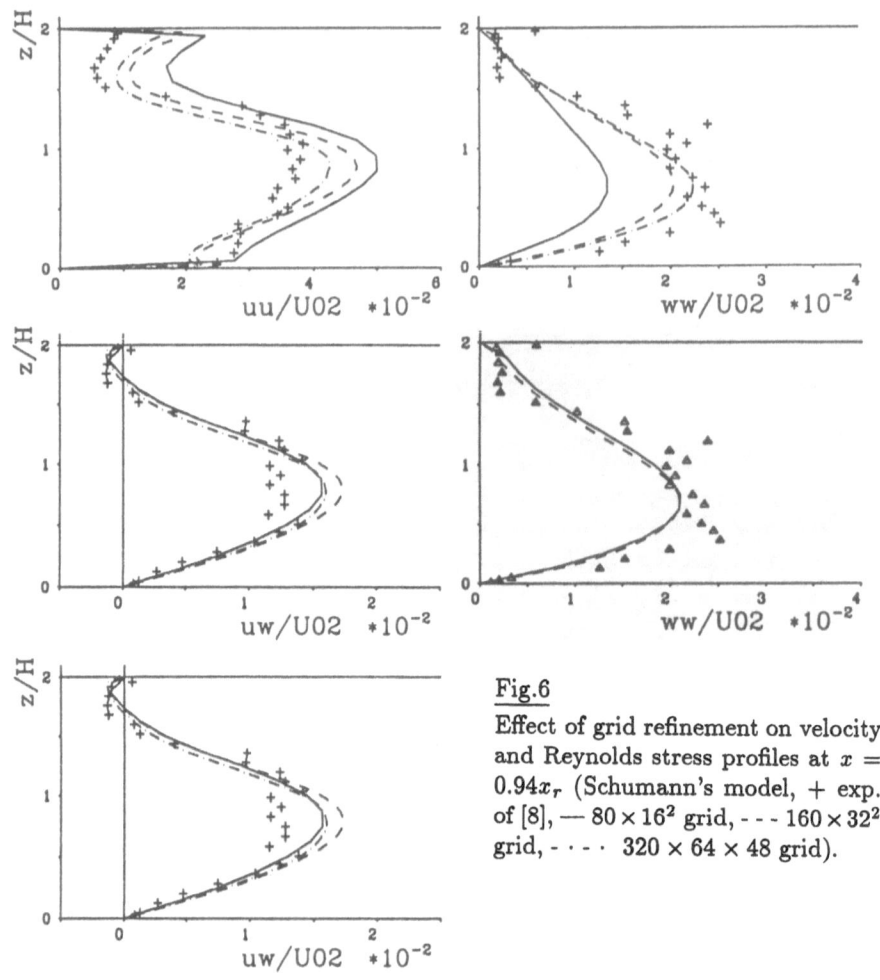

Fig.6
Effect of grid refinement on velocity and Reynolds stress profiles at $x = 0.94x_r$ (Schumann's model, + exp. of [8], — 80×16^2 grid, - - - 160×32^2 grid, - · · · $320 \times 64 \times 48$ grid).

Grid anisotropy

All the presented results have been obtained with equidistant grids and a relation between the three mesh sizes of $\Delta x = \Delta y = 2\Delta z$. Such grids are weakly anisotropic and thus have only a minor effect on the turbulence structure which can be explained by different filter widths in the x- and z-directions. Strong grid anisotropy leads to a spurious exchange of fluctuating energy in the three directions. In fig.7 results with Schumann's model for an isotropic grid with $\Delta x = \Delta y = \Delta z = 1/8H$ (solid line) and an anisotropic grid with $\Delta x = \Delta y = 2\Delta z = 1/8H$ (dashed line) are compared and a non-negligible effect is found. The computational domain is $8H$ wide.

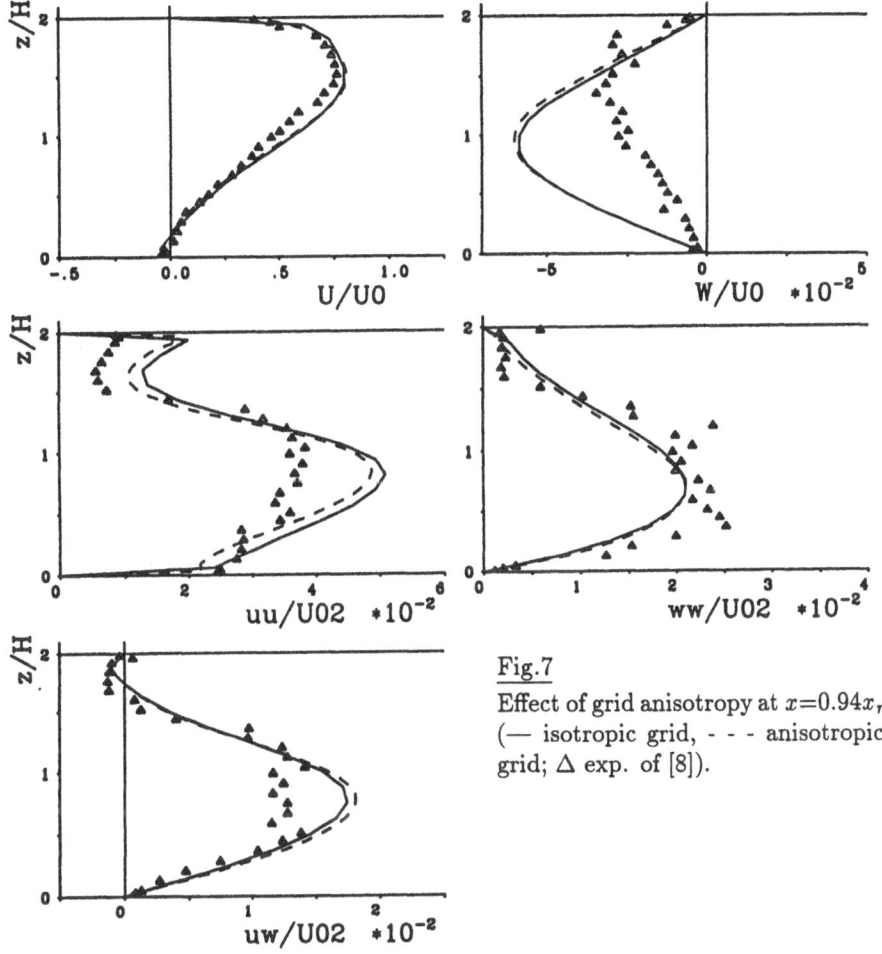

Fig.7
Effect of grid anisotropy at $x=0.94x_r$
(— isotropic grid, - - - anisotropic
grid; $\triangle$ exp. of [8]).

Reynolds-stress transport

The turbulence structure of the bfs flow considered is statistically 2D. Its dynamics is controlled by 4 nonzero components of the Reynolds-stress tensor, namely $\langle u^2 \rangle$, $\langle v^2 \rangle$, $\langle w^2 \rangle$ and $\langle -uw \rangle$. They follow from a transport equation which in the case of statistically stationary high Reynolds number flow reads:

$$
\underbrace{\langle u_k \rangle \frac{\partial \langle u_i u_j \rangle}{\partial x_k}}_{\text{Conv}} + \underbrace{\langle u_k u_j \rangle \frac{\partial \langle u_i \rangle}{\partial x_k} + \langle u_k u_i \rangle \frac{\partial \langle u_j \rangle}{\partial x_k}}_{\text{Prod}} - \underbrace{\frac{1}{\rho} \left\langle p \left(\frac{\partial u_i}{\partial x_j} + \frac{\partial u_j}{\partial x_i} \right) \right\rangle}_{\text{P}-\text{S}}
$$

$$
+ \underbrace{2\nu \left\langle \frac{\partial u_i}{x_k} \frac{\partial u_j}{x_k} \right\rangle}_{\text{Diss}} + \underbrace{\frac{\partial}{\partial x_k} \left(\langle u_i u_j u_k \rangle + \frac{1}{\rho} \left\langle (\delta_{jk} u_i + \delta_{ik} u_j) p \right\rangle \right)}_{\text{Diff}} = 0 \qquad (2)
$$

13

The terms describe convection, production, redistribution, dissipation and diffusion. Except in the dissipation term the dummy index k represents a summation from 1 to 2. Viscous diffusion has been neglected since it is only important in the near wall region where it is not resolved with the present equidistant grid ($\Delta z = 1/16H$). Most turbulence models, e.g. [9,10] assume local isotropy in the turbulence fine scale structure and model the dissipation rate in the following way

$$2\nu \left\langle \frac{\partial u_i}{\partial x_k} \frac{\partial u_j}{\partial x_k} \right\rangle = \frac{2}{3}\delta_{ij}\varepsilon \quad , \tag{3}$$

where ε is the rate of dissipation appearing in the turbulence energy equation. While convection and diffusion are neglected in some algebraic stress models (ASM) which means that there is a balance between production, redistribution and dissipation in (2), Rodi [10] relaxes this assumption by relating convection and diffusion of $\langle u_i u_j \rangle$ to the corresponding terms in the kinetic energy (k) equation. The present LES results will show that none of these models is valid in the bfs flow. At the same time we find strong anisotropy in the dissipation rates which we computed as differences of all the other terms in (2). This indicates that statistical turbulence models for bfs flows must contain the full Reynolds-stress equations and transport equations for the components of the dissipation rate tensor.

In fig.8 we have plotted profiles of the transport terms of eq.(2) for the 4 Reynolds stress components. It is obvious that the turbulence dynamics are especially pronounced in the free shear layer and the reattachment zone. At $x = 0.94x_r$ the vertical fluctuations are strongly affected due to the presence of the wall which leads to important diffusion, redistribution and dissipation effects. A comparison of two different algebraic stress models is provided in fig.9. It shows that models which assume the vanishing of convection plus diffusion produce errors of the order of the terms themselves. Rodi's ASM [9] seems to behave better especially in planes close to reattachment.

Conclusions

Using statistical data and the comparison with the experiment we have tried to demonstrate the reliability of LES for complex flows. Through inspection of the Reynolds stress transport we found that simple algebraic turbulence models are incapable of correctly predicting bfs flow. There is a need for solving the full Reynolds-stress equations along with the transport equations for the components of the dissipation rate tensor. This makes the statistical approach very laborious apart from the remaining uncertainties in modeling the dissipation rate equations. For the moment being LES can hardly contribute to this problem. A detailed analysis of the instantaneous turbulent bfs flow has been given earlier [11,12,13,14].

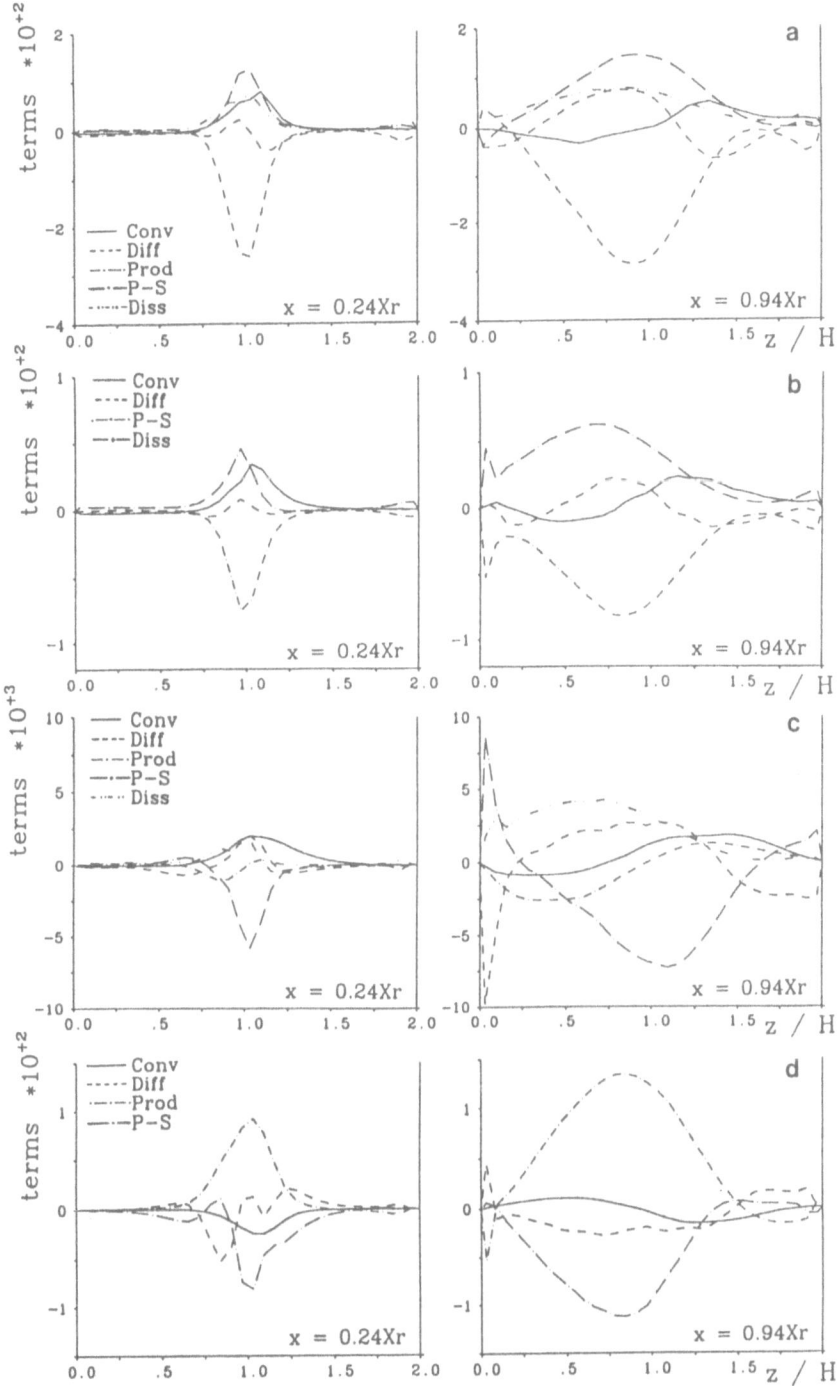

Fig.8 Transport terms of the Reynolds stress equations.
a) $\langle u^2 \rangle$, b) $\langle v^2 \rangle$, c) $\langle w^2 \rangle$, d) $\langle uw \rangle$

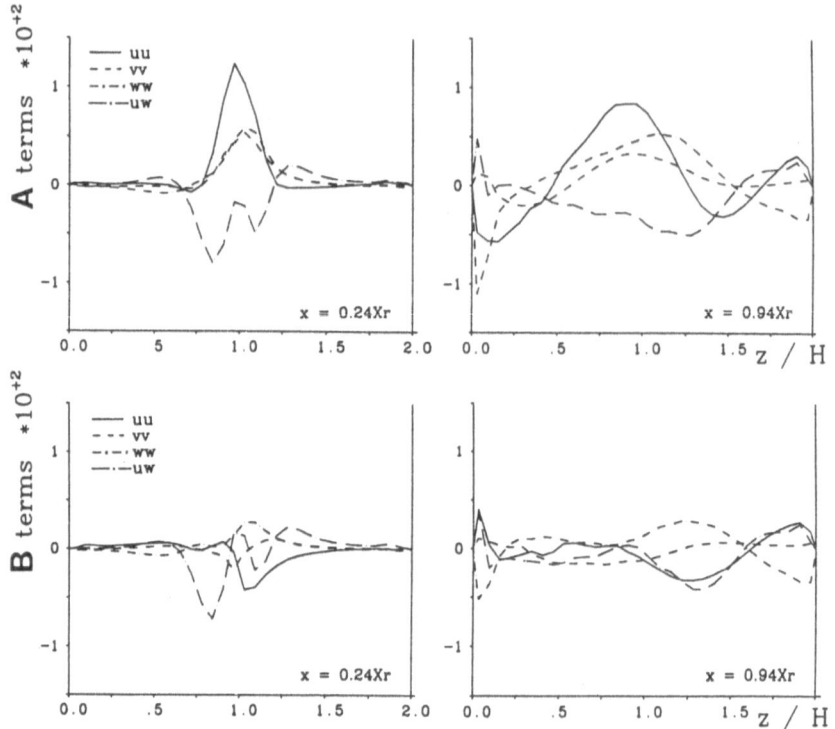

Fig.9 Test of algebraic stress models. $A = CONV(\langle u_i u_j \rangle) + DIFF(\langle u_i u_j \rangle)$, $B = A - \langle u_i u_j \rangle (CONV(k) + DIFF(k))/k$.

Literature

[1] Schumann,U.: "Subgrid scale model for finite difference simulations of turbulent flows in plane channels and annuli", J.Comp.Phys., <u>18</u> (1975) pp. 376-404.

[2] Schmitt,L.: "Grobstruktursimulation turbulenter Grenzschicht-, Kanal- und Stufenströmungen", Dissertation, TU München (1988).

[3] Arnal,M., Friedrich,R.: "On the effects of spatial resolution and subgrid-scale modelling in the large-eddy simulation of a recirculating flow ."Proc. 9th GAMM-Conference on Numerical Methods in Fluid Mechanics, Lausanne, 1991. - To appear in: Notes on Numerical Fluid Mechanics,(1992).

[4] Richter,K., Friedrich,R., Schmitt,L.: "Large-eddy simulation of turbulent wall boundary layers with pressure gradient". - In: Proc. 6th Symp. on Turbulent Shear Flows, Toulouse (1987), pp. 22.3.1-7.

[5] Smagorinsky,S.S.: "General circulation experiments with the primitive equations", Mon. Weather Rev., <u>91</u> (1963) pp. 99-164.

[6] Friedrich,R.: "On large- eddy simulation of turbulent flows". - In: H.Niki, M.Kawahara (eds.), Computational Methods in Flow Analysis, Okayama University press, Okayama (1988), Vol.2, pp. 833-843.

[7] Roshko,A., Lau,I.C.: "Some observations on transition and reattachment of a free shear layer in incompressible flow". Proc. of the Heat Transfer and Fluid Mechanics Institute, E.F. Chorwat (ed.), Stanford Univ. Press (1965).

[8] Durst,F., Schmitt,F.: "Experimental study of high Reynolds number backward-facing step flow", in Proc. 5th Symp. on Turbulent Shear Flows, Aug. 7-9 (1985), Cornell University, Ithaca, NY.

[9] Launder,B.E., Reece,G.J., Rodi,W.: "Progress in the development of a Reynolds stress turbulence closure". J. Fluid Mech. 68 (1975), pp. 537-566.

[10] Rodi,W.: "A new algebraic relation for calculating the Reynolds stresses". ZAMM 56 (1976), T 219-T 221.

[11] Friedrich,R., Arnal,M.: "Analysing turbulent backward- facing step flow with the lowpass-filtered Navier-Stokes equations". J.Wind Engg. & Industrial Aerodyn., 35 (1990), pp.101-128.

[12] Arnal,M., Friedrich,R.: "The instantaneous structure of a turbulent flow over a backward-facing step". - In: V.V.Kozlov., A.V.Dovgal (Eds.), Separated Flows and Jets, Springer-Verlag (1991), pp.709-717.

[13] Arnal,M., Friedrich,R.: "Investigation of the pressure and velocity fields in a turbulent separated flow using the LES technique." 29th Aerospace Sciences Meeting, Reno, Jan. 7-10 (1991). AIAA paper 91-0251.

[14] Arnal,M., Friedrich,R.: "Large-eddy simulation of a turbulent flow with separation". - To appear in: Turbulent Shear Flows 8, Springer-Verlag (1992).

17

Turbulent Flow over Surface-Mounted Obstacles in Plane and Axisymmetric Geometries

M. Perić and C. Tropea

Lehrstuhl für Strömungsmechanik, University of Erlangen-Nürnberg
Cauerstraße 4, 8520 Erlangen, Germany

Summary

Numerical and experimental investigations of the flow over surface-mounted obstacles in fully developed channel and pipe flow have been performed. The numerical work was restricted to two-dimensional geometries and results obtained with the k-ϵ turbulence model were compared with those obtained using a second-moment turbulence closure model. The experimental work included obstacles of finite length in the channel flow, with very detailed velocity measurements (LDA) being performed for the cube obstacle. Unique characteristics of two and three dimensional flow fields have been identified. Most recent results include the time dependent behaviour of these flow fields, studied using flow visualization and two-component velocity measurements.

1. Introduction

The phenomenon of flow separation in internal flows caused by sudden changes in cross-sectional area is a subject of particular relevance to numerous practical engineering applications due to the associated influence on the pressure loss, heat and mass transfer and the resulting effects of erosion and/or corrosion. Experimental and numerical investigations of these flow fields have often been restricted to simplified, two-dimensional flow geometries, for instance ribs, fences or the backward facing step, to concentrate first on the basic physics of separated flows and to evaluate turbulence models and numerical solution procedures. Only more recently, with the growing capabilities of numerical codes, are three-dimensional geometries being studied in more detail. Such geometries often have much more relevance to practical flow fields rather than being only of fundamental importance. Common to all obstacle flows however, is a flow separation upstream of the obstacle which is a distinguishing feature of all the flows investigated in the present work.

This report covers research performed over a period of seven years on various flow geometries and by numerous researchers and students, as summarized in Table 1 and documented fully in the list of references. Some of the more immediate questions and goals addressed and completed in the course of this work are summarized below in chronological order:

- influence of obstacle length (streamwise direction) and height on global flow field parameters (x_s, x_R) [5,4,22]
- documentation of time-averaged velocity field for flow over selected plane and axisymmetric geometries: data suitable for detailed comparison with numerical results [19,24,26]
- comparison of turbulence quantities with other separated flows, eg. backstep flow [2]
- computation of 2D flows with k-ϵ turbulence model [1,7]
- development of a second moment turbulence closure model and comparison with results obtained using k-ϵ model and with experiments [16,17]

Table 1: Overview of experimental and numerical work performed on surface-mounted obstacles

Researcher	Geometry	Work	'85	'86	'87	'88	'89	'90	'91
Dimaczek, G.		E	⊢———————————————⊣						
Kessler, R.		N	⊢——————————⊣						
Obi, S.		N		⊢———————————————⊣					
Zeigler, T.		E		⊢————⊣					
Eh, C.		E			⊢————⊣				
Foss, J.		G	⊢————⊣						
Martinuzzi, R.		E + N			⊢————————————————————————⊣				
König, J.		E						⊢————⊣	
Simpson, R.J.		G						⊢————⊣	
Theisinger, J.		N						⊢————⊣	
Larousse, A.		E							⊢—⊣
Volkert, J.		E							⊢—⊣
Wang, A.B.		E + N		⊢—————————————————————————————⊣					
Founti, M.		E	⊢—————————⊣						
Pollard, A.		G	⊢————⊣						

E: experimental N : numerical G : guest researcher

- influence of obstacle width (plane geometry) on global flow parameters [10,15]
- detailed measurements of time-averaged velocity field over cube obstacle; comparison of turbulence quantities with 2D geometries [14,12]
- change of flow topology between 2D and 3D flow geometries
- study of time-dependent flow field for cube obstacle (flow visualization, LDA) [11,23]
- comparison of time-dependent measurements with (third party) numerical computations [13]

In the following sections a brief overview of the work performed is given. Some of the more important results and conclusions are also summarized.

2. Axisymmetric Obstacle Flow

Four experimental and nine numerical obstacle geometries were investigated at various Reynolds numbers for the axisymmetric case, as summarized in Table 2.

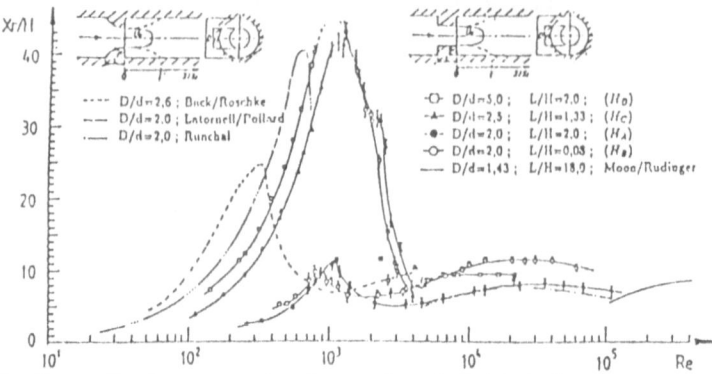

Figure 1: Normalized mean reattachment length behind the obstacle as a function of Reynolds number and diameter ratio

Table 2: Axisymmetric geometries investigated

D/d	L/H	Exp. and/or Num.
2	0.08	E + N
2	0.5	N
2	1	N
2	2	N
2	3	E + N
2	4	N
2	5	N
2	6	N
2	∞	N (pipe expansion)
2	∞	N (pipe contraction)
5	2	E
2.5	1.33	E

Detailed velocity measurements were performed at the Reynolds numbers 13,000 and 22,000 (based on maximum pipe velocity and pipe diameter) and measurements of the separation zone lengths were obtained for the Reynolds number range 100 - 10^5. Velocity measurements were obtained using a one-component LDA. Refractive index matching was used to allow measurement of all velocity components and to also permit near wall measurements.

In Fig. 1 the results of Reynolds number variations on the primary mean reattachment length (x_R) are shown for various obstacles and compared with those of other authors. A close correlation between x_R and the Reynolds number at which x_R reaches fully turbulent conditions can be made to the the the diameter ratio D/d. In Fig. 2 the Reynolds number similarity of the mean velocity and normal stresses on the centerline is shown. It is also apparent from this diagram that the obstacle has an upstream influence on the flow of at most one pipe diameter. Further details of the flow similarity can be found in [6]. A complete documentation of work performed on axisymmetric geometries, including tabulated measurement results, can be found in [24].

3. Plane, Two-Dimensional Obstacle Flow

The experiments for the plane, two-dimensional obstacle were performed in a 5×60 cm² water channel, the obstacles spanning the section at a downstream position at which the turbulent channel flow was fully developed. Velocity measurements were obtained with a one-component frequency-shifted LDA operated in forward scatter. The crystal violet technique [3] of surface flow visualization was used to deduce general flow patterns and mean reattachment lengths.

The mean velocity field is characterised by the mean separation and reattachment positions, which are summarized for various obstacles in Table 3. This table includes numerical

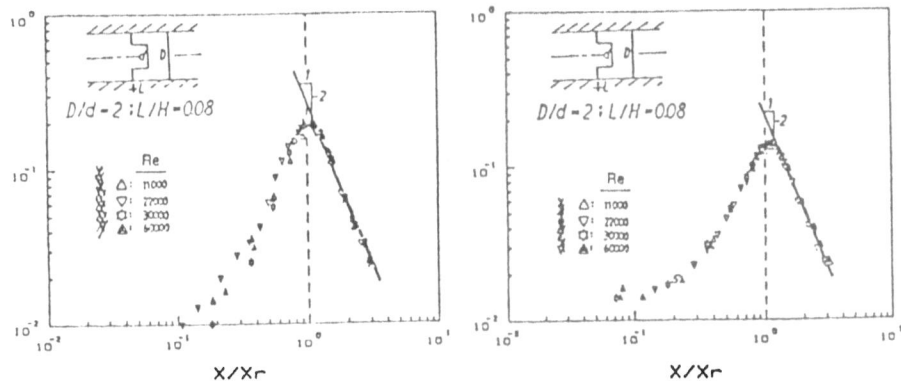

Figure 2: Dimensionless a) $\sqrt{\overline{u^2}}$ and b) $\sqrt{\overline{v^2}}$ distributions along the centerline for various Reynolds numbers

Table 3: Summary of mean separation and reattachment locations ($Re_H = 42000$)

Obstacle H × L (mm)	L/H	X_R/H			X_F/H			X_T/H		
		num	vis	LDA	num	vis	LDA	num	vis	LDA
25 × 25	1.00	7.1	7.1	7.1	0.65	0.50	0.55	N/A	N/A	N/A
25 × 50	2.00	6.9	6.4	-	0.65	0.50	-	0.47	0.8	-
37.5 × 25	0.67	6.5	6.5	-	0.54	0.50	-	0.60	0.8	-
32.5 × 25	2.00	6.5	6.5	-	0.70	0.50	-	0.75	1.0	-
25 × 5	0.20	8.2	9.4	-	0.69	0.36	-	N/A	N/A	N/A
X_R - reatt. behind obst.; X_F - sep. in front of obst.; X_T - reatt. on top of obst.										

results (num) obtained using the k-ϵ turbulence model, results from the flow visualization (vis) and LDA measurements (LDA). Increasing either L/H or the channel blockage favours formation of a recirculation region on top of the obstacle, already at very modest values of L/H. The length of the main recirculation region also shows a decrease with increasing L/H. Both high and low blockage ratios lead to a decrease in the mean reattachment length.

The primary separation zone behind the rib fluctuates in time as evident from the reverse flow indicator shown in Fig. 3. These fluctuations are larger than in backstep or sudden expansion flows and lead also to larger measured turbulent kinetic energy, as shown in Fig. 4. This figure also includes computations obtained with the high Reynolds number k-ϵ turbulence model. Both experiments and computations indicate two maxima of k, one immediately above the obstacle and one downstream. This differs from simple sudden expansion geometries and is a result of the forward edge separation from the obstacle.

Details of these investigations can be found in [1,26]. Further comparisons between plane and axisymmetric geometries are given in [2].

In further studies using the same experimental apparatus, the influence of obstacle width was investigated. In particular the length of the separation zones in front of and behind the obstacle were measured as a function of obstacle width. Fig. 5 indicates that a width

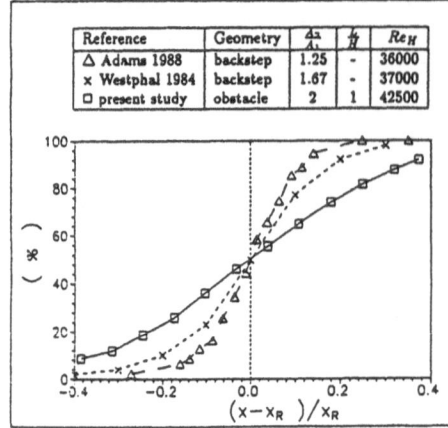

Reference	Geometry	$\frac{\Delta_2}{\Delta_1}$	$\frac{y}{H}$	Re_H
△ Adams 1988	backstep	1.25	-	36000
× Westphal 1984	backstep	1.67	-	37000
□ present study	obstacle	2	1	42500

Figure 3: Forward flow fraction ($y/H = 0.02$).

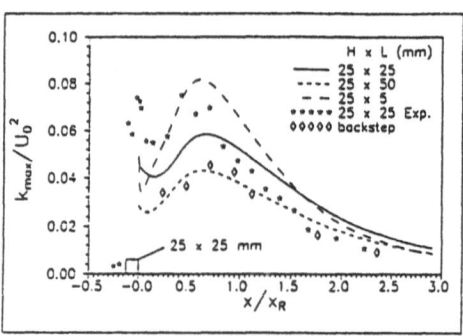

Figure 4: Development of k_{max} in separated shear layer from computations and experiments : $Re_H = 42000$ (Backstep data from [21]).

independent flow is achieved in front of the obstacle already at aspect ratios of $W/H=3$, whereas behind the obstacle the flow is sensitive to obstacle width up to $W/H=9$. These results are confirmed by extensive flow visualization studies performed using the crystal violet technique [11].

Furthermore a distinct pattern of node and saddle points was observed on the front face of the obstacle for $W/H > 6$. This indicates a three-dimensionality of the flow which is also documented in profiles of $\overline{w^2}$ in front of the obstacle. These observations and a topologically consistent interpretation of the flow field is given in Fig. 6. More details of work can be found in [10,15,20].

4. Numerical Investigations

The test case investigated here is a separating flow over a square rib mounted in a plane channel for which detailed measurements are performed [1]. Initial computational studies using the standard k-ϵ model [9] have shown that this is a very severe test case for the numerical method as well as the turbulence model, because of the strong streamline curvature immediately upstream of the rib, followed by a remarkable flow accerlation, separation at the sharp corner, an abrupt area expansion behind the obstacle and flow reattachment associated with an adverse pressure gradient. The k-ϵ turbulence model has been shown to be incapable of representing this flow field in detail, especially around the obstacle. It was also demonstrated that the maximum numerical solution error for turbulent kinetic energy exceeded 30% even on a very fine grid consisting of 190 × 120 control volumes.

A second-order Reynolds stress turbulence model is applied to the calculation of the flow over a square rib. Accurate solutions are obtained by using a finite-volume computation procedure and a very fine numerical grid, consisting of 240 × 120 control volumes. Numerical errors are estimated by comparing solutions on three systematically refined grids, and are of the order of 1%. In order to stabilize the iterative solution procedure, time marching is used in addition to under-relaxation. Convergence is monitored by evaluating the sum of absolute residuals for every equation at the beginning of each time step. These are required to fall below 0.1% of reference inlet fluxes.

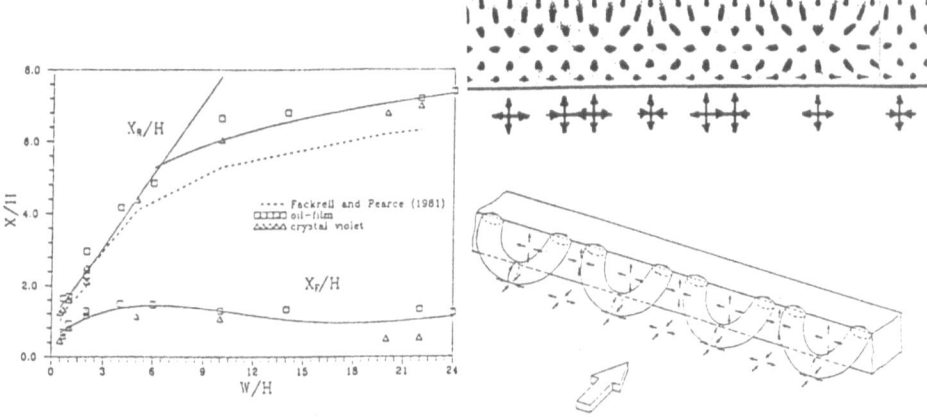

Figure 5: Separation x_F/H and reattachment lengths x_R/H as a function of W/H.

Figure 6: Visualization of flow on obstacle front face using crystal violet technique and interpretation of flow field.

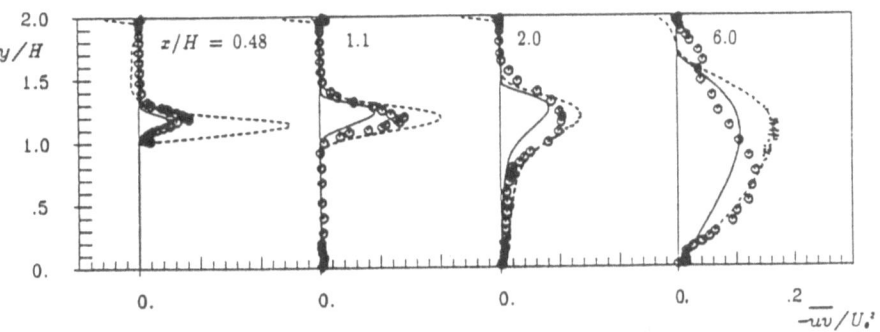

Figure 7: Turbulent shear stress: ⊙-experiment; ——— second-moment closure model; – – – – k-ϵ model (finest grid).

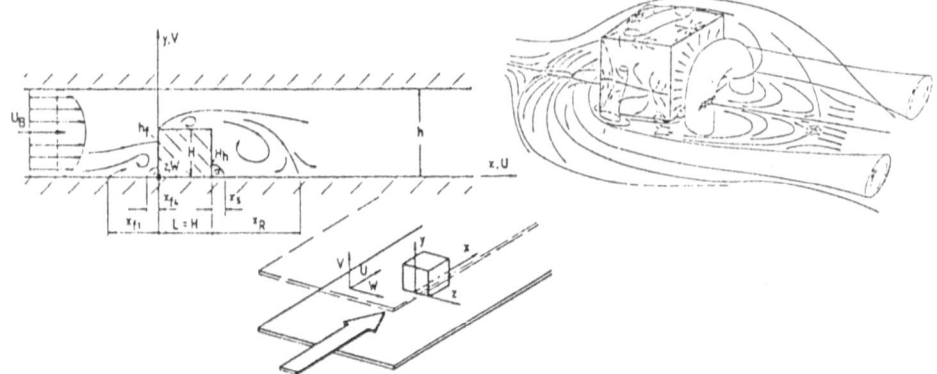

Figure 8: Sketch of the obstacle geometry in a channel flow

Figure 9: Sketch of the flow around the cube

Good agreement of the second-moment closure model and the data is demonstrated for the mean velocity distributions above and just downstream of the rib. The k-ϵ model predicts a too small and attached separation zone on top of the rib; the blockage of the cross section is thus smaller, resulting in an underestimation of the maximum velocity. Good agreement between the computed and measured values of $\overline{uv}$ immediately on top of the rib are seen in Fig. 7. The rather poor representation of the flow field by the k-ϵ model in this region is attributed to inappropriate modelling of the production rates. Because of the strong flow acceleration immediately upstream of the obstacle, the dominating term in the production rate of k, namely $-\overline{u^2}(\partial U/\partial x)$ is *negative*; in the k-ϵ model however, this is modelled as $\nu_t(\partial U/\partial x)^2$, which is always *positive*, thus resulting in the observed high values of k.

Use of the second-moment turbulence closure leads to more realistic simulation of the mean velocity and turbulence field around the obstacle as compared to k-ϵ model calculations. However, there is still substantial disagreement with experiments. As the data show, three dimensional effects are present even for this geometry and it remains unclear as to whether this disagreement can be blamed on the turbulence model or on the two-dimensional formulation of the equations. Further details of the numerical work performed can be found in [1,9,16].

5. Cube Obstacle

The flow field around a surface-mounted cube placed in a channel was investigated using a single and a two-component LDA, by performing wall pressure measruements as well as by conducting oil-film and laser sheet flow visualization experiments. All experiments were performed in a fully developed channel flow in air at a Reynolds number of 80000, based on the channel height and the mean channel velocity. A schematic representation of the geometry is given in Fig. 8.

Extensive velocity measurements and topological considerations lead to a mean flow field representation as shown in Fig. 9. In the mean, the separated shear layer does not reattach on the top side of the cube. A large primary recirculation vortex dominates the local structure of the flow field. The separated flow regions around the cube do not form a closed recirculation bubble.

The flow field in front and on top of the cube is instationary. Typically, the velocity probability distribution function contains a double peak for at least one of the velocity

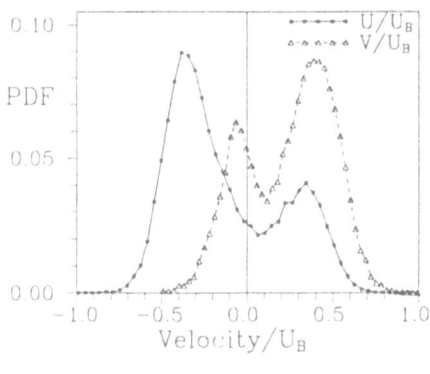

Figure 10: Velocity PDF at x = -0.44H, y=0.10H, z = 0.

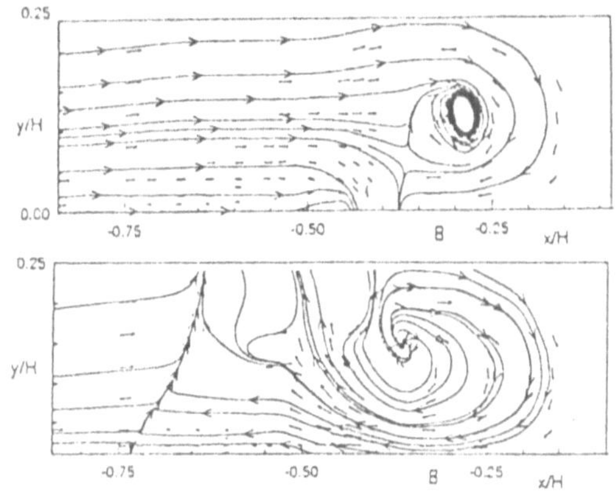

Figure 11: Mode segregated flow field in front of cube.

components measured in these regions, implying fluctuations of the corresponding velocity component about two preferred values. This behaviour, illustrated in Fig. 10, is interpreted to indicate that the local flow field can alternately occupy one of two states or modes. By reprocessing the velocity data and discriminating in amplitude between the two possible velocity states, mode-segregated velocity fields can be reconstructed yielding two probable streamline patterns for each region of the flow. An example of this mode segregation for the region in front of the obstacle is given in Fig. 11. Such decompositions of the velocity field compare remarkably well with video visualizations of the flow and with time dependent computations using large eddy simulations [25].

Complete documentation of the investigations of the flow around a surface mounted cube is given in [12,13,14]

References

[1] Dimaczek, G., Kessler, R., Martinuzzi, R., Tropea, C. *The Flow Over Two-Dimensional, Surface-Mounted Obstacles at High Reynolds Numbers* 7th Symp. on Turb. Shear Flows, Stanford Univ., Aug. 1989.

[2] Dimaczek, G., Tropea, C., Wang, A.B. *Turbulent Flow Over Two-Dimensional, Surface-Mounted Obstacles: Plane and Axisymmetric Geometries* 2nd European Turb. Conf., Berlin, 1988.

[3] Dimaczek, G., Eh, C., Tropea, C. *Stromungssichtbarmachung von Wasserströmungen mit Hilfe des Kristallviolettverfahrens* DGLR Workshop 2D Meßtechnik, Markdorf, 1988.

[4] Durst, F., Founti, M., Pereira, J.C.F., Gackstatter, R., Tropea, C. *The Wall-Reattaching Flow Over Two-Dimensional Obstacles* 2nd Int. Symp. on Appl. of Laser Anemom. to Fluid Mech. Lisbon July 2-4, 1984.

[5] Durst, F., Founti, M., Wang, A.B. *Experimental Investigation of the Flow through an Axisymmetric Constriction* 6th Symp. of Turb. Shear Flows, Berlin, Heidelberg, 1989, 338-350.

[6] Durst, F. Founti. M., Wang, A.B. *Similarity Phenomena and Computations of the Flow through an Axisymmetric Ring-type Obstacle Attached to a Pipe Wall* Hydrocomp 89, Elsevier Applied Science Publishers 1989, 484-499.

[7] Durst, F., Wang, A.B. *Experimental and Numerical Investigations of the Axisymmetric, Turbulent Pipe Flow over a Wall-Mounted thin Obstacle* 7th Symp. on Turb. Shear Flows, Stanford Univ., Aug. 1989.

[8] Eh, C. *Stromungssichtbarmachung von Wasserstromungen mit Hilfe eines Naßdiffusionsverfahrens* Studienarbeit, LSTM, Univ. Erlangen-Nürnberg, 1987.

[9] Kessler, R., Perić, M., Scheuerer, G. AGARD Conf. Proc. No. 437, P9.1-P9.12, 1988.

[10] König, J., Martinuzzi, R., Tropea, C. *Die Umströmung dreidimensionaler Hindernisse* 7th DGLR-Fach-Symposium: Strömungen mit Ablösung, Aachen, 7-9 Nov. 1990.

[11] König, J. *Anwendung des Kristallviolettverfahrens zur Untersuchung der dreidimensionalen Hindernisströmung* Studienarbeit, Lehrstuhl für Strömungsmechanik, Univ. Erlangen-Nürnberg, 1991.

[12] Larousse, A., Martinuzzi, R., Tropea, C. *Flow Around Surface-Mounted Obstacles* 8th Symp. on Turb. Shear Flows, München, 1991.

[13] Martinuzzi, R., Tropea, C. *Flow Around a Surface-Mounted Cube* 6th Symp. on Appl. of Laser Techniques to Fluid Mech., Lisbon, 1992.

[14] Martinuzzi, R. *Experimentelle Untersuchungen der Umströmung wandgebundener, rechteckiger, prismatischer Hindernisse* Dissertation LSTM, Univ. Erlangen-Nürnberg, 1992.

[15] Martinuzzi, R., Tropea, C. *The Flow Around Surface-Mounted, Prismatic Obstacles Placed in a Fully Developed Channel Flow* Trans. ASME, J. Fluids Engg., accepted for publication.

[16] Obi, S., Perić, M., Scheuerer, G. *Finite-Volume Computation of the Flow Over a Square Rib Using a Second-Moment Turbulence Closure* Int. Symp. on Eng. Turb. Modelling and Meas. Dubrovnik, Sept. 24-28, 1990.

[17] Obi, S. *Berechnung komplexer turbulenter Strömungen mit einem Reynolds-Spannungs-Modell* Dissertation, LSTM, Univ. Erlangen-Nürnberg, 1991.

[18] Obi, S., Perić, M., Scheuerer, G. *Second-Moment Calculation Procedure for Turbulent Flows with Collocated Variable Arrangement* AIAA J. **29** 585-590, 1991.

[19] Stieglmeier, M., Tropea, C., Weiser, N., Nitsche, W. *Experimental Investigation of the Flow Through Axisymmetric Expansions* Trans. ASME, J. Fluids Engg. **111** 464-471, 1987.

[20] Theisinger, J. *Untersuchungen der Dreidimensionalität einer Rippenströmung im Flachkanal mittels LDA* Diplomarbeit, LSTM, Univ. Erlangen-Nürnberg, 1990.

[21] Tropea, C. *Die turbulente Stufenströmung in Flachkanälen und offenen Gerinnen* Dissertation, Univ. Karlruhe, 1982.

[22] Tropea, C., Gackstatter, R. *The Flow Over Two-Dimensional Surface-Mounted Obstacles at Low Reynolds Numbers* J. Fluids Eng. ASME **107** 1985, 489-494.

[23] Volkert, J. *Anwendung des Laser-Lichtschnittverfahrens zur Untersuchung der dreidimensionalen Hindernisströmung* Studienarbeit, LSTM, Univ. Erlangen-Nürnberg, 1991.

[24] Wang, A.B. *Strömungen in Rohren mit Ringförmigem Hindernis* Dissertation, LSTM, Univ. Erlangen-Nürnberg, 1991.

[25] Werner, H., Wengle, H. *Large-eddy Simulation of Turbulent Flow Around a Cube in a Plane Channel* 8th Symp. Turb. Shear Flows, Munich, 1991.

[26] Ziegler, T. *Experimentelle Untersuchung der Strömung über ein zweidimensionales Hindernis* Diplomarbeit, LSTM, Univ. Erlangen-Nürnberg, 1987.

CALCULATION OF SEPARATED FLOWS WITH A TWO-LAYER TURBULENCE MODEL

J. Cordes, W. Rodi, N.H. Cho

Institute for Hydroemchanics
University of Karlsruhe
D-7500 Karlsruhe, F.R. Germany

SUMMARY

A two-layer turbulence model is described which uses the standard k-ε model in the bulk of the flow and resolves the viscosity-affected regions near walls with a one-equation model based on empirical length-scale prescriptions. Applications of the two-layer model to the following separated flows are presented: flow over a backward-facing step, over a T-profile, over a channel bed with periodic dunes and past an airfoil at 14° angle of attack. In all cases with larger separation regions, the two-layer model yielded a significant improvement of the predictions compared with the standard k-ε model using wall functions; in particular the length of the separation region is predicted in better agreement with the experiments.

1. INTRODUCTION

Separated flows occur in many engineering devices, and hence there is a great need for reliable methods to predict such flows. Turbulence processes exercise a considerable influence on the development of separated flows so that adequate turbulence models are essential for a realistic calculation of these flows. Unlike in attached or free shear layers, the distribution of the turbulent length scale is not easy to prescribe empirically in a general way in separated flows, whence simple eddy-viscosity models such as mixing-length-type models or one-equation models based on empirical length-scale prescriptions are not very suitable for flows with larger separation regions. The simplest models with which such flows can be calculated with some success are two-equation models which determine the length-scale distribution from a differential equation, and the most popular of these is the k-ε model. This has become the most widely used model for practical flow calculations, but when applied to separated flows its success was often only moderate. One reason for the limited success is the fact that most calculations carried out so far with the k-ε model were done by bridging the viscosity-affected near-wall layers by wall functions. This approach was used mainly because the numerical resolution of the thin near-wall layers with steep gradients was beyond the available computing resources but also because the standard k-ε model is not applicable in this viscosity-affected region. Instead of resolving this region, the first grid point away from the wall was placed outside the viscous sublayer. Basically, wall functions relate the velocity as well as the turbulence parameters k and ε at the first grid point to the friction velocity and lean heavily on the assumption of a logarithmic velocity distribution and the validity of local equilibrium of turbulence (production = dissipation) at this point. These assumptions are generally not valid for separated flows, as was shown in [1]. Wall functions involving the friction velocity are of course particularly unsuited for separation and reattachment regions where the friction velocity changes sign.

Recent increases in computing power have opened up the possibility of resolving the viscosity-affected near-wall region, and various low-Reynolds-number models for simulating the turbulent processes in this region were developed. A variety of low-Re versions of the k-ε model involving different near-wall functions and additional terms have been proposed, and the pre-1984 models have been reviewed by Patel et al. [2]. More recent proposals are compiled in [3]. All versions have been tested mainly in boundary-layer flows and were found to require rather high numerical resolution near the wall, mainly because of the steep gradient of the dissipation rate ε, whose distribution is determined by solving a transport equation for this quantity. Typically, 60 to 80 grid points across the boundary layers are required for proper numerical resolution. In spite of the recent advances in computing power, this means that in flows involving several walls, the limits of available computer resources are reached quickly. Further, k-ε models have been found to perform rather poorly in decelerating boundary layers approaching separation [4]. Also, the low-Re k-ε models were all developed for attached boundary layers, and initial applications to separated flows indicated that the near-wall damping functions are not always well behaved in these flows.

In order to save grid points and hence computer storage and time and to increase the robustness of the method, a recent trend has been to use the k-ε model only away from the wall and to resolve the viscosity-affected near-wall layer with a simpler model involving an empirical length-scale prescription. Iacovides and Launder [5] employed as simpler model the van Driest version of the mixing-length model while other authors [6, 7] used the one-equation model due to Wolfshtein [8]. The purpose of the work reported here was to develop and test a two-layer model suitable for separated flows. This paper describes briefly the model and presents applications to a number of separated flows. A more detailed account is given in Cordes [9].

2. TWO-LAYER TURBULENCE MODEL

The two-layer model employs the standard high-Reynolds-number k-ε model in the bulk of the flow not too close to walls. This model is based on the eddy-viscosity concept and calculates the distribution of the eddy viscosity from

$$\nu_t = c_\mu \frac{k^2}{\varepsilon}. \tag{1}$$

The distributions of the turbulence parameters k (turbulent kinetic energy) and ε (dissipation rate) are calculated from the following model equations governing these quantities:

$$U_j \frac{\partial k}{\partial x_j} = \frac{\partial}{\partial x_j} \left[(\nu + \frac{\nu_t}{\sigma_k}) \frac{\partial k}{\partial x_j} \right] + \underbrace{\nu_t \left(\frac{\partial U_i}{\partial x_j} + \frac{\partial U_j}{\partial x_i} \right) \frac{\partial U_i}{\partial x_j}}_{P_k} - \varepsilon \tag{2}$$

$$U_j \frac{\partial \varepsilon}{\partial x_j} = \frac{\partial}{\partial x_j} \left(\frac{\nu_t}{\sigma_\varepsilon} \frac{\partial \varepsilon}{\partial x_j} \right) + c_{\varepsilon 1} \frac{\varepsilon}{k} P_k - c_{\varepsilon 2} \frac{\varepsilon^2}{k}. \tag{3}$$

In the k-equation, the molecular viscosity ν is negligible compared with ν_t/σ_k in regions where the standard k-ε model is applied. The standard values are used for the constants appearing in the model ($c_\mu = 0.09$, $c_{\varepsilon 1} = 1.44$, $c_{\varepsilon 2} = 1.92$, $\sigma_k = 1.0$, $\sigma_\varepsilon = 1.3$).

The viscosity-affected regions near walls are resolved with a one-equation turbulence model due to Norris and Reynolds [10]. This was chosen because it proved to be successful in a variety of boundary layer calculations, in particular for boundary layers with adverse pressure gradient [4]. In the one-equation model, the eddy viscosity is calculated from

$$\nu_t = c_\mu k^{1/2} \ell_\mu. \tag{4}$$

k is again determined from the k-equation (2), but the dissipation rate ε appearing in this equation is not determined from a separate transport equation but from the following formula:

$$\varepsilon = \frac{k^{3/2}}{\ell_\varepsilon}. \tag{5}$$

ℓ_μ and ℓ_ε are length scales which are prescribed empirically in the one-equation model. When the coefficient c_μ in (4) is chosen as the square of the structure parameter uv/k under local equilibrium conditions in shear layers, the length scales ℓ_μ and ℓ_ε are the same in the log-law region, where they vary linearly ($\ell_\mu = \ell_\varepsilon = \kappa c_\mu^{-3/4} y$, κ = von Karman constant). Very close to the wall, deviations from the linear distribution occur, and these are different for ℓ_μ and ℓ_ε. The observed damping of the eddy viscosity very near walls is effected by a reduction of ℓ_μ with the aid of an exponential function similar to the van Driest damping function used in the mixing-length model. In particular, the following ℓ_μ-distribution is used in the Norris-Reynolds model in regions near walls:

$$\ell_\mu = 2.495\,(1 - \exp(-0.0198\,Re_y)) \quad , \quad Re_y = \frac{k^{1/2}y}{\nu} \, . \tag{6}$$

In contrast to the original van Driest damping function, the argument Re_y does not involve the friction velocity U_τ and hence the function (6) is also applicable to separated flow. For the length scale ℓ_ε, Norris and Reynolds chose the following distribution:

$$\ell_\varepsilon = \frac{2.495y}{1 + 5.3/Re_y} \tag{7}$$

involving the same argument Re_y. This function reduces the ℓ_ε-distribution below the linear one very close to the wall which is in contrast to recent observations from direct numerical simulations (DNS) which indicate an increase above the linear distribution near the wall [11]. Hence, in this region, the ℓ_ε-distribution (7) leads to an ε-distribution which is in disagreement with that resulting from the DNS data, but this is of little consequence for the prediction of flow quantities of engineering interest. There is, of course, room for further development.

In two-layer modelling, the two model components have to be matched at some location, and this should be placed near the edge of the viscous sublayer. Various matching criteria have been tested [9], and the one chosen is to match the models at a location where the ratio of eddy viscosity to molecular viscosity has a certain (relatively high) value. ν_t/ν in the range of 12 to 48 have been tested [9] and the calculation results were found to be rather insensitive to the exact matching criterion when the ratio was beyond 30. The matching criterion used here appears to be more general than matching at a preselected grid line, a practice adopted by previous authors [5 - 7].

3. APPLICATION TO SEPARATED FLOWS

The new two-layer model was tested first by application to developed channel flow. The results are compared in [9] with experiments and direct simulation data. In general, there is good agreement, except that the peak of k near the wall ($y^+ \approx 15$) is underpredicted and that the distribution of ε very near the wall is not in accord with the DNS data due to the particular ℓ_ε-function used, as was mentioned already. Similar observations can be deduced from independent calculations of boundary-layer flow in zero pressure gradient [12]. The one-equation model must be blamed for the discrepancies mentioned, and they can be remedied with an improved one-equation model.

In the following, four examples of model applications to separated flows are presented, namely to the flow over a backward-facing step, over a T-profile, over periodic dunes and over an airfoil. In each case, the calculations are compared with results obtained with the standard k-ε model and with experiments. The first two flow examples have rectangular geometry and the governing flow equations were solved with a computer program based on the TEACH code [13] employing rectangular numerical grids. However, the hybrid central/upwind differencing scheme in the original TEACH code was replaced by the third-order accurate QUICK scheme for solving the momentum equations. The last two flow examples were calculated with a finite-volume method developed by Majumdar [14] which allows the use of boundary-fitted, curvilinear, non-orthogonal grids. Also in this code, the higher-order QUICK scheme is incorporated as well as the hybrid linear parabolic approximation (HLPA) method of Zhu [15], and for the example of the flow past an airfoil the influence of the discretization scheme on the results will be demonstrated.

29

Flow over backward-facing step. The flow over a step forming a channel expansion is probably the separated flow studied most extensively in experiments and is hence the prime flow example for testing the suitability of turbulence models for separated flows. The two-layer model was therefore applied first to this flow, and a situation was chosen which was studied experimentally in detail by Driver and Seegmiller [16] and which was a test case at the 1980/81 Stanford Conference on Turbulent Flows and serves again as test case for the current Collaborate Testing of Turbulence Models effort. The flow geometry is given in Fig. 1a. Fig. 1b displays the distribution of the friction coefficient along the bottom wall, and the zero crossing marks the reattachment point. As is well known, the standard k-ε model underpredicts the separation length by about 20% and it also produces too small negative velocities and hence c_f-values in the reverse-flow region. The two-layer model brings the reattachment point close to the observed one, and the negative friction coefficient in the reverse-flow region is now considerably larger and is in fact somewhat overpredicted. The two-layer model calculation

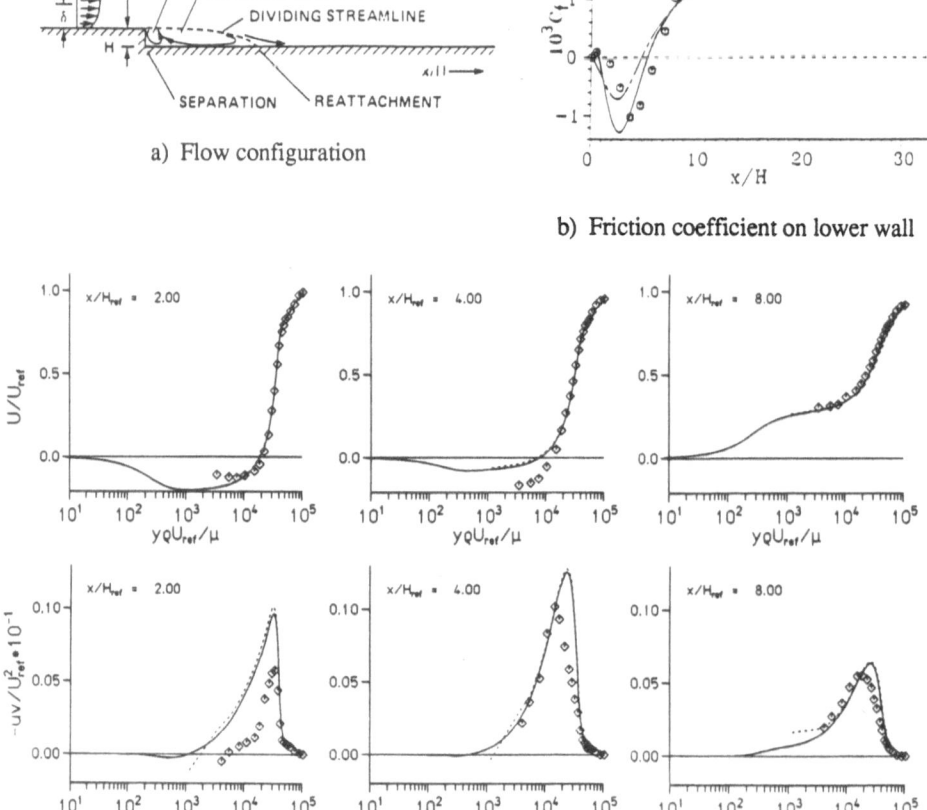

a) Flow configuration

b) Friction coefficient on lower wall

c) Velocity (U) and shear stress ($\overline{uv}$) distribution

Fig. 1: Flow over backward-facing step, calculations with QUICK scheme:
—— two-layer model, —·— ---- k-ε model; experiments [16]: o, ◊

30

also produced a small second-corner eddy which is absent in the calculation with the standard k-ε model [9]. Fig. 1c exhibits the velocity and shear-stress profiles at three downstream locations. There is relatively little difference between the two-equation and the k-ε model predictions. The velocity in the reverse-flow region is first over- and later underpredicted somewhat, but in the outer region and also in the redevelopment region there is good agreement. The shear stress $\overline{uv}$ is overpredicted by 20 to 40% in the shear layer bounding the separation region, and this can be remedied to a major extent by introducing a curvature correction to the coefficient c_μ in (1) proposed by Leschziner and Rodi [17]. This also brings the reattachment length even closer to the measured one. Cordes [9] presents further results and comparisons with experiments; for example the velocity profile plotted in the reverse-flow region in wall coordinates shows clearly that the calculations obtained with the two-layer model are far from the standard logarithmic distribution in this region.

Flow over T-configuration. The second example concerns the flow over a T-configuration (see Fig. 2a) which was studied experimentally by Jaroch and Fernholz [18]. Although these authors point out the basically three-dimensional nature of the flow in their experiment, two-dimensional calculations were carried out for the flow in the symmetry plane. Fig. 2b shows the calculated streamlines in the separation region and also gives the reattachment points; it demonstrates clearly that in this case the two-layer model predicts the reattachment length in much better agreement with experiments than does the standard k-ε model employing wall functions. On the other hand, the velocity profiles given in Fig. 2c indicate that the separation zone is predicted too thin by both models. This then leads to a shift in the shear-stress distribution (also Fig. 2c) towards the wall. When the calculated shear-stress profiles were compared with the measurements [9] the shear-stress level was underpredicted considerably. In the meantime it became apparent that the measured high shear-stress values were due to vibrations of the hot-wire probe [10] and hence no data are included here. Overall it can be

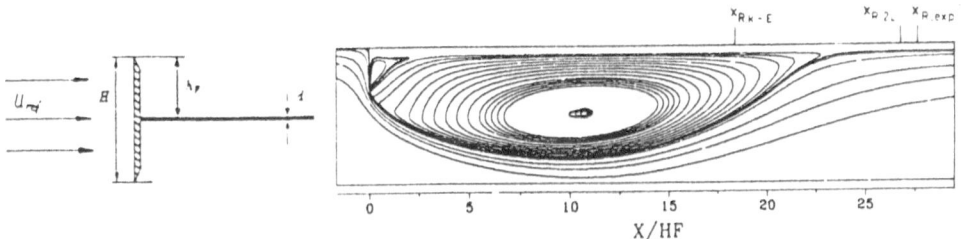

a) Flow configuration   b) Calculated streamlines in separation region

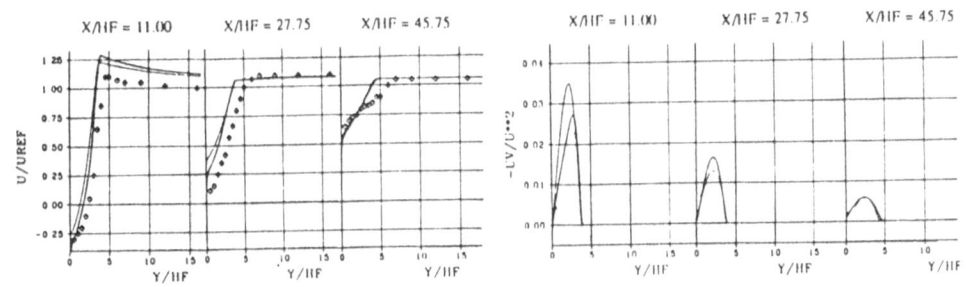

c) Velocity (U) and shear stress ($\overline{uv}$) profiles

Fig. 2: Flow over T-configuration; calculations by Bührle [25] with QUICK scheme:
—— 2-layer model, - - - k-ε model; ◊ experiments [18]

31

observed that, for practical purposes, the main features of the flow are reasonably well predicted by the two-layer model.

Flow over periodic dunes. The next example has non-rectangular geometry, and the separation point is not fixed a priori but is determined by the flow calculation. The flow considered is that in an open channel over a bed with periodic, fixed dunes which was investigated at the Delft Hydraulics Laboratory [20] with an LDA and served as test case for the 13th meeting of the IAHR Working Group on Refined Modelling of Flows. In the experiment, the channel bed

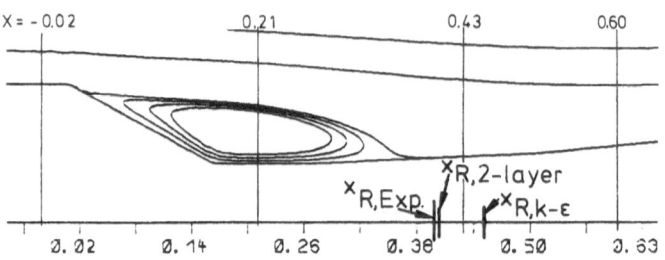

a) Calculated streamlines in separation region

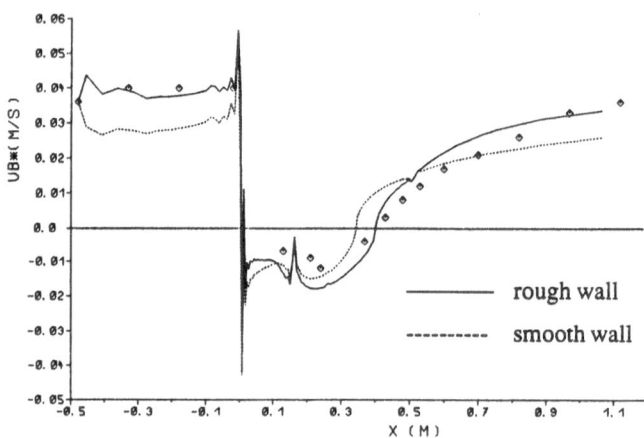

b) Friction velocity along dune wall, lines are calculations, ◊ experiments [20]

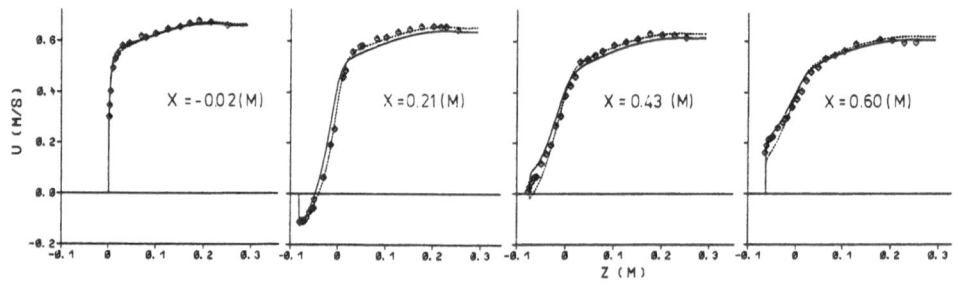

c) Velocity profiles; calculations: —— two-layer model, - - - k-ε model; experiments [20] ◊

Fig. 3: Flow over periodic dunes; calculations are with QUICK scheme

consisted of 33 dunes of equal geometries so that periodic flow developed and only one dune section needed to be covered in the calculation. The length of such a section was 1.6 m and the average water depth was 0.334 m. Fig. 3a shows the dune geometry in the vicinity of the crest and trough. In the calculations, the water surface was treated as symmetry plane. The concrete bed in the experiment was roughened with glued-on sand, and hence roughness effects needed to be accounted for in the calculations. Scheuerer [21] developed a method for this in the context of a low-Reynolds-number k-ε model. He found that the roughness effects could be simulated satisfactorily by adding a production term near the wall to the k-equation, introducing extra turbulence due to the roughness elements. This model was incorporated into the one-equation model near the wall [9]. Fig. 3b shows the influence of this roughness correction on the distribution of the friction coefficient along the dune bed. Calculations obtained with the assumption of a smooth bed and experimental data are included for comparison. Clearly, upstream of the dune crest the smooth-bed calculations yield too low a friction coefficient. At the separation point just downstream of the crest, the friction coefficient drops sharply to negative values in the separation zone. Further downstream c_f becomes positive again, and the zero crossing, indicating the reattachment point, as well as the redevelopment further downstream are well predicted. Also in this region, the predictions with the roughness model switched on are better than those obtained by assuming the wall to be smooth. Fig. 3a exhibits the streamlines in the vicinity of the separation region. It indicates that separation occurs shortly after the crest, as was observed in the experiments. Fig. 3a also shows the reattachment points predicted both with the two-layer model and the standard k-ε model in comparison with the experimental location. In this case, the k-ε model yields too long a separation region, and the two-layer model is again in much better agreement with the experiment. Finally, the velocity profiles are displayed at various sections in Fig. 3c which shows that their development is well predicted. Because of the somewhat excessive separation length produced by the k-ε model this also yields a belated redevelopment of the profile. Cordes [9] compares calculated uv- and k-profiles with the measurements and finds that the shapes are in good agreement but the peak values in the separated shear layer are too low by up to 40%. The reasons for this underprediction are not known.

Flow over NACA-4412 airfoil. As last example, calculations are presented for the flow over a NACA-4412 airfoil at an angle of attack of 13.87° and Re = cU_0/v = 1.5 x 10[6], where c is the chord length. This flow was investigated by Coles and Wadcock [22] with a flying hot-wire. In the experiments, transition was forced by transition strips placed at x/c = 0.025 and x/c = 0.103 on the upper and lower surface, respectively. The calculations were carried out with a body-fitted C-grid, and the original computer code due to Majumdar [14] had to be modified in order to account for the branch cut present in this case [9]. The two-layer model calculations were carried out with a 146 x 83 grid and the k-ε model calculations with a 146 x 37 grid. In the latter calculations, the turbulence model was switched on everywhere so that the initial laminar boundary layers were not accounted for. With the two-layer model, various approaches were tried including a strong damping of the length-scale ℓ_μ upstream of the transition strips so that the boundary layer remained laminar in this region. However, the various measures had little effect on the calculation results. The calculation domain extended 2.2c upstream of the leading edge and 3c downstream of the trailing edge. The side boundaries were walls roughly 1.7c above and below the airfoil at the locations of the wind tunnel walls. Wall functions were used here. Because of the relatively short working section of the wind tunnel causing a Venturi effect, the free-stream was not entirely parallel to the approach velocity [23]. In his calculations, Rhie [24] took this into account and optimised his calculations by matching the free-stream velocity to the measured one. This was not attempted in the present calculations.

Fig. 4c shows that the discretization scheme used for the convective terms in the momentum equations has considerable influence on the results (The hybrid central/upwind scheme was always used for the k- and ε-equations). When a low-order scheme (hybrid central/upwind scheme) is used, the flow separates far too early (at x/s ≈ 0.4), leading to unrealisticly wide velocity profiles, and there is little difference between k-ε and two-layer model predictions [9].

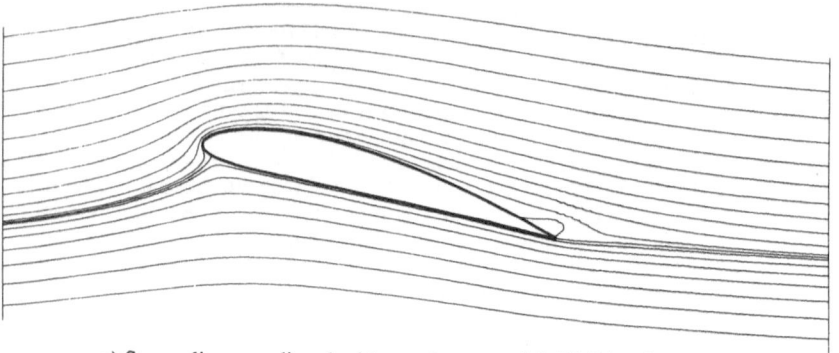

a) Streamlines predicted with two-layer model, HLPA scheme

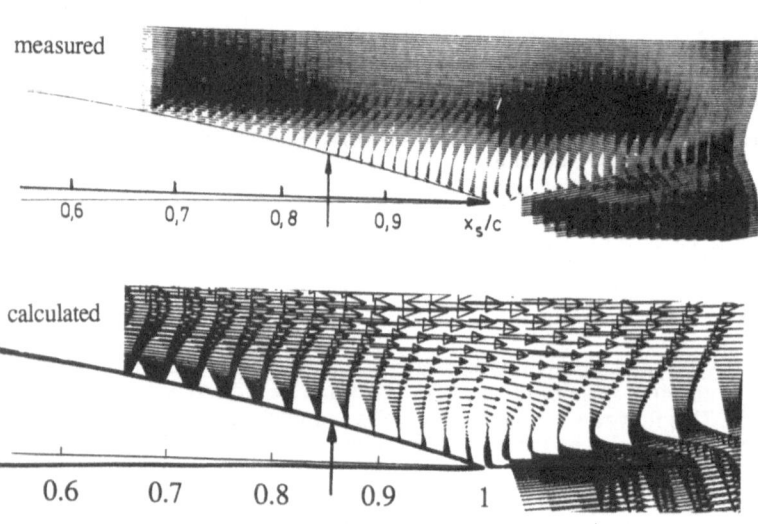

b) Velocity vectors near trailing edge

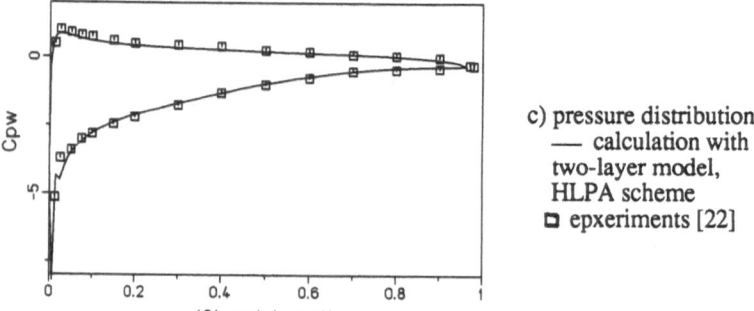

c) pressure distribution,
— calculation with
two-layer model,
HLPA scheme
□ epxeriments [22]

Fig. 4: Flow past NACA 4412 airfoil

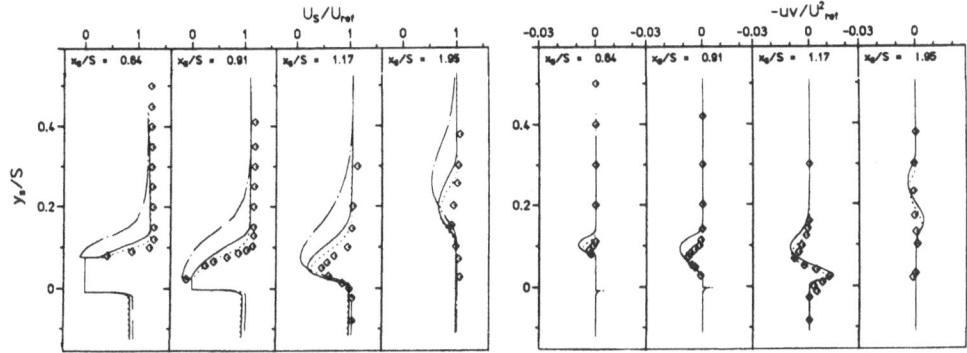

d) Velocity (U_s) and shear stress ($\overline{uv}$) distribution; calculations:
— two-layer model with HLPA, - - - k-ε model with QUICK,
— . — k-ε model with hybrid central/upwind, ◊ epxeriments [22]

Fig. 4 continued

With the more accurate QUICK or HLPA schemes, a much smaller separation region is predicted (see Fig. 4a). k-ε model calculations with the most accurate QUICK scheme yield velocity profiles in good agreement with the measurements (Fig. 4c). Properly converged two-layer model calculations could not be obtained with this unbounded scheme. Fig. 4d includes two-layer model calculations obtained with the HLPA scheme which is bounded but a little more diffusive than the QUICK scheme so that the results are in slightly worse agreement with the measurements. Fig. 4d also includes a comparison of predicted and measured shear-stress distributions showing that both model versions yield fairly realistic results. The streamlines calculated with the two-layer model are displayed in Fig. 4a, showing a small separation zone on the upper side near the trailing edge. Calculated and measured velocity vectors in the trailing-edge region are compared in Fig. 4b and show good agreement. An arrow indicates the separation point which in both calculations and experiments is around x/c ≈ 0.85. The pressure distributions on the upper and lower surface of the airfoil predicted with the two-layer model are compared with measurements in Fig. 4c. The agreement is very good. As a consequence, the lift coefficient C_L is also predicted in good agreement with the measurments: the two-layer model yields C_L = 1.62 and the k-ε model C_L = 1.69 while the experimental value is C_L = 1.67. The drag coefficient (including the friction part) calculated with both models is C_D ≈ 0.065 which compares with C_D = 0.072 quoted in [26] for the measurements.

4. CONCLUSIONS

A two-layer turbulence model was developed which allows the calculation of separated flows without the use of wall functions; instead the near-wall region is resolved with a one-equation model. Applications to a variety of separated flows have demonstrated that, for situations with larger separation zones, the results obtained with the two-layer model are clearly improved over those obtained with the k-ε model using wall functions; in particular the length of the separation region is predicted with higher fidelity. For the airfoil flow considered, which has only a small separation zone, superiority of the two-layer model could not be established. The model should be tested on further airfoil flows with larger separation regions and more clearly defined free-stream conditions. Test calculations should also be extended to three-dimensional separated flows, where the use of wall functions is even more questionable. With the computing power now available, such calculations with 10 to 15 grid points in the viscosity-affected near-wall region should be possible so that the model is a promising tool for solving complex practical flow problems.

REFERENCES

[1]   Rodi, W.: "Experience with two-layer turbulence models combining the k-ε model with a one-equation model near the wall", AIAA Paper 91-0216 (1991).

[2]   Patel, V.C., Rodi, W., Scheuerer, G.: "Turbulence models for near-wall and low-Reynolds number flows: A review", AIAA J., Vol. 23, pp. 1308-1319 (1985).

[3]   Shih, T.H., Mansour, N.N.: "Modelling of near-wall turbulence", in Eng. Turbulence Modelling and Experiments, eds. W. Rodi, E.N. Ganic´, Elsevier, pp. 13-22 (1990).

[4]   Rodi, W., Scheuerer, G.: "Scrutinizing the k-ε model under adverse pressure gradient conditions", J. Fluids Eng., 108, pp. 174-179 (1986).

[5]   Iacovides, H., Launder, B.E.: "The numerical simulation of flow and heat transfer in tubes in orthogonal-mode rotation", Proc. 6th Symp. on Turbulent Shear Flows, Toulouse (1987).

[6]   Chen, H.C., Patel, V.C.: "Near-wall turbulence models for complex flows including separation", AIAA J., Vol. 26, pp. 641-648 (1988).

[7]   Yap, C.L.: "Turbulent heat and momentum transfer in recirculating and impinging flows", Ph.D. Thesis, UMIST, Manchester (1987).

[8]   Wolfshtein, M.: "The velocity and temperature distribution in one-dimensional flow with turbulence augmentation and pressure gradient", Int. J. Heat Mass Transfer, 12, pp. 301-318 (1969).

[9]   Cordes, J.: "Entwicklung eines Zweischichten-Turbulenzmodells und seine Anwendung auf abgelöste, zweidimensionale Strömungen", Dissertation Universität Karlsruhe (1992).

[10]   Norris, L.H., Reynolds, W.C.: "Turbulent channel flow with a moving wavy boundary", Rept. No. FM-10, Stanford University, Dept. Mech. Eng. (1975).

[11]   Rodi, W., Mansour, N.N.: "One-equation near-wall turbulence modelling with the aid of direct simulation data", Proc. 1990 CTR Summer Program, Stanford University (1990).

[12]   Fujisawa, N., Rodi, W., Schönung, B.: "Calculation of transitional boundary layers with a two-layer model of turbulence", Proc. 3rd Int. Symp. on Transport Phenomena in Dynamics of Rotating Machinery, Honolulu (1990).

[13]   Gosman, A.D., Pun, W.M.: "Calculation of recirculating flows", Lecture Notes for Course entitled "TEACH", Imperial College, London (1973).

[14]   Majumdar, S.: "Development of a finite-volume procedure for prediction of fluid flow problems with complex irregular boundaries", Rept. SFB 210/T/29, Universität Karlsruhe (1986).

[15]   Zhu, J.: "A low-diffusive and oscillation-free convection scheme", Commun. in Appl. Num. Meth., 7, pp. 225-232 (1991).

[16]   Driver, D.M., Seegmiller, H.J.: "Features of reattaching turbulent shear layer in divergent channel flow", AIAA J., 23, pp. 163-171 (1985).

[17]   Leschziner, M.A., Rodi, W.: "Calculation of annular and twin parallel jets using various discretization schemes and turbulence model variations", J. Fluids Eng., 103, pp. 352-360 (1981).

[18]   Jaroch, M., Fernholz, H.-H.: "The three-dimensional character of a nominally two-dimensional separated turbulent shear flow", J. Fluid Mech., 205, pp. 523-552 (1989).

[19]   Fernholz, H.-H., private communcation (1992)

[20]   IAHR Working Group on Refined Modelling of Flows, Test Problem for 13th meeting, Delft Hydraulics (1986).

[21]   Scheuerer, G.: "Entwicklung eines Verfahrens zur Berechnung zweidimensionaler Grenzschichten an Gasturbinenschaufeln", Dissertation Universität Karlsruhe (1983).

[22]   Coles, D., Wadcock, A.J.: "Flying-hot-wire study of flow past an NACA 4412 airfoil at maximum lift", AIAA J., 17, pp. 321-329 (1979).[23]   Wadcock, A.J.: "Investigation of low-speed turbulent separated flow around airfoils", NASA CR 177450 (1987).

[24]   Rhie, C.M.: "A numerical study of the flow past an isolated airfoil with separation", PH.D. Thesis, University of Illinois at Urbana Champaign (1981).

[25]   Bührle, P. "Numerische Berechnung der abgelösten Strömung über eine zur Anströmung senkrechte ebene Platte mit einer zur Anströmung parallelen Trennplatte in der Symmetrieebene", Studienarbeit Institut für Hydromechanik, Universität Karlsruhe (1989).

[26]   Kline, S.J., Cantwell, B.J., Lilley, G.M.: Proc. 1980-81 AFOSR-HTTM-Stanford Conference on Complex Turbulent Flows, Stanford University (1982).

AN EVALUATION OF THE DISCRETE VORTEX METHOD AS A MODEL FOR THE FLOW PAST A FLAT PLATE NORMAL TO THE FLOW WITH A LARGE WAKE SPLITTER PLATE

H.H. Fernholz, M.P.G. Jaroch, J.M.R. Graham

Hermann-Föttinger-Institut für Thermo- und Fluiddynamik,
Technischen Universität Berlin,
Straße des 17. Juni 135, D-1000 Berlin 12

ABSTRACT

A discrete vortex model of the Biot Savart type is proposed for the flow past a flat plate normal to the flow with a long wake splitter plate in its plane of symmetry. A very basic model, incorporating potential vortices, removal of vortices far downstream and near the wall and a two-equation model for the introduction of new vortices into the flow field did not allow calculations beyond the level of determining the drag of the plates.

Incorporating a more elaborate Rankine vortex model and a temporal decrease of the circulation of the individual vortices produced consistent results for the pressure and the velocity field. The qualitative agreement with experiments was quite good since experimental results were used to select empirical parameters, the quantitative agreement was acceptable only for the time mean values. For the Reynolds stresses the quantitative agreement was as bad as was expected from the calculation of other authors.

A simulation of the creation of vorticity of opposite sign at walls showed, that this is only one of the main physical effects which were simulated by temporal decrease of circulation and not the cause for the discrepancy between numerical and wind-tunnel results. The physical effects of three-dimensional vortex-vortex interaction and three-dimensional transport of vorticity cannot be modelled in the simple two-dimensional model used here and are the probable causes for the inability of discrete vortex model to predict turbulent intensities and correlations.

REFERENCES

[1] *Jaroch, M.P.G.; Graham, J.M.R.*: An evaluation of the discrete vortex method as a model for the flow past a flat plate normal to the flow with a long splitter plate. J. Mecanique theor. appl. 7, pp. 105-134, 1988

[2] *Jaroch, M.P.G.*: Oil flow visualization experiments in the separated and re-attachment regions of the flow past a transverse flat plate with a long splitter plate. ZFW 11, S. 230-236, 1987

[3] *Jaroch, M.P.G.*: Eine kritische Betrachtung der Methode disketer Wirbel als Modell für eine abgelöste Strömung mit abgeschlossener Ablöseblase auf der Basis experimenteller Erkenntnisse. Diss. TU Berlin, 1987

SKIN FRICTION MEASUREMENTS IN TWO- AND THREE-DIMENSIONAL HIGHLY TURBULENT FLOWS WITH SEPARATION

H.H. Fernholz, P. Dengel, M. Hess

Hermann-Föttinger-Institut für Thermo- und Fluiddynamik,
Technischen Universität Berlin,
Straße des 17. Juni 135, D-1000 Berlin 12

ABSTRACT

The results reported in [1] present comparative skin-friction measurements obtained by a Preston tube, a surface fence, a single film and a McCroskey hot-film probe, and a wall pulsed wire in two- and three-dimensional highly turbulent flows (near wall turbulence level > 40%) with separation and instantaneous reverse flow. Fluctuating values of skin friction were obtained by the wall pulsed wire and the hot films in addition to the mean values and the direction of the wall stream line. Differences in the results of the measuring techniques are explained and the failure of devices under these severe conditions is displayed.

REFERENCE

[1]   *Dengel, P.; Fernholz, H.H.; Hess, M.*: Skin-frictin measurements in two- and three-dimensional highly turbulent flows with separation, Advances in Turbulence, Proc. of the first European Turbulence Conference, Lyon, France, 1.-4. July 1986, ed.: G. Comte-Bellot, J. Mathieu, Springer-Verlag Heidelberg, pp. 470-479, 1987

Near-wall structure in separated flows

P.M. Wagner & H.H. Fernholz

Hermann-Föttinger-Institut für Thermo- und Fluiddynamik
Technische Universität Berlin, D-1000 Berlin 12, Germany

Abstract

Results obtained from flow-visualisation experiments and skin-friction measurements in the near-wall region of a separation bubble indicate the existence of large coherent structures. The observed patterns can be divided into large low-velocity "pockets" and smaller high-velocity streaks. The existence of these structures can be explained partly by high-velocity material from the free shear layer penetrating to the wall from above. Both the form and the frequency of occurence of these structures differ considerably from those observed in turbulent boundary layers.

1.Introduction

The structure of turbulence in free and wall-bounded shear-flows has been the object of a large research effort over the past decades. It is well known that these structures contribute substantially to momentum, heat, and mass transfer as described e.g. by Kline et al. [1] or Cantwell [2].

Only little information is available so far on the appearance of coherent structures in separated flow regions. The free shear layer separating from a wall due to an adverse pressure gradient or from a sharp edge interacts with the reverse-flow region near the wall. Since the fluctuations are not small compared with the mean flow and the instantaneous and mean velocity gradients are of the same order of magnitude, calculation methods with conventional closure schemes fail to predict the behaviour of separated flows. Velocity profiles scaling with the mean skin friction (log-law profiles) do not exist in the reverse-flow region since there appears to be no balance between energy production and dissipation. A good review of some of the phenomena involved was given by Simpson [3].

This investigation aims at describing the turbulence structure in the reverse-flow region behind a vertical fence mounted in a flat-plate boundary layer. Flow visualisation experiments and measurements were carried out in order to explore the spatial character of the turbulent structures. Since the near-wall phenomena dominate the momentum, heat, and mass transfer, near-wall quantities, such as mean and fluctuating values of skin friction and its higher moments, will be considered first. Wall pulsed-wire probes were used to determine accurate long-time statistics, while near-wall hot-wires provided time-resolved skin-friction data. The time-resolved data were used to obtain spacial correlations and ensemble averages of signal patterns associated with so-called turbulence generated "events". A good review on the conditional sampling and ensemble averaging techniques for turbulent flows is provided e.g. by Antonia [4].

2. Equipment

The experiments were conducted in a wind tunnel and in a water channel.

2.1 Wind tunnel

The wind tunnel was of the open-return type with a digitally controlled fan. The test section is shown in Fig. 1. It is made of a flat plate consisting of several elements of different length which could be easily exchanged, two fixed sidewalls and a removeable roof. The boundary layer developing on the flat plate was tripped 45 mm downstream of the leading edge. The flow separates upstream and downstream of a 17.0 mm high vertical fence mounted normal to the streamwise direction and spanning the width of the test section. The fence had an aspect ratio ($width/height$) of 25.9 and the blockage of the test section ($height\ of\ the\ section/height\ of\ the\ fence$) was 12.5 %. A zero pressure gradient turbulent boundary layer could be generated by removing the fence.

Almost all experiments were carried out in the near-wall region of the large reverse-flow region downstream of the fence. In order to detect coherent structures in the reverse-flow region the flow had to be optimized so that there exists a large area, where the near-wall streamlines are normal to the fence, and that the reattachement line — the line, where the mean skin friction is zero — is symmetric and normal to the streamwise direction. In the first tests large spanwise variations of the skin-friction were observed, which were apparently due to the screens in the settling chamber. Replacing all screens by only one precisely machined perforated metal screen (blockage ratio 64 %) as described by Dengel & Fernholz [5] resulted in a uniform separation line upstream and a nominally two-dimensional over two-thirds of the span.

2.2 Water channel

A small closed-circuit water channel was used. Both sidewalls of the test section were made of plexiglass. The flat plate used for both the turbulent boundary layer and the separation experiments was similiar to the one used in the wind tunnel. Only qualitative results were drawn from flow visualisations in the water channel.

2.3 Flow-visualisation techniques

In the wind tunnel, the flow was visualised by means of both smoke injection and by evaporating paraffin oil from a heated wire. Good results were obtained by using the modified smoke-wire streak technique developed by Choi [6]. The smoke wire was spray-painted to produce marks at regular intervals. The illumination was provided by a 6 W Argon-Laser. A prism was used to split the laser beam into two different colours. The diverging colour beams passed through a second prism and through a two-dimensional lens to form parallel light sheets. The light sheets were adjusted to be parallel with the flat plate and the smoke wire was mounted 1 mm from the wall, thereby visualising both the horizontal and the vertical motion of the smoke particles.

In the water channel, the flow was visualised by hydrogen bubbles generated at a 50 μm diameter platinum wire mounted 1 mm above the wall. Illumination was provided by a 150 W slide projector. Coloured light sheets were produced by fixing coloured plastic foil at the plexiglass sidewalls. A pulse-generator with an adjustable duty cycle was used to produce the hydrogen bubble lines at a rate from 0.01 to 10 Hz.

2.4 Instrumentaion and data acquisition

A 16 bit microcomputer with a 4 $MByte$ contiguous memory and a 105 $MByte$

disk drive was used for the data acquisition. An industrial bus interface provided all neccessary analog and digital input and output connections.

Pulsed-wire measurements were taken with a dual-channel pulsed-wire anemometer. The probes were calibrated against a Preston tube using a third order polynomial on the mean-sqared time-of-flight readings. In the reverse-flow region 10000 samples taken with a sampling rate of 25 Hz, were used to derive the probability density distributions.

The hot-wire sensors were operated using a modified 10 channel AA LAB Systems Model AN-1003 anemometer. The bridge outputs passed through signal-conditioners and low-pass filters and were read from the microcomputer through a 12 Bit A/D converter with 16 synchronous sample-and-hold inputs. The system was able to acquire up to a total of 1.8 $million$ samples with a sampling rate of up to 180 kHz. The data presented here are derived from time records of 50.000 $samples$ per wire per data point using a sampling rate of 8.0 kHz with the lowpass filters set to 3.8 kHz. The velocity probes were calibrated in the free stream against velocities obtained from Prandtl-tube readings. The near-wall hot-wire probes were calibrated in a turbulent boundary layer against the mean skin friction obtained from Preston-tube readings.

The near-wall hot-wire technique was developed since no other technique was available which could provide time-resolved skin-friction measurements in air within the necessary frequency range. The problems arising with surface hot films in various flows are discussed in detail by Alfredsson et al. [7], an analytical discussion of the arising difficulties is provided by Tardu et al. [8], and our own measuring technique including details on probe fabrication and calibration was described by Wagner [9].

Since the hot wires were operated at an overheat ratio of 1.6 in an air-conditioned laboratory and since all instruments were allowed to heat up for at least two hours before measurements were taken, no compensation for the temperature drift was needed. After each series of measurements the calibration was checked and the data were discarded if the deviation was larger than 1.5 %.

All pulsed-wire and hot-wire probes were manufactured in house. Details on their fabrication and error estimates are described by Dahm & Vagt [10] and Dengel et al. [11].

3. Results

3.1 Flow visualisation

A major part of this study was devoted to flow visualisation of the reattaching flow field in order to obtain at least qualitative information on the characteristics of the turbulent structure. Visualisations with light sheets parallel to the wall (Fig. 2) exhibit comparatively large lumps of unmarked fluid displacing the visualised particles thereby revealing the black colour of the wall underneath. These spotty "black holes" are lumps of energetic turbulent fluid which were transported from the free shear layer almost vertically down into the near-wall region where they tend to spread in all directions. Between these dark areas the near-wall fluid moving in the reverse-flow direction is forming streak-like patterns with high shear. The fluid in the streaky regions is lifted away from the wall. Visualisations with light sheets normal to the wall (Fig. 3) show the upward motion at both sides of a "black hole". It was found, that these "lifted backflow" events occur randomly over a large part of the reverse flow region $(0.3 < x/x_R < 1.0)$ and not only for $x/x_R > 0.7$ as reported by Pronchik [12] who visualised the flow behind a backward facing step. Video sequences taken from the hydrogen bubble visualisations show, that upstream velocities occur only at the upstream front of the stagnation areas. The high-shear streaks are wider and shorter and break up much faster when compared with the slim patterns observed in a zero pressure gradient boundary layer.

3.2 Measurements

Fig. 4 and Fig. 5 show the development of the pressure coefficient $\overline{c_p}$, the skin friction coefficient $\overline{c_f}$, the skin friction fluctuation coefficient c'_f, the reverse flow factor χ_w, and the higher moments of the wall shear stress along the symmetry line downstream of the vertical fence obtained from wall-pulsed wire measurements plotted against the dimensionless distance x/x_R where x_R is the length of the reverse-flow region. The data are qualitatively very similiar to measurements in a bluff-plate/splitter-plate configuration reported e.g. by Ruderich & Fernholz [13] and Jaroch & Fernholz [14]. The higher moments show a strong non-Gaussian behaviour up to at least $x/x_R = 1.6$. The flatness F_w reaches a maximum of at least 10 indicating a "spotty" behaviour of wall shear stress. This is consistent with the dual nature observed in the visualization experiments.

In the region $0.35 < x/x_R < 0.65$ the reverse flow factor χ_w reaches values above 0.95. In this region time records of the wall-shear stress were obtained from near-wall hot-wire measurements. Since hot-wire probes cannot discriminate between positive and negative velocities, these time-records are not completely correct. A comparision of wall shear stress histograms obtained from pulsed-wire and hot-wire measurements (Fig. 6) reveals, that the errors are restricted, however to samples with relatively low positive velocities. Therefore all samples with values lower than the negative of the maximum can be used with confidence (i.e. they are within the error range for hot-wire measurements). Fig. 7 shows time records of the wall shear stress in the reverse-flow region ($Re_{h_F} = 14000$) and in a zero pressure gradient boundary layer with the same mean skin friction. The extreme intensity of the "peaks" and the long duration of the "valleys" indicate the "dual behaviour" of the wall shear stress in the reverse-flow region. While the signals in the boundary layer are not correlated, the signals in the reverse-flow region exhibit a high correlation at least for low skin-friction values. Spanwise measurements (Fig. 8) reveal that the crosscovariance does not vanish for distances of several hundred viscous units.

Ensemble averages obtained with conditional sampling techniques indicate that the high shear events in the reverse-flow region contain much more energy than those in the boundary layer.

4. Conclusions

Flow visualisation experiments and measurements of the instantaneous skin friction using pulsed-wire probes and arrays of near-wall hot-wire probes in the separation region downstream of a fence mounted in a turbulent boundary layer reveal the existence of large coherent structures in the reverse-flow region. Conditional averages indicate that the observed patterns can be divided into large low-velocity "pockets" and smaller high-velocity "streaks". It is suggested that the low-velocity patterns originate from ejections of the free shear layer. The downward moving fluid is strongly decelerated but occasionally maintains its positive velocity thereby preventing the forward flow fraction from reaching zero. The fluid in the near-wall region is displaced and forms streaky upward moving patterns with strong negative velocities.

Acknowledgements

The present study was undertaken with financial support from the German Research Society (DFG), which we gratefully acknowledge. The authors would like to thank A. Dahm and T. Lange, who manufactured the probes, and P. Dengel who had always time for fruitful discussions.

References

1. S.J. Kline, W.C. Reynolds, F.A. Schraub & P.W. Runstadler (1967), The structure of turbulent boundary layers. *J. Fluid Mech.* **30,** *741*

2. B.J. Cantwell (1981), Organized Motion in turbulent flow. *Ann. Rev. Fluid Mech.* **13,** *457*

3. R.L. Simpson (1981), A review of some phenomena in turbulent flow separation. *J. Fluid Eng.* **103,** *520*

4. R.A. Antonia (1991), Conditional sampling in turbulence measurement. *Ann. Rev. Fluid Mech.* **13,** *131*

5. P. Dengel & H.H. Fernholz (1989), Generation of and measurements in a turbulent boundary layer with zero skin friction. In *Advances in Turbulence 2* (ed. H.H Fernholz & H. Fiedler),*432.*

6. K.S. Choi (1985), Near-wall structure of a turbulent boundary layer with riblets. *J. Fluid Mech.* **208,** *417*

7. P.H. Alfredsson, A.V. Johansson, A.V. Haritonidis & H. Eckelmann (1987), The flucutating wall-shear stress and the velocity field in the viscous sublayer. *Phys. Fluids* **31,** *1026*

8. S. Tardu, C.T. Pham & G. Binder (1991), Effects of longitudinal diffusion in the fluid and of heat conduction to the substrate on the response of wall hot-film gages. In *Advances in turbulence 3* (ed. A.V. Johansson & P.H. Alfredsson)

9. P.M. Wagner (1991), The use of near-wall hot-wire probes for time-resolved skin-friction measurements. In *Advances in turbulence 3* (ed. A.V. Johansson & P.H. Alfredsson)

10. A. Dahm & J.D. Vagt (1981), Entwicklung und Herstellung interferenzarmer Hitzdrahtsonden. *HFI Institutsbericht IB-01/77,* TU Berlin.

11. P. Dengel, H.H. Fernholz & M. Hess (1987), Skin-friction measurements in two- and three-dimensional highly turbulent flows with separation. In *Advances in turbulence 1* (ed. G. Compte-Bellot & J. Mathieu)

12. S.W. Pronchik (1983), An experimental investigation of the structure of a turbulent reattaching flow behind a backward-facing step. *Ph.D. Thesis, Stanford University.*

13. R. Ruderich & H.H. Fernholz (1986), An experimental investigation of a turbulent shear flow with separation, reverse flow and reattachment. *J. Fluid Mech.* **163,** *283*

14. M.P. Jaroch & H.H. Fernholz (19899), The three-dimensional character of a nominally two-dimensional separated turbulent shear flow. *J. Fluid Mech.* **205,** *523*

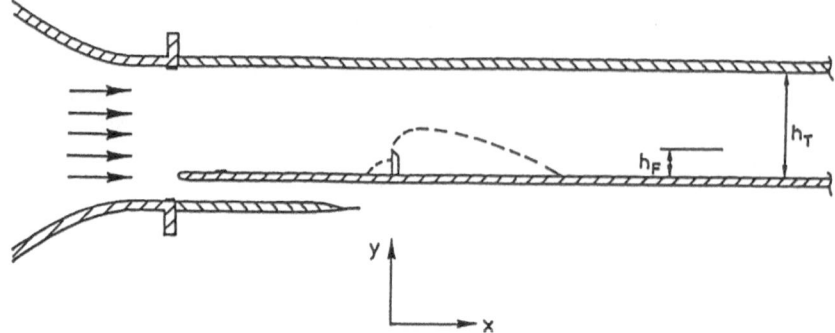

1. Test section used for the experiments in the separation bubble and in a zero pressure gradient boundary layer (fence removed).

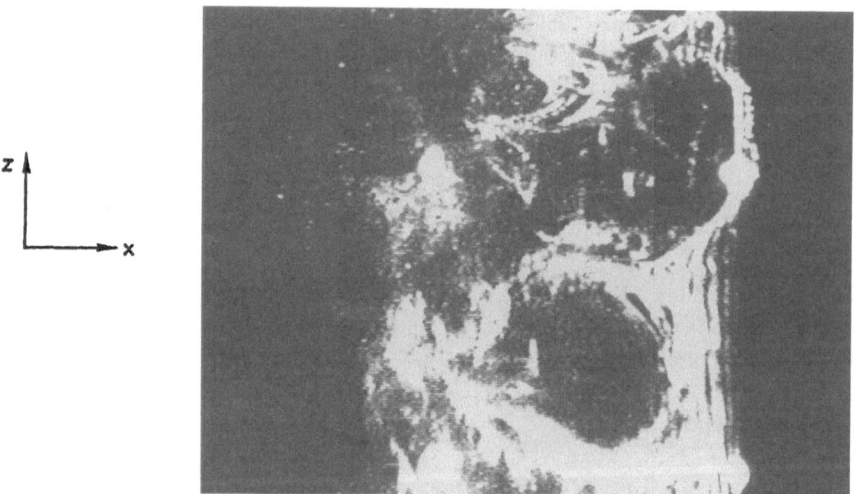

2. Hydrogen bubble visualisation of the near-wall region behind the fence illuminated by a light sheet parallel to the wall.

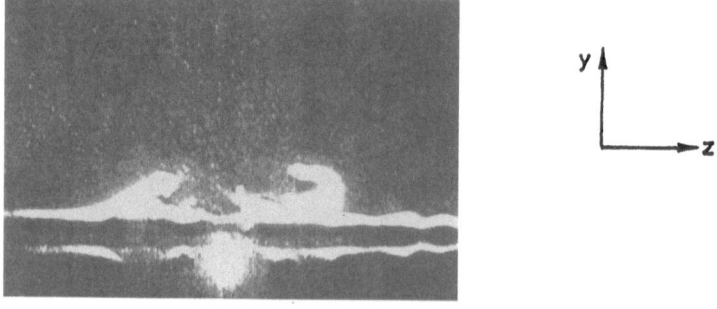

3. Hydrogen bubble visualisation of the near-wall region behind the fence illuminated by a vertical light sheet seen from downstream.

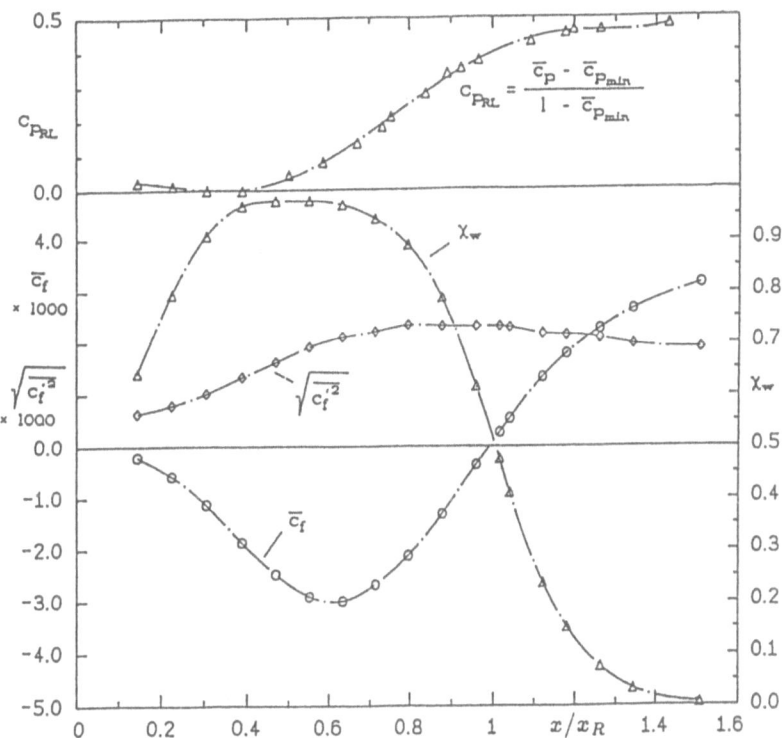

4. Development of the pressure coefficient $\overline{c_p}$, the skin friction coefficient $\overline{c_f}$, the skin friction fluctuation coefficient c'_f, and the reverse flow factor at the wall χ_w along the symmetry line downstream of the fence at $Re_{h_F} = 14000$.

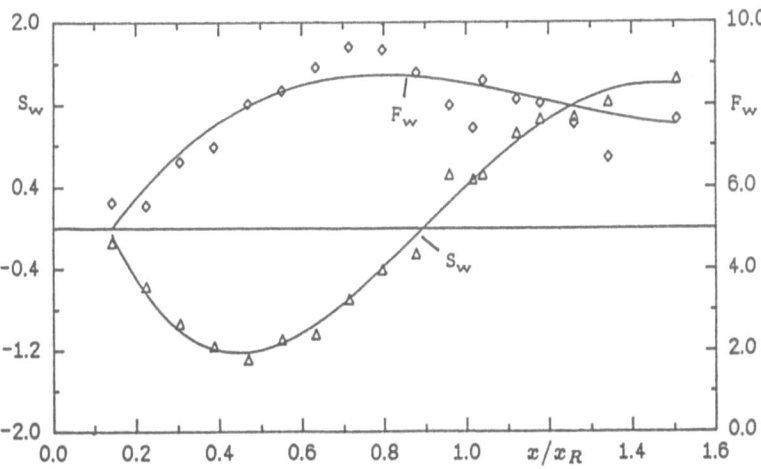

5. Development of the skewness S_w and the flatness F_w of the wall shear stress along the symmetry line downstream of the fence at $Re_{h_F} = 14000$.

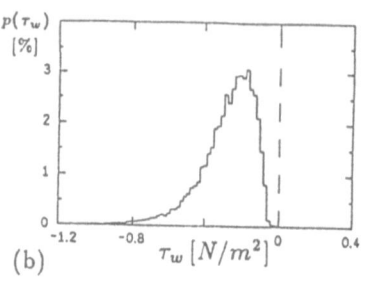 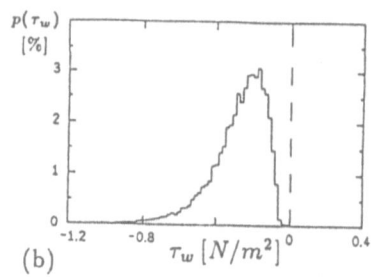

6. Comparision of wall shear stress histograms obtained from (a) wall pulsed wire and (b) near-wall hot-wire measurements.

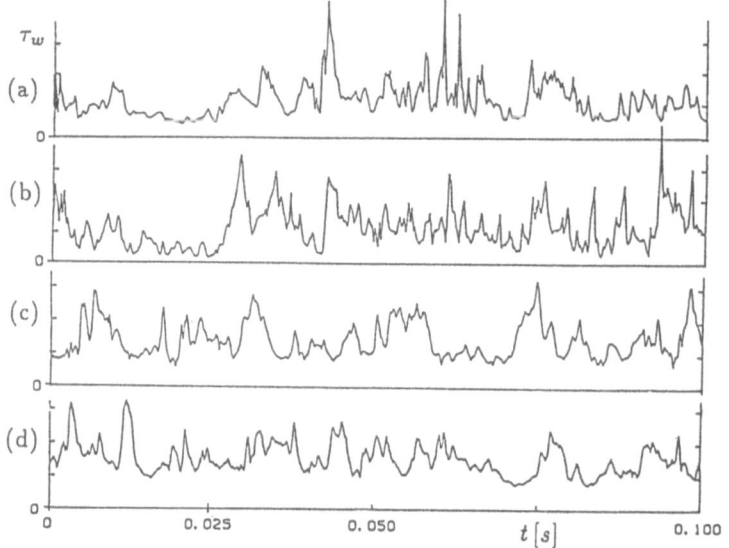

7. Time records of the wall shear stress obtained simultanously with two near-wall hot-wire probes separated 75 viscous units in the spanwise direction. (a) and (b) reverse flow region ($x/x_R = 0.58$), (c) and (d) zero pressure gradient boundary layer ($Re_\theta = 950$) at the same mean skin friction ($u_\tau = 0.457$ m/s).

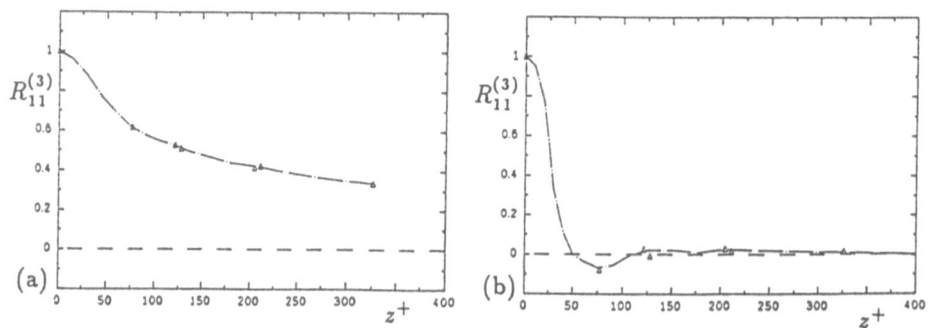

8. Spanwise crosscovariance of the wall shear stress (a) in the reverse flow region ($x/x_R = 0.58$) and (b) in a zero pressure gradient boundary layer.

46

Flow Separation over the Inclined Step

B. Ruck & B. Makiola

Forschungsgruppe Zweiphasenströmungen
Institut für Hydromechanik
Universität Karlsruhe
Kaiserstr. 12, 7500 Karlsruhe 1, Germany

Summary

Flow field investigations were carried out in a backward-facing single-sided step flow with variation of the wall inclination angle. The aim of this investigation was to determine the differences in turbulent flow field quantities when compared to the 90° step geometry. The inclination angle was varied between 10° and 90°. Reynolds numbers (based on step height H) were realized up to 64000. Additionally, the influence of the expansion ratio on the flow field was to investigate with three different ratios 1.48, 2.0, 3.27. Velocity data were measured by laser Doppler anemometry allowing to determine integral flow quantities as e.g. separation streamlines or reattachment lengths. The experiments should both, contribute to the understanding of the phenomenology of separated flows and establish a comprehensive data base for the validation of numerical codes.

Introduction

The flow over a single-sided backward-facing step provides a classic example of flow separation. The simple geometry and the easily attainable two-dimensionality predestinates the step geometry to study separation phemomena. Whereas the 90° step geometry has often been investigated in the past, see

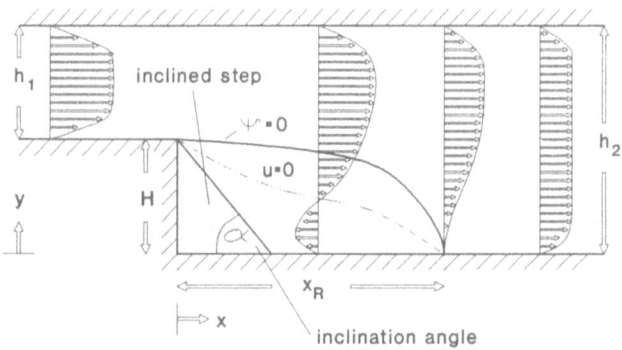

Figure 1: Sketch of the single-sided backward-facing step flow

e.g. [1,2,3,4], the flow over the inclined step has attracted less attention. Some experimental studies exist for rotational symmetrical diffusors [5,6], however, systematic investigations for single-sided step flows with different inclination angles are mostly lacking. The purpose of the investigations presented in this paper is to supply experimental data for technically relevant flow cases with inclined step geometries involved, see Figure 1. For the experimental study, the angle graduations were 10°, 15°, 20°, 25°, 30°, 45° and 90°. Reynolds numbers up to Re_H=64.000 were realized and the expansion ratio was varied ER = 1.48, 2.0, 3.27.

Experiments

For the investigations a closed-loop wind tunnel was used. The test section was made of glass to ensure optical accessibility for a laser Doppler anemometer. The inlet section had a length of 100 step heights (ER = 2) to realize fully developed flow condition at the location of the step. The channel width of 40*H ensured a sufficient two-dimensionality of the flow in the middle plane (measuring plane). To vary the step geometry, differently shaped wedges were inserted into the test section, see also Figure 1.

The LDA measurements were carried out by a dual beam anemometer system including a 25 mW HeNe laser. The LDA system was equipped with double Bragg cells and worked in the forward light scatter mode. A transient recorder (Iwatsu DM-901) in combination with a microcomputer was used to process the LDA

Table 1: Parameter configurations investigated

ER	α [°]	Reynolds number					
		5000	8000	11000	15000	47000	64000
1,48	90	x	x	x	x	x	x
1,48	45	x			x	x	
1,48	30	x	x	x	x	x	x
1,48	25	x	x	x	x	x	x
1,48	20	x	x	x	x	x	x
1,48	15	x	x	x	x	x	x
1,48	10	x	x	x	x	x	x
2,0	90				x	x	x
2,0	45				x	x	
2,0	30				x	x	x
2,0	25				x	x	x
2,0	20				x	x	x
2,0	15				x	x	x
2,0	10				x	x	x
3,27	90	x	x	x	x		
3,27	25	x	x	x	x		
3,27	20	x	x	x	x		
3,27	15	x	x	x	x		

75 configurations

signals. The data evaluation was performed by a software discrimination scheme based on the comparison of the cycle length within signal bursts. Table 1 gives an indication of all the parameter configurations investigated.

Results

Figure 2 fits the measurements of the reattachment length in the whole laminar and turbulent range. The experiments were performed in the turbulent right part of Figure 2 for Re > 5000. In Figure 2, the behaviour for laminar flow condition is taken from [7,8]. Considering the results for turbulent flow condition reveals that for constant Reynolds numbers and an expansion ratio of ER = 3,27, a decrease in inclination angle shortens the reattachment length, see also [9,10,11]. The same findings can be inferred from Figure 3 in a non-logarithmic scale. The changes in reattachment length seem not to be significant in the inclination range from 90° to 45° , however, in the range below 45°, the reattachment length is shortened rapidly with step angle. This indicates that the spreading of the free shear layer is damped for small step angles due to the presence of the nearer coming step wall. It can also be seen that for Re > 10.000 the reattachment length remains almost constant with increasing Reynolds number, which allows to identify fixed reattachment lengths for specific step angles and a given expansion ratio, see Figure 4, where the curves for different Reynolds numbers coincide.

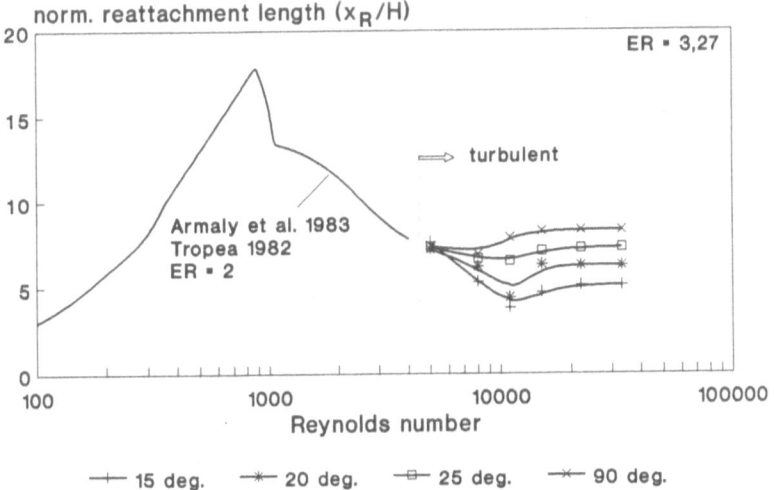

Figure 2: Reattachment length as a function of step angle with different Reynolds numbers; logarithmic scale

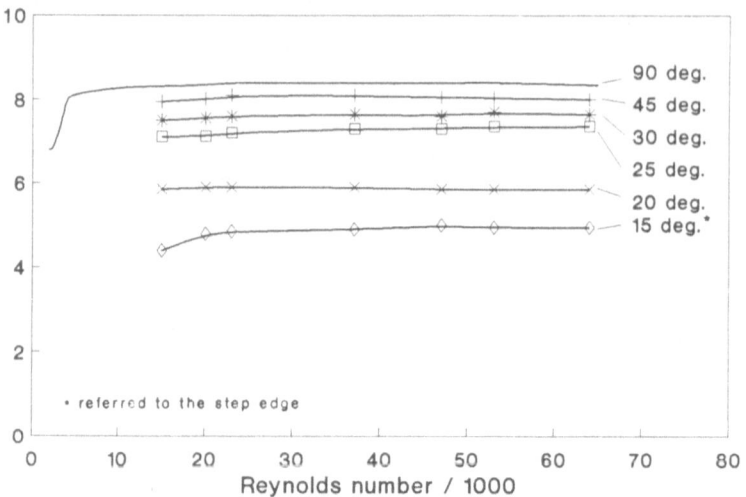

Figure 3: **Reattachment length for different step angles as a function of Reynolds number for Re > 15.000**

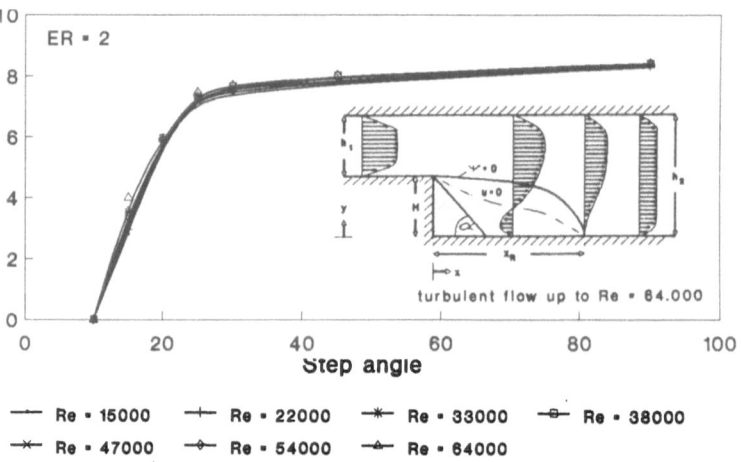

Figure 4: Reattachment length as a function of step angle for Re > 15.000

Considering the measurements of the normal shear stress, see e.g. Figure 5 where a comparison is given for a 10° and 30° step geometry, it could be shown that for a step angle of 20° - 25° a maximum occurs in the cross section under investigation as might be seen in Figure 6. It could be proved that in this particular range of angles, turbulent intensity becomes a

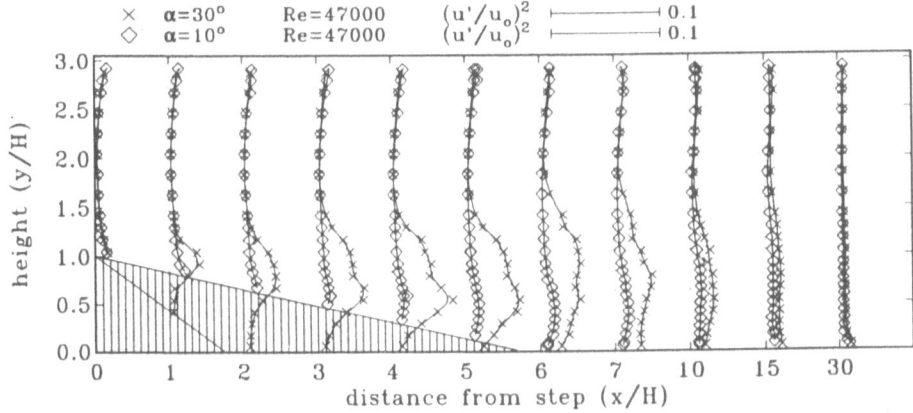

Figure 5: Turbulence intensity profiles behind the step edge
for $\alpha = 10°$ and $30°$ at Re = 47.000

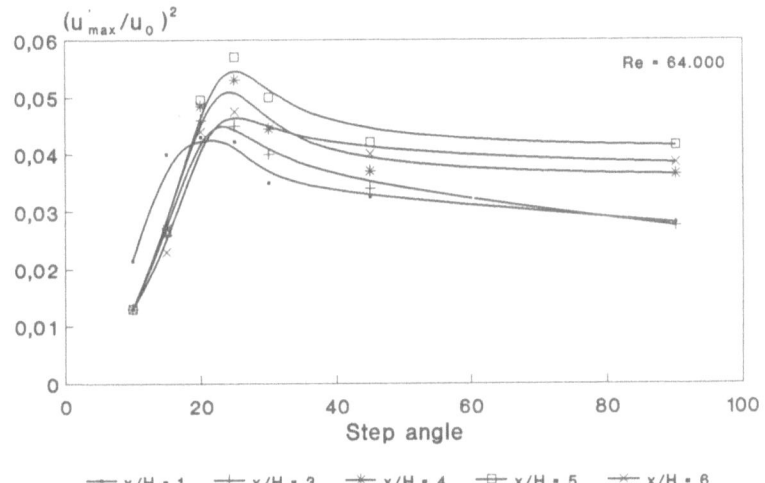

Figure 6: Maximum turbulence intensity in cross section as a
function of step angle at different positions
downstream of the step

maximum and the turbulent data exceed the values of the 90°
step flow. The plot of the lines of constant turbulence
intensity for different step angles reveals that the position
(not the amount) of the cross sectional maximum turbulence
intensity is only little affected by the step angle inclina-

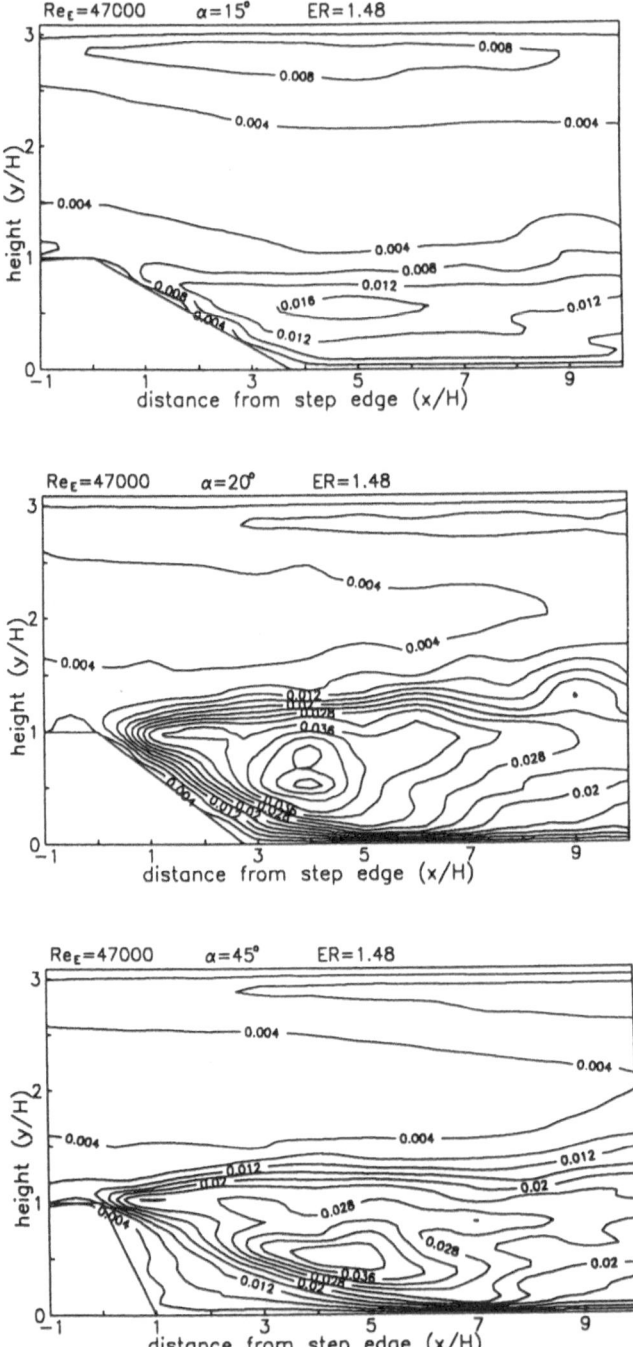

Figure 7: Lines of constant turbulence intensity for different
step angles at Re = 47.000

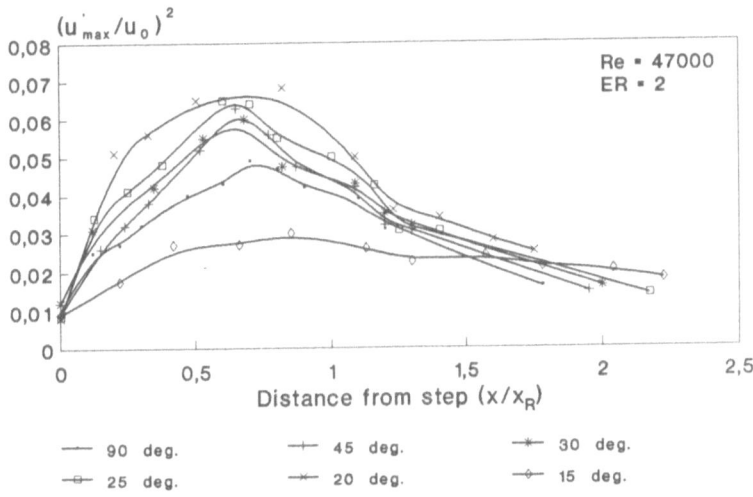

Figure 8: Maximum turbulence intensity as a function of distance from step (normalized with the reattachment length)

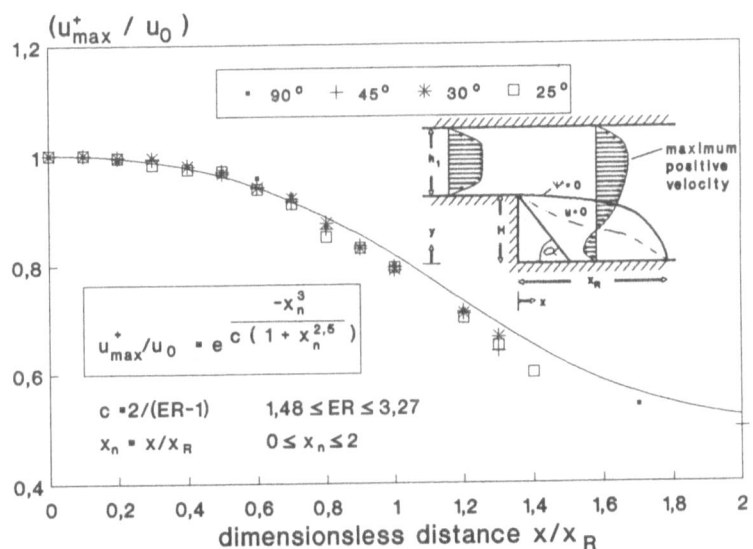

Figure 9: Velocity decay behind the step with different step angles and expansion ratios

tion, see Figure 7. Thus, when normalized with the reattachment length of the individual step geometry, the turbulence
intensity is peaked for the different angle variations in the
cross section $x/H \approx 0.6$, as might be seen in Figure 8.

Considering the velocity decay in the streamwise portion of
the velocity profile behind the step, the experimental data
allow to indicate a formula which accounts for fully turbulent
flow condition (Re > 15.000), step angle variation (25° - 90°)
and expansion ratio differences (ER = 1.48 - 3.27), see Figure
9. The formula represents a good approximation for the velocity decay having an accuracy of 4% in the parameter range investigated.

Acknowledgements

This work was supported by the Deutsche Forschungsgemeinschaft
under Grant No. Ru 345/6. The support is gratefully acknowledged.

Literature

[1]   Abbott, D.E., Kline, S.J., 1962: "Experimental Investi-
      gation of Subsonic Turbulent Flow over Single and
      Double Backward-Facing Steps", Trans. ASME J. Basic
      Engng., 317-325

[2]   Bradshaw, P., Wong, F.Y., 1972: "The Reattachment and
      Relaxation of a Turbulent Shear Layer", Part 1, J.
      Fluid Mech., 52, 1, 113-135

[3]   Schmitt, F., 1987: "Untersuchung der turbulenten
      Stufenströmung bei hohen Reynoldszahlen", Fort-
      schritts-Berichte VDI (Dissertation), 117, 7

[4]   Adams, E.W., Johnston, J.P., 1988: "Effects of the
      Separating Shear Layer on the Reattachment Flow Struc-
      ture, Part 2: Reattachment Length and Wall Shear
      Stress", Experiments in Fluids, 6, 493-499

[5]   Kline, S.J., Abbot, D.E., Fox, R.W., 1959: "Optimum
      Design of Straight Walled Diffusors, J. Basic Eng.,
      Vol. 81, pp. 321

[6]   Chaturvedi, M.C., 1963: "Flow Characteristics of Axi-
      symmetric Expansions", J. of the Hydraulics Division,
      89, HY3, 61-92

[7]   Armaly, B.F., Durst, F., Pereira, J.C.F., Schönung, B.,
      1983: "Experimental and Theoretical Investigation of
      Backward-Facing Step Flow", J. Fluid Mech. 127,
      473-496

[8]   Tropea, C., 1982: "Die turbulente Stufenströmung in
       Flachkanälen und offenen Gerinnen", Dissertation, Uni-
       versität Karlsruhe

[9]   Ruck, B., Makiola, B., 1989: "Backward-Facing Step
       Flow with Inclined Step Geometries", Proc. 4th Asian
       Congress of Fluid Mechanics, Hong Kong, C, 236-239

[10]   Ruck, B., Makiola, B., 1989: "Flow over Single-Sided
       Backward- Facing Steps with Angle Variations", Proc.
       3rd Int. Symp. on Laser Anemometry, Swansea, Wales,
       40.1-40.10

[11]   Makiola, B., Ruck, B., 1990: "Experimental Investiga-
       tion of a Single-Sided Backward-Facing Step Flow with
       Inclined Step Geometries", in Engineering Turbulence
       Modelling and Experiments, Rodi, W., Ganic, E.N.
       (eds.), Elsevier Science Publ. Co., Inc., 487-496,
       ISBN: 0-444-01563-9

Dynamics of Micrometer-Sized Particles in a Separated Step Flow

B. Ruck & B. Makiola

Institut für Hydromechanik
Forschungsgruppe Zweiphasenströmungen
Universität Karlsruhe
Kaiserstr.12, 7500 Karlsruhe 1, FRG

Summary

Particles in the diameter range from 1 to 70 μm were suspended in an air flow and the particle motion over a backward-facing step was measured by means of laser-Doppler anemometry. Thus, the local and integral flow quantities, i.e. the mean and turbulent velocity data could be measured precisely. In the experiments, monodispersed particle size distributions were used to exclude particle size related information ambiguity, known as triggering or size bias. The results of this study show qualitatively and quantitatively the difference in time-averaged particle dynamics for selected particle sizes in a backward-facing step flow. The experiments show the changes in the particle velocity field when compared with the velocity field of the continuous phase deduced from the 1 μm particles. Additionally, the results imply the strong influences which different particle sizes have on flow data measurement when size effects are not taken into account with particle-related optical measuring techniques.

Introduction

Particle transport and deposition in turbulent flows has been intensively investigated in experiments [1,2,3,4,5,6]. Compared to the great number of experimental studies in particle-laden channel flows, detailed experimental investigations of the particle size-resolved dispersion in separated step flows are rare. This holds especially for the technically most relevant particle diameter range of 1-100 μm. The paucity of knowledge in this field is probably due to experimental and financial restraints. On the one hand, experimental on-line equipment to measure simultaneously individual particle sizes and velocities with a polydispersed size distribution in the flow [5,7,8,9] is not commonly in use and not trivial to apply. On the other hand, the costs for monodispersed particles, which can be advantageously used for size-dependent dispersion studies, are substantial.

To measure particle velocities, laser-Doppler anemometry [10,11,12] is an advantageous measuring method because the velocity of the individual particles suspended in the flow can be measured directly. Normal LDA applications infer the velocity information of a continuous fluid phase from the velocity data of small suspended particles. This can be done with a sufficient accuracy, as long as the particles are very small and ideally follow the flow fluctuations. If the particles increase in size, it is not the velocity of the continuous phase but the velocities of the individual particle phases which are measured. Thus, using monodispersed particles, one can measure the size-dependent particle dynamics, e.g. in a separated flow regime.

The investigations in this paper describe the dynamics of particles of different sizes in a single-sided backward-facing step flow, see also [13,14]. The particle diameters under investigation were 1, 15, 30 and 70 μm. It seemed reasonable to carry out the experiments with a 90°-step geometry, which represents the most commonly investigated and understood flow separation regime, see Figure 1. The experiments were carried out for turbulent flow conditions.

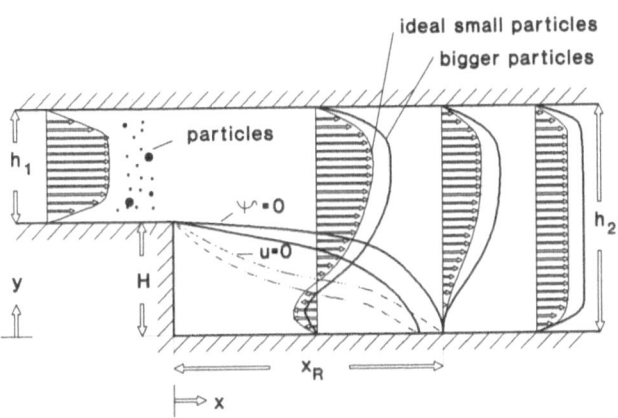

Figure 1: 90° backward-facing step geometry; x_R reattachment length; H step height; u_0 max. velocity in inlet profile

Experiments

The step test section was a part of a closed loop wind tunnel. The test section was made of glass to ensure optical access. An expansion ratio of 1 : 2 and a length of 116 step heights were realized. The inner width of the test section was 500 mm, the step height was 25 mm and the channel height after the step was 50 mm. The dimensions of the inlet length of the test section were calculated to ensure fully-developed flow conditions at the step with respect to vertical and lateral profile symmetry, mean flow and turbulent quantities. The wind tunnel was driven by a fan, which allowed us to realize flow velocities up to 50 m/s and which is equivalent to $Re_H \leq 10^5$.

The local particle velocity was measured using a conventional one-dimensional LDA. The LDA system (15 mW He-Ne laser) was working in the forward light scattering direction. The system was equipped with a double Bragg cell arrangement to discriminate the flow direction and to resolve regions of recirculation. The diameter of the LDA-measuring volume was chosen as 500 μm. The half-angle of the crossing laser beams was 2.86°, which corresponds to a fringe spacing of 6.33 μm. Figure 2 shows the LDA mounted on the traversing unit.

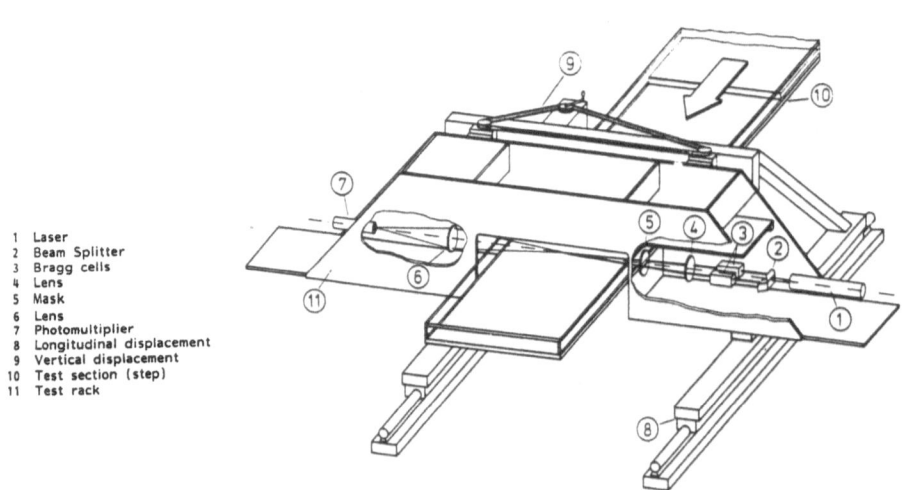

```
 1  Laser
 2  Beam Splitter
 3  Bragg cells
 4  Lens
 5  Mask
 6  Lens
 7  Photomultiplier
 8  Longitudinal displacement
 9  Vertical displacement
10  Test section (step)
11  Test rack
```

Figure 2: Laser Doppler anemometer mounted on a traversing unit

The measurements were based on a single particle velocity evaluation. The data evaluation was performed by a 100 MHz transient recorder with software processing in an associated microcomputer. To trace the continuous fluid phase in the experiments, small oil droplets of a median dia $d_p \approx 1$ μm were used. The oil droplets were generated by an atomizer. For the investigation of the dynamics of differently sized particles, starch particles of almost uniform size were used. The density of the particle medium is $\rho = 1500$ kg/m^3. The particles show a spherical shape and are not soluable in cold water. Figure 3 shows a photograph of 50 μm dia starch particles, taken by electron microscopy. Three different size classes 15, 30 and 70 μm dia were used. The particles were injected into the wind tunnel after the radial fan. The injection was performed by a

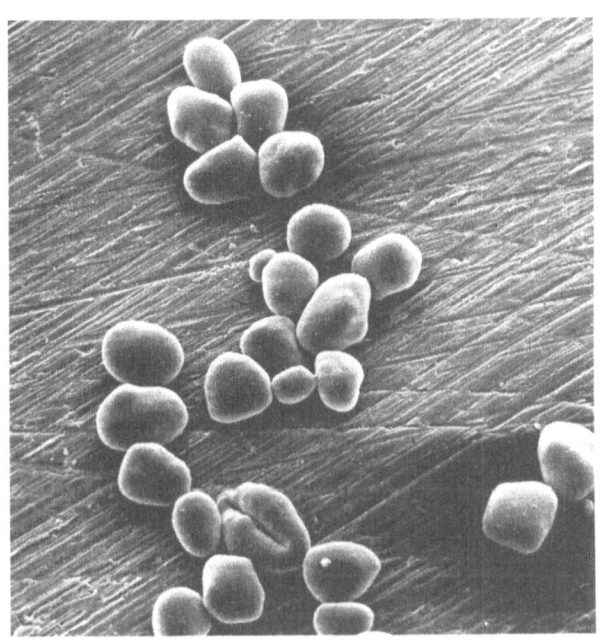

Figure 3: Starch particles of 50 μm dia under the electron microscope

Table 1: Particles suspended in the step flow

particle media	Reynolds number			
	15000	30000	47000	64000
oil (1 μm)	$2,16 \ 10^{-5}$	$4,34 \ 10^{-5}$	$6,97 \ 10^{-5}$	$9,9 \ 10^{-5}$
starch (15 μm)	$9,0 \ 10^{-3}$	$1,8 \ 10^{-2}$	$2,9 \ 10^{-2}$	$4,13 \ 10^{-2}$
starch (30 μm)	$3,6 \ 10^{-2}$	$7,23 \ 10^{-2}$	$1,16 \ 10^{-1}$	$1,65 \ 10^{-1}$
starch (70 μm)	$1,96 \ 10^{-1}$	$3,94 \ 10^{-1}$	$6,3 \ 10^{-1}$	$8,99 \ 10^{-1}$

particle media	diameter d (μm)	relaxation time $\tau \cdot 10^6$ (sec)	density ratio s	settling velocity u_E (m/sec)
oil	1	2,51	660	$2,44 \ 10^{-5}$
starch	15	1047	1200	$1,02 \ 10^{-2}$
starch	30	4190	1200	$4,08 \ 10^{-2}$
starch	70	22811	1200	$2,22 \ 10^{-1}$

diameter	number concentration	volume concentration
d = 1 μm	$0,97 \ 10^3/cm^3$	$0,508 \ 10^{-9} cm^3/cm^3$
d = 15 μm	$0,34 \ 10^3/cm^3$	$0,60 \ 10^{-6} cm^3/cm^3$
d = 30 μm	$0,13 \ 10^3/cm^3$	$1,84 \ 10^{-6} cm^3/cm^3$
d = 70 μm	$0,07 \ 10^3/cm^3$	$1,26 \ 10^{-5} cm^3/cm^3$

special device based on premixing of the particles in a pressurized tank. The normal loss of suspended particles due to wall deposition was compensated by a permanent low seeding rate. Degradation of the particles was not observed. Table 1 gives a compilation of the particles used in the experiments. The particle number concentration in the experiments was relatively low (10^2-10^3/cm^3), so that the influence measured was that of the flow on the particle movement and not that of the particles on the flow. In this concentration and size range, the distortion of the flow field by particles is negligible.

Results

The following results represent time-averaged flow velocities and turbulence data. It should be remarked that the actual local mean data are ensemble-averaged quantities, which rely on 1000 single LDA bursts. As mentioned before, the signal processing was performed by a transient recorder-microcomputer combination. The processable data rate was about 10 readings/s which is rather low and predominantly caused by the data handling between recorder and computer. Together with a sufficiently high particle number concentration, a periodic sampling of the information was induced.

Figure 4 shows two typical series of measured mean velocity profiles after the single-sided step for oil particles of 1 μm dia and starch particles of 70 μm dia. It can be inferred that the 70μm particles show smaller recirculating velocities than the 1 μm particles. Furthermore, a smaller recirculation zone

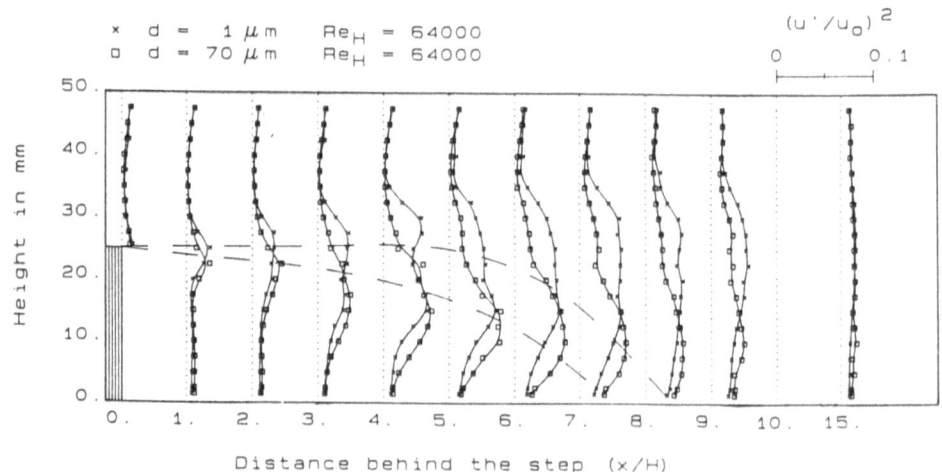

Figure 4: Mean velocity profiles for 1 and 70 μm dia particles behind the backward-facing step with Re_H = 64.000

would be inferred from the velocity profile of the larger particles. As a trend, all the experiments deliver higher positive values for bigger particles in all the cross-sections
under investigation. Figure 5 gives the corresponding r.m.s.
velocities to Figure 4. It can be inferred that the cross-
sectional bimodal r.m.s. velocity distribution in the free
shear layer is damped significantly for bigger particles. The
maximum r.m.s. values of bigger particles are shifted more
towards the lower wall.

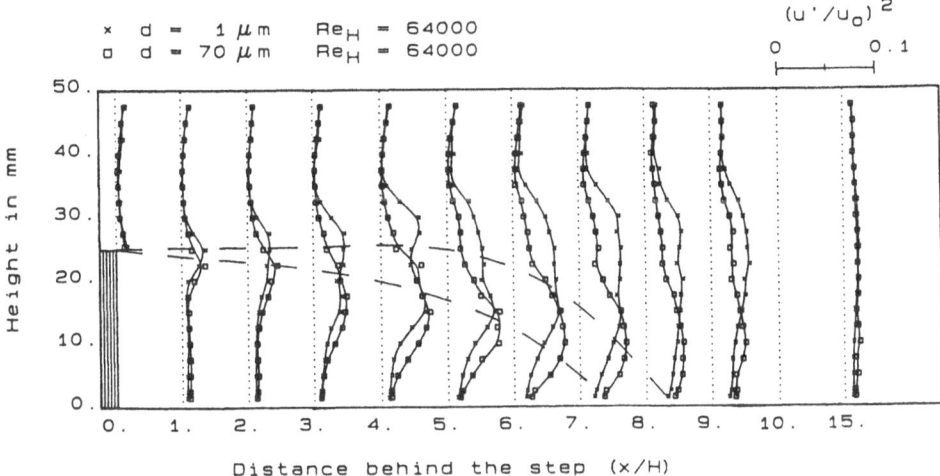

Figure 5: r.m.s. velocity profiles for 1 and 70 μm dia particles behind the step, see Figure 4

Bigger particles suspended in a flow have a higher momentum,
which leads to a longer stopping distance when the continuous
flow is slowed down, e.g. by an expansion of the channel
geometry. This can be seen in figure 6, where the streamwise
local mean velocities stay at a higher level longer for the
70μm particles than for the smaller ones. For u = 40 m/s the
corresponding stopping distances were 10^{-4} m for 1 μm particles and 0.9 m for 70 μm particles respectively, see Table 1.

In figures 6 and 7 the maximum positive streamwise and negative counterstreamwise velocity for the velocity profile considered is given. The aforementioned particle dynamics are
reflected well in these graphs. It can be inferred that up to
position x/H ≈ 3 no significant changes between the particle
velocities in the streamwise and recirculation direction can
be observed. The differences in particle velocity data begin
to grow behind x/H ≈ 3. The position of the maximum negative
velocity in the separation zone is shifted to slightly smaller
values of x/h with increasing particle size, but still remains
around x/H ≈ 4.

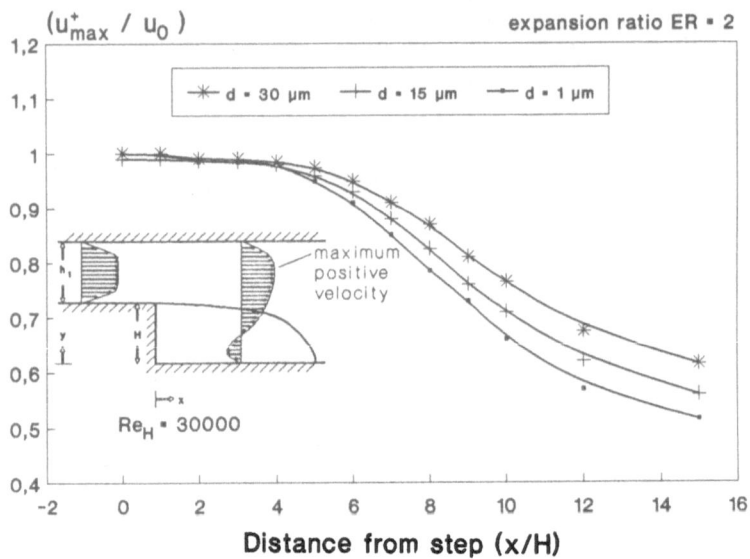

Fig. 6: Maximum positive velocity u^+_{max} (normalized with u_0) in the streamwise portion of the velocity profile of particles of different sizes.

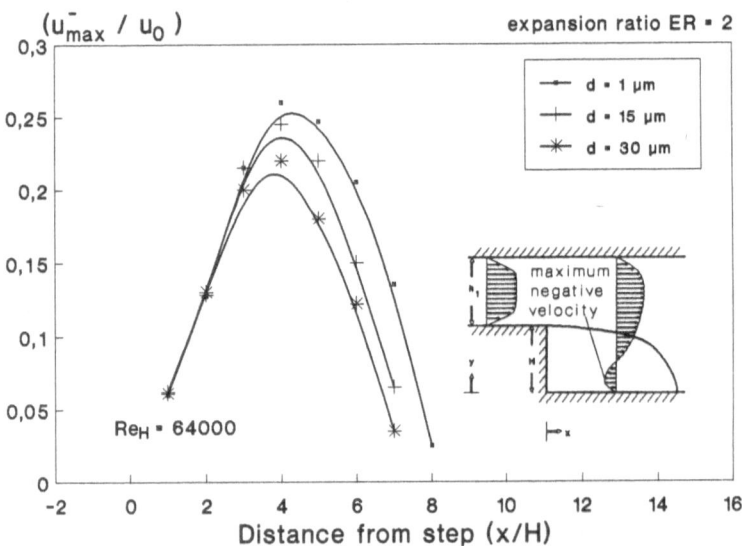

Figure 7: Maximum negative velocity u^-_{max} (normalized with u_0) in the recirculation zone of particles of different sizes.

As shown, the particle velocity data of different particle classes deviate from each other. If the particle information is used to describe the continuous phase flow field, different velocity profiles will result. In figure 8, the dividing streamline and line of streamwise zero velocity in the recirculation region are shown for all particle size classes investigated and a Reynolds number of $Re_H = 15,000$. The reattachment length of the particle velocity field is effectively shortened with increasing particle diameter. This suggests that in separated flow regions the use of particle-related measuring techniques can lead to erroneous measurements, when the particulate phase is not controlled. The experiments show clearly that a distinction between flow velocity field and particle velocity field has to be made in separated flow regions even in the particle size range of micrometers.

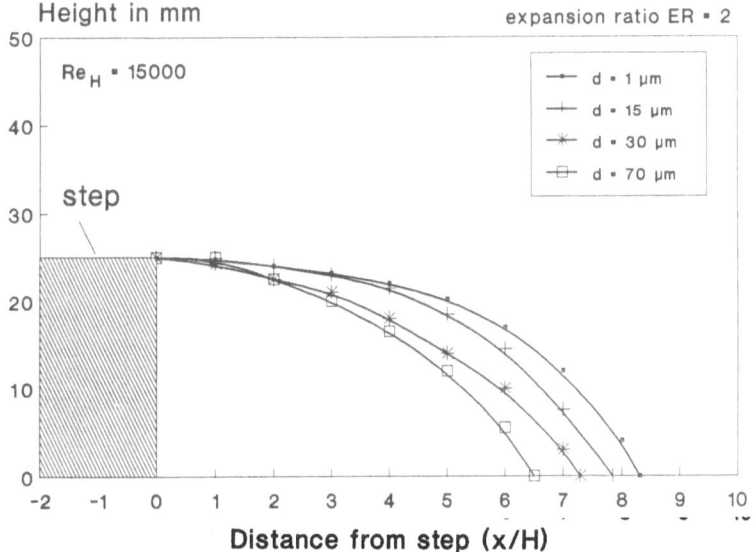

Figure 8: Dividing streamline derived from velocities of particles of different sizes

Conclusions

The results show that with increasing particle size, the particle velocity field differs increasingly from the flow velocity field of the continuous phase. For bigger particles, the velocity fluctuations in all the cross-sections under investigation decrease. In the zone of recirculation, big particles give a lower velocity than small ones. The opposite holds for the streamwise portion of the velocity field, where big particles have a higher velocity in the cross-sections near the step. If derived from particle velocity information, the dimensions of the recirculation zone decrease with increasing particle size. This is due to the reduction of negative counter-streamwise particle velocities in the separation region, which effectively shortens the measured reattachment length. Comparing the results of differently sized particles in the flow, it can be inferred that particle diffusion differs from the eddy diffusion even in the range of micron-sized particles. Additionally, the results demonstrate the dependence of particle sizes on flow data, when particle-related measuring techniques are used. Due to a delayed momentum decay of bigger particles, erroneous continuous phase flow measurements can result.

Acknowledgements

This work was supported by the Deutsche Forschungsgemeinschaft under Grant No. Ru 345/2. The support is gratefully acknowledged.

References

[1]   Farmer, R., Griffith, P., Rohsenow, W.M., 1970: "Liquid Droplet Deposition in Two-Phase Flow", J. of Heat Transfer, 587-594

[2]   Simpson, H.C., Brolls, E.K., 1974: "Droplet Deposition on a Flat Plate from an Air/Water Mist in Turbulent Flow over the Plate", Symp. on Two-Phase Flow Systems (A3), University of Strathclyde, Glasgow, 1

[3]   Hetsroni, G., Sokolov, M., 1971: "Distribution of Mass, Velocity, and Intensity of Turbulence in a Two-Phase Turbulent Jet", J. Applied Mech., 315-327

[4]   Popper, J., Abuaf, N., Hetsroni, G., 1974: "Velocity Measurements in a Two-Phase Turbulent Jet", Int. J. Multiphase Flow, 1, 715-726

[5]   Ruck, B., Pavlowski, B., 1984: "Kombinierte optische Messung von Teilchengrößen- und Teilchengeschwindigkeitsverteilungen im Rohr", tm-Technisches Messen, 51. Jahrg., 2, 61-67

[6]   Lee, S.L., Börner, T., 1987: "Fluid Flow Structure in a Dilute Turbulent Two-Phase Suspension Flow in a Vertical Pipe", Int. J. Multiphase Flow, 13, 2, 233-246

[7]   Bauckhage, K., Flögel, H.H., 1984: "Simultaneous Measurement of Droplet Size and Velocity in Nozzle Sprays", Proc. 2nd Int. Symp. on Applications of Laser Anemometry to Fluid Mechanics, Lisbon, Portugal, 18.1

[8]   Hishida,   K.,   Tajima, K., Maeda, M., Kano, H., 1984:
      "Measurements of Two-Phase Turbulent Flow by LDA with
      Particle Size Discrimination", 2nd Int. Symp. on
      Applications of Laser Anemometry to Fluid Mechanics,
      Lissabon

[9]   Ruck, B., Schmitt, F., Loy, T., 1986: "Particle Dyna-
      mics in a Separated Step Flow", 3rd Int. Symp. on
      Application of Laser- Anemometry to Fluid Mechanics,
      Lissabon, Proc., chapter 2.1

[10] Durst, F., Melling, A., Whitelaw, J.H., 1976: "Princi-
      ples and Practice of Laser-Doppler Anemometry", Aca-
      demic Press, London

[11] Ruck, B., 1987: "Laser-Doppler-Anemometrie", AT-Fach-
      verlag Stuttgart, ISBN: 3-921 681-00-6

[12] Ruck, B., (Hrsg.), 1990: "Lasermethoden in der Strö-
      mungsmeßtechnik", AT-Fachverlag Stuttgart, ISBN:
      3-921 681-01-4

[13] Ruck, B., Makiola, B., 1988: "Particle Dispersion in
      a Single-Sided Backward-Facing Step flow", Int.
      Journal Multiphase Flow, Vol. 14, No. 6, pp. 787-800

[14] Ruck, B., Makiola, B., 1990: "The Dispersion of Par-
      ticles in a Separated Backward-Facing Step Flow",
      Proc. of IUTAM Symp. 20.-24.8.90, La Jolla, Califor-
      nia/USA

The Structure of the Drag Reduced Turbulent Flow over a Backward Facing Step - an experimental study

J. Dohmann
Universität Dortmund, Fachbereich Chemietechnik, Lehrstuhl Energieprozeßtechnik,
Postfach 500 500, W-4600 Dortmund 50

1. Summary

Aqueous solutions of polyacrylamide (Separan AP45, 50 ppm) or some cationic surfactant additives (mixtures of tetradecyl-trimethyleammoniumbromide with sodium salicylate) exhibit the effect of drag reduction in turbulent pipe flows. From previous investigations it could be concluded that this effect is based on the change of the structure of turbulent motion. The aim of this work is the experimental investigation on the influence of drag reducing additives on the features of turbulence.
The description of turbulent structures leads to a better understanding of the effect of drag reduction and offers deeper insights to turbulence in general. The chosen geometry of a backward facing step supplements investigations of other turbulent flows, e.g. the plane mixing layer (Kwade 1982) or pipe flow (Bewersdorff et al. 1986).

2. Experimental Set-Up

A solid body with an unsteady expansion was introduced in a rectangular channel, forming a flow field as indicated in figure 1.

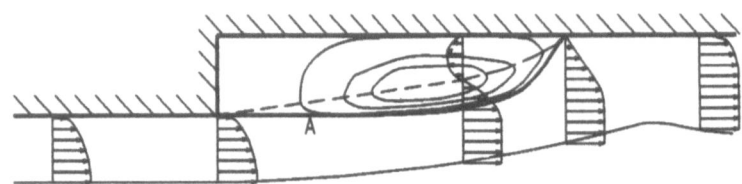

Figure 1: Sketch of the backward facing step. Line A indicates the separating streamline dividing the free shearlayer and the recirculation vortex.

The flow over this backward facing step was examined by two independent laser-Doppler-velocimeters (LDV). In addition to the determination of the mean velocity and the components of Reynolds stress tensor as a function of space a conditional sampling was performed to obtain the spatial structure of coherent turbulent motions. This conditional sampling method was described by Kaplan (1973) or Antonia (1981). The signal of one LDV at a reference point x_a was analyzed for fluctuations being characteristic for a coherent structure. If a coherent structure was detected the signal of the second LDV at a point x_b was used to calculate an ensemble average. Blackwelder (1977) defined a correlation function which can be written as

$$< u_i(x_b) >= \frac{\overline{c(t) \cdot u_i(x_b,t)}}{\overline{c(t)}} \quad , \tag{1}$$

where $u_i(x,t)$ is the velocity signal of the second LDV and $c(x,t)$ is a detection function indicating the presence of a coherent structure at the reference point. $< u_i(x) >$ is called the ensemble average. By variation of the location x_b of the LDV the spatial velocity distribution could be scanned. (For more details see Riediger 1989).

3. Non-Newtonian Testfluids

The first testfluid is a high dilute solution of a long chain polyacrylamide ($M=1.6 \cdot 10^6$ kg/mole) with a concentration of 50 wppm. The solution shows a non-Newtonian behaviour, e.g. shear thinning and non zero normal stress differences. The other testfluid is an aqueous solution of a cationic surfactant $C_{14}TASal+NaBr$ (950 ppm) with a complicate rheological behaviour. If a critical shear strain is exceeded the solution shows a steep increase in viscosity and changes to a shear induced state with anisotropic and viscoelastic properties. For comparison all experiments are performed using water as well.

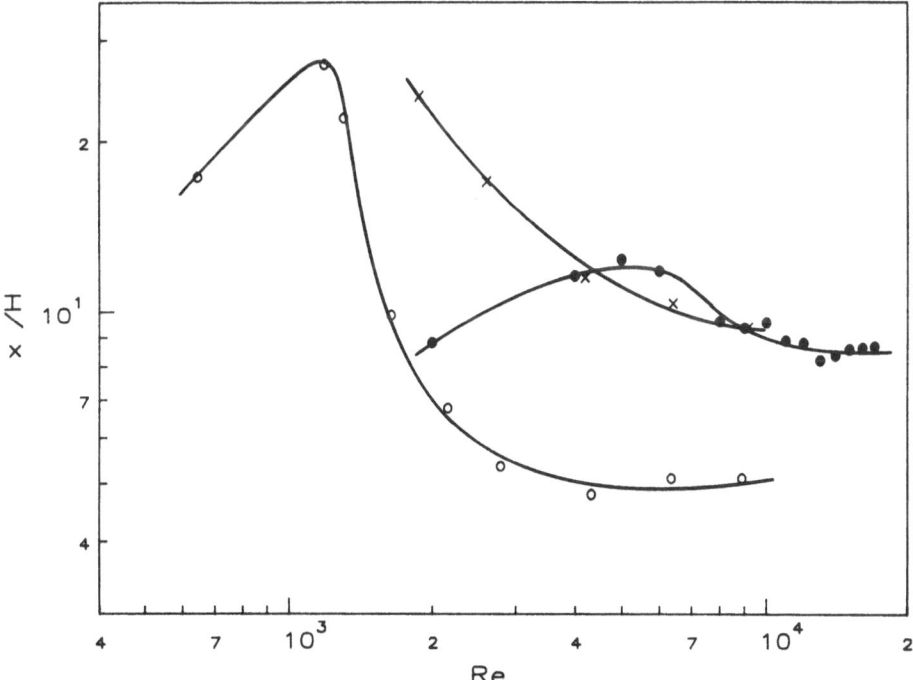

Figure 2: Separation length x_R/H vs. Reynolds number which is formed with the height H of the step. o - Water, x -polymer solution, - surfactant solution.

67

4. Selected Results

The drag reducing additives affect the mean velocity profiles and the shape of the main vortex. The latter is represented e.g. by its separation length, shown in figure 2 as a function of Reynolds number. In all the cases the separation length is increased. Both, the Newtonian and the non-Newtonian fluids show an asymptotic separation length. The asymptotic level is increased by a factor of 2. It can be shown that below $Re_H = 2000$ the separation length of both fluids, water and surfactant solution coincide because the low shear strain does not induce non-Newtonian behaviour.

The velocity profiles (figure 3) indicate a shear layer which grows with the distance from the separation point. The shear strain is attenuated, so this process is comparable to the plane mixing layer studied by Kwade or Riediger. Differences between the three testfluids occur. The Newtonian fluid exhibits a fast attenuation of the shear strain, the velocity profile gets smooth. At large distances no influence of the backward facing step could be seen: the flow seems to have become a developed turbulent channel flow. In contrast the surfactant solution exhibits even in the largest distances a behaviour comparable to that of a Newtonian but laminar flow with a prolongated recirculation vortex. The polymer solution exhibits a behaviour between that one shown by the testfluids water and surfactant solution.

This behaviour can be better understood when the Reynolds stress is taken into concideration (see figure 4). In most of the cases the location of the maximum of the Reynolds stress coincide with the location of maximum velocity gradient. In the Newtonian case the most intense stresses are obtained, followed from that obtained in the polymer solution. In the latter case the spatial distribution of the Reynolds stresses is changed significantly. The maximum is found closer to the recirculation area. The high level of the Reynolds stresses yields a developed turbulent channel flow at large distances from the separation point. In contrast the surfactant additive attenuates the Reynolds stress. At x=220 the stress approximates the zero-level which lead to the described mean-velocity profile.

Using the conditional sampling technique the spatial distributions of the vorticity could be calculated from the "momentary" ensemble averaged spatial velocity field. The following figure 5 give an impression of this property. It is shown that in the non-Newtonian case the vortex less rough and only exhibits a single maximum in vorticity. In the Newtonian case the vortex covers an area exceeding that one of the non-Newtonian case. The reduced size of the vortex corresponds well with the reduced width of the Reynolds- stress profile due to the polymer additive. As found in the plane mixing layer (for further details see Riediger 1989) the small scale turbulence seems to be reduced. The position of the maximum in vorticity corresponds with the position of maximum shear strain too. At large distances from the separation point it was not possible to measure coherent vorticity. In literature the effect of vortex "tearing" is proposed to be responsable (Hussain 1986). Other effects such as the phase scrambling effect can be made responsable too.

The flow of the surfactant solution does not indicate any occurance of turbulent coherent structures. This effect was subject of a following investigation (see Dohmann 1992).

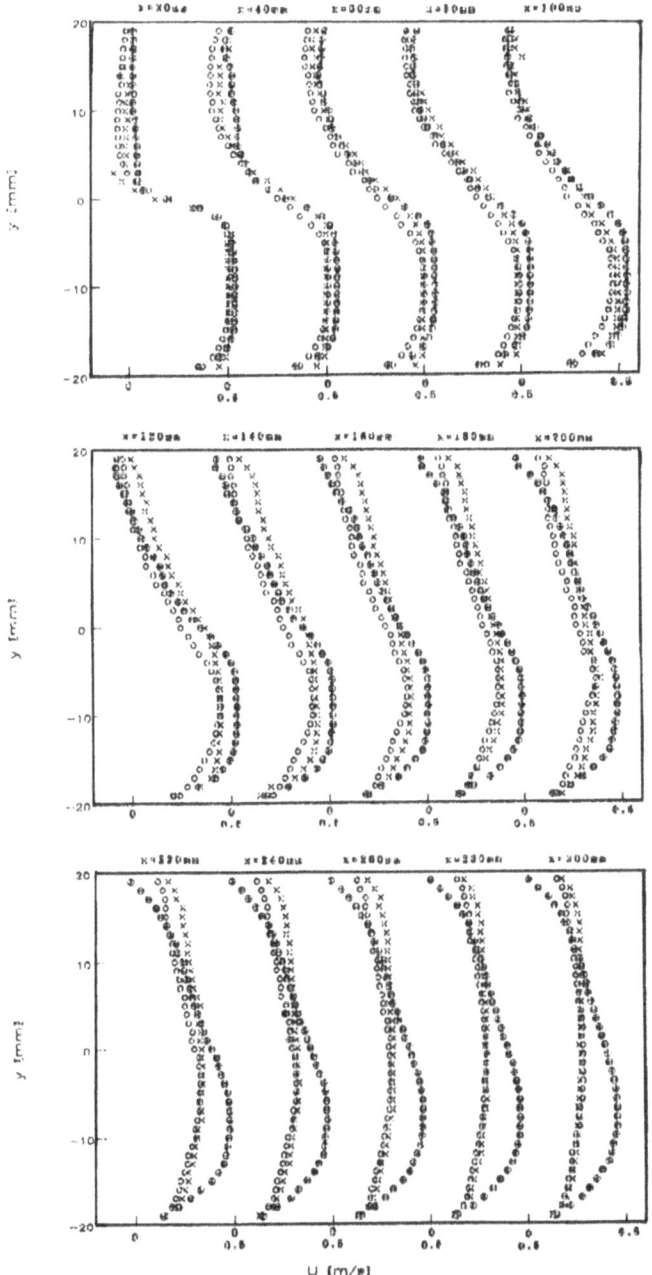

Figure 3: Mean velocity profiles at different downstream positions.
Re=5000, o - water, x- polymer solution, •- surfactant solution.

69

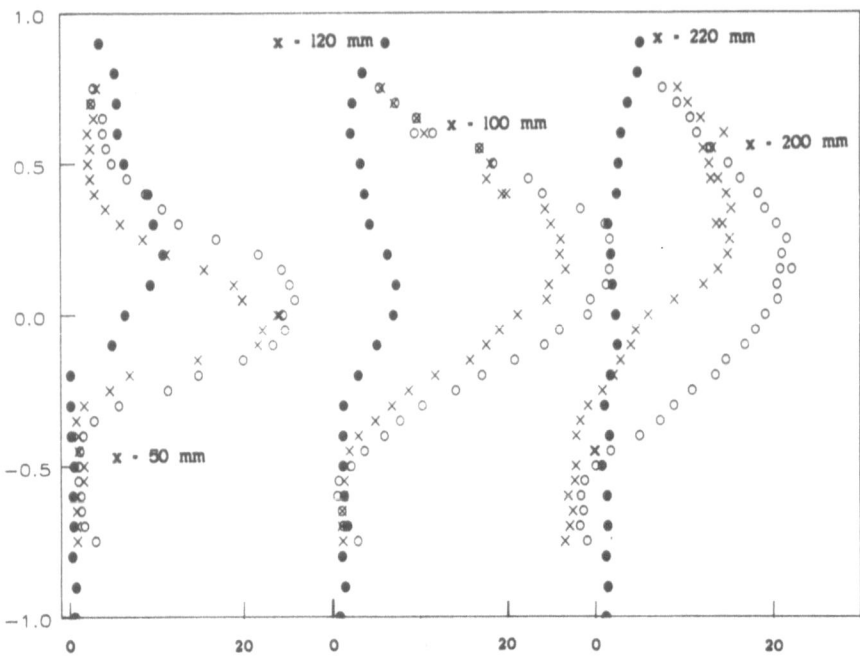

Figure 4: Reynolds stress $[\text{cm}^2/\text{s}^2]$ at different positions

The described results lead to the knowledge of important differences between the features of the Newtonian and the non-Newtonian flows. Some remaining questions concerning the behaviour of the non-Newtonian fluids in turbulent flows will be subject of future investigations due to the difficult description of the fluid properties.

5. Restricting conclusions

- As can be concluded from results obtained in Newtonian fluids (Kottke 1983) the condition of the boundary layer located upstream the separation point affect the separation length and the stability (growth of disturbances) of the free shear layer. The additives exhibit a strong influence on these boundary conditions. This fact reduces the comparability of the flows of the different fluids.

- It is noteworthy to mention that the high turbulence intensity level in regions of the flow with small mean velocities lead to less significant informations about the local velocity. The unsatisfactory quality of the LDV signals requires an advanced technology for signal processing, e.g. burst spectrum analyzers.

- From velocity measurements a three-dimensional motion could be concluded. It could be shown that this effect is emphasized by the used additives.

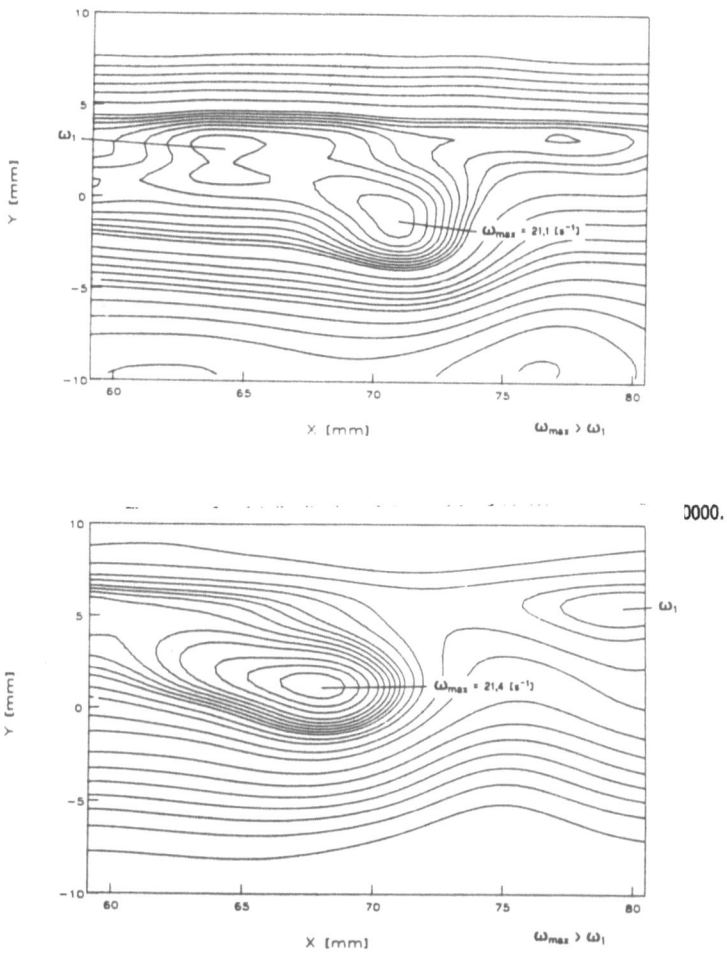

)000.

Figure 5b: Spatial distribution of the vorticity field. Polymer solution. x=70 Re=10000

6. Acknowledgement

The reported subject was investigated between 1983 and 1988 at the department "Chemietechnik" by Prof. Dr. rer. nat. H. Giesekus (Lehrstuhl Strömungsmechanik). This report is based on the work of his Co-workers.

The financial support by the DFG is greatfully acknowledged.

7. References

Antonia, R. A.
    Conditional sampling in turbulence measurement.
    Ann. Review of Fluid Mech., 13 (1981) 131-156.

Bewersdorff, H.-W.; Frings, B.; Lindner, P.; Oberthür, R. C.
    The conformation of frag reducing micelles from small-angle-neutron-scattering
    experiments. Rheol. Acta, 25 (1986) pp. 642-646.

Blackwelder, R.
    On the role of phase information in conditional sampling.
    Physics of Fluids, 20 (1977) 232-242.

Dohmann, J.
    Anisotrope Turbulenz der Kármanschen Wirbelstraße - eine phänomeno-
    logische Untersuchung. Universität Dortmund, Dissertation 1992.

Hussain, A.K.M.F.
    Coherent structures and turbulence. J. of Fluid Mech. 173 (1986) 303-356.

Kaplan, R. E.
    Conditional sampling techniques. 3rd Sympoium on Turbulence in Liquids
    (1973) University of Missouri-Rolla.

Kwade, M.
    Beeinflussung der Turbulenzstruktur in der ebenen Mischungsschicht zweier
    Ströme durch Polymerzusätze. Rheol. Acta, 21 (1982) pp. 120-149.

Riediger, S.
    Untersuchung der Turbulenzstruktur in einer ebenen Mischungsschicht und
    hinter einer Kanalerweiterung bei der Strömung zweier nicht-newtonscher
    Flüssigkeiten. Universität Dortmund, Dissertation 1989.

DIRECT NUMERICAL SIMULATION OF UNSTEADY SEPARATED
BOUNDARY-LAYER FLOWS OVER SMOOTH BACKWARD-FACING STEPS

H. Bestek[*], K. Gruber[‡], H. Fasel[+]

[*] Institut für Aerodynamik und Gasdynamik
Universität Stuttgart, Pfaffenwaldring 21, 7000 Stuttgart 80, FRG
[‡] Currently at Mercedes-Benz AG, 7032 Sindelfingen, FRG
[+] Aerospace and Mechanical Engineering Department
University of Arizona, Tucson, Arizona 85721, USA

SUMMARY

Direct numerical simulations based on finite-difference solutions of the unsteady
Navier-Stokes equations are used for investigating two-dimensional laminar separation bubbles in incompressible boundary-layer flows over smooth backward-facing
steps. An attempt has been made to clarify the mechanism of natural unsteadiness of the separation bubble that appears if the step height exceeds a certain
limit. It was found that this phenomenon is caused by hydrodynamic instability
of the separated boundary layer.

INTRODUCTION

Laminar and transitional separation bubbles have been subject of numerous studies
[1, 2]. In the past several years interest in airfoils operating at low chord Reynolds
numbers has focussed attention on the various phenomena involved in the formation of transitional separation bubbles. A better understanding of the structure
and behavior of transitional separation bubbles is necessary for improved design
and analysis techniques of low Reynolds number airfoils [3].

Different approaches and numerical procedures have been developed for the calculation of two-dimensional boundary-layer flows in the presence of transitional separation bubbles. Empirical models of transitional separation bubbles with characteristic
parameters determined from experimental observations are often used in airfoil
design and analysis methods (see for example [4]). Other methods, such as viscous-
inviscid interaction methods, methods based on triple-deck-theory, and Navier-
Stokes solvers, aim at calculating the flow field in the vicinity of a laminar or transitional separation bubble. With many of these methods the formation of a laminar
separation bubble, i.e. laminar at separation and laminar at reattachment, can be
accurately predicted. This has been demonstrated by many investigators for a laminar flow over a flat plate with a separated region caused by a trough in the wall [5].
This flow geometry has become a test case for interacting-boundary-layer and triple-
deck modelling [6, 7]. Another model problem to test newly-developed methods is
the retarded flow on a flat plate, which was proposed by Briley [8]. However, for
laminar bubbles many authors report steady solutions only up to certain limits of
bubble size or Reynolds number. For larger Reynolds numbers the numerical methods
often fail to converge or are otherwise unable to yield meaningful solutions.
Attempts have been made to apply these methods for the calculation of transitional
separation bubbles [9,10] using empirical transition models to switch the solution
from laminar to turbulent flow. However, these calculations have shown that the
numerical solution is strongly dependent on the transition criteria used.

73

Transition in separation bubbles is not yet understood. In previous studies (see for example [11, 12]), transition was thought to occur in the highly unstable separated shear layer resulting in a turbulent shear layer that grows rapidly due to turbulent mixing with the external stream, and, as a consequence, would lead to reattachment of the layer. However, in recent detailed experimental investigations of two-dimensional separation bubbles [13], it was found that transition is initiated by the amplification of laminar instability waves, similar to the Tollmien-Schlichting waves observed in attached boundary layers. The amplification of these waves proceeds from a critical point, which is generally upstream of separation, according to the well-known instability of the boundary layer profiles approaching separation.

Numerical investigations of laminar separated flows using finite-difference methods for solving the two-dimensional Navier-Stokes equations have been initiated at the University of Stuttgart more than 15 years ago. For a boundary layer flow over a backward-facing step Fasel et al. [14] obtained steady separation bubbles up to a certain step height, while for larger step heights an unsteady development of the separated flow was observed which was initially interpreted as vortex shedding. However, the phenomenon was not been fully understood at that time. Thereupon, to better understand the underlying mechanisms, unsteady laminar separation phenomena in two-dimensional boundary layers were investigated in great detail by Gruber [15-17] using direct numerical simulations. For these investigations a fully implicit finite-difference method for the solution of the incompressible Navier-Stokes equations was employed, that was developed previously by Fasel [18] for the numerical simulation of spatially amplified Tollmien-Schlichting waves in boundary layer flows.

In the present paper, results for laminar separation bubbles in boundary layer flows over smooth backward-facing steps are presented. The results clearly indicate a connection between hydrodynamic instability and the self-excited unsteadiness of the separated flow. These investigations also demonstrate that the numerical method is applicable for realistic simulations of highly unsteady, viscous flow phenomena in laminar separated flows.

NUMERICAL MODEL

The numerical method is based on the two-dimensional Navier-Stokes equations in vorticity-velocity formulation. For the numerical solution, a fully implicit finite-difference method of fourth order accuracy in space and second order in time was employed. For the numerical model, a finite integration domain is selected to represent a certain region of a boundary layer flow over a flat plate containing a smooth backward-facing step. In Fig. 1a, the integration domain is shown in real dimensions. At the inflow boundary the Blasius profiles corresponding to the chosen x-position are specified. Separation is induced by the surface geometry depending on the step height S. The length of the integration domain is chosen to be large enough so that the inflow and outflow boundaries are sufficiently far away from the generated bubble. The vertical extent of the integration domain covers several boundary layer thicknesses. In the numerical model, the calculations are performed on a rectangular grid in a rectangular domain A-B-C-D, as sketched in Fig. 1b. For this, the interaction model of Veldman [19] was implemented into the numerical method and used to prescribe an external velocity distribution $U_e(x)$ at the free-stream boundary C-D of the integration domain according to the pressure distribution generated by surface geometry. Details of the numerical method can be found in [16].

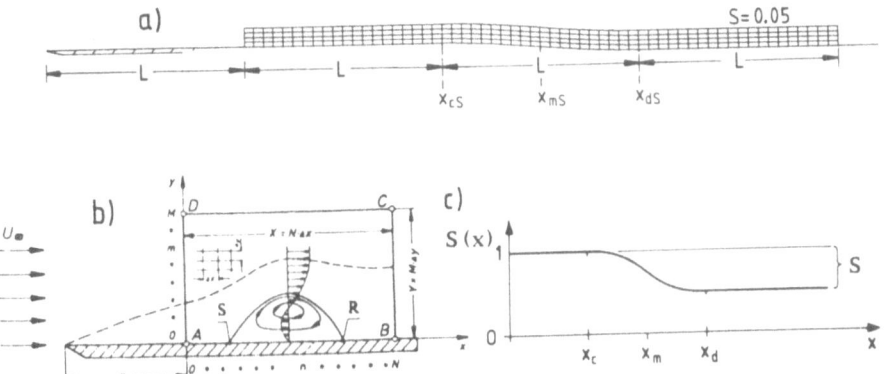

Fig. 1   Numerical model: a) Integration domain in real dimensions.
b) Integration domain in the numerical method. c) Wall contour.

With the numerical model the transient formation of a steady laminar separation
bubble is simulated. First, as initial conditions at t=0, a steady boundary-layer flow
over a flat plate (i.e. for S=0) is calculated, solving iteratively the steady Navier-
Stokes equations with U_e=1 prescribed at C-D. Then, for t >0, a surface contour
according to Fig. 1c is imposed impulsively (i.e. within several time steps of the
time-dependent calculation), and the reaction of the flow to the suddenly imposed
pressure gradient is then calculated solving the unsteady Navier-Stokes equations.
With this approach the time-dependent formation of a steady laminar separation
bubble can be simulated, if the step height S is small enough. In contrast to other
pseudo-unsteady calculations of steady flows, with our numerical method the
characteristic phenomena of the transient viscous flow development are also
simulated accurately. For larger step sizes the flow field becomes unsteady.

RESULTS AND DISCUSSION

For the calculations presented in this paper, if not otherwise specified, the length
of the integration domain extends from $Re(x_o)$=0.8 *10^5 (A-D) to $Re(x_N)$=3.2*10^5
(B-C), with the Reynolds Re based on U_∞ and reference length L (see Fig. 1a).
The Reynolds number Re_{δ_1} based on displacement thickness δ_1 of the Blasius boundary layer was Re_{δ_1}=486 at A-D. The global reference Reynolds number was Re =0.8*10^5.

Steady laminar separation bubbles were obtained for step heights up to S=0.05.
For S=0.04, in Fig. 2 the local values of the flow variables v and ω are plotted in
perspective representation versus x and y at time-level t=800Δt. The view is in the
direction away from the wall, looking slightly in upstream direction. For a rough
estimate of the amplitudes, the numbers given above and beneath each box at the
left hand side can be used. They indicate the maximal and minimal values of
the linear vertical scale. The distribution of the vorticity at the wall (proportional
to the wall shear stress) indicates the appearence of a laminar separation bubble
with a slightly asymmetrical distribution of the negative wall shear stress. The
normal velocity v is strongly affected; the large positive values far outside the
boundary layer indicate the displacement effect of the bubble. When the time-
dependent calculation was continued until t=1600Δt, the flow field did not change.
Apparently, a steady laminar separation bubble has been calculated.

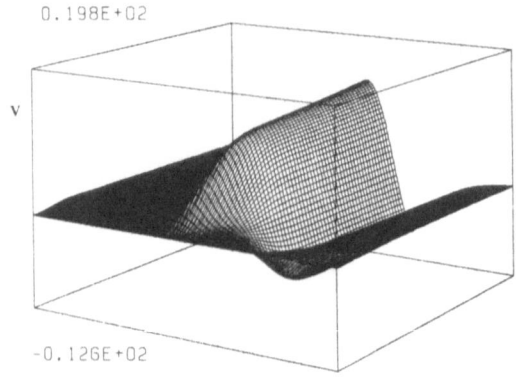

0.198E+02

v

-0.126E+02

0.498E+00

ω

-0.819E-01

x → y

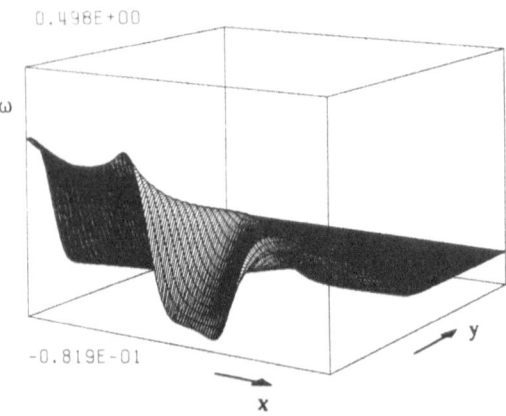

Fig. 2   Steady laminar separation bubble for S=0.04. Flow variables v and ω versus x and y in perspective representation

Likewise, for S=0.05 a steady solution was obtained from the transient flow development. The instantaneous streamlines plotted in Fig. 3 for time step t=300Δt show that a steady separation bubble has formed. Continueing the calculation in time the bubble slightly grows in size until t=550Δt, but then does not change as time increases further.

However, for a slightly larger step size S=0.052 no steady state solution was obtained, as can be seen from Figs. 4 and 5. The instantaneous streamlines in Fig. 4 show an unsteady development of the separation bubble that might be interpreted as unsteady separation: the separation bubble seems to split and to shed vortices in a nearly periodical fashion. However, from the instantaneous vorticity field, plotted in perspective representation in Fig. 5, it is obvious that these vortex—like structures are associated with wave-like vorticity fluctuations that appear in the rear part of the separated region and spread in downstream direction. For the wall vorticity, the fluctuation amplitudes are so large that negative values of the total vorticity appear periodically, which in the

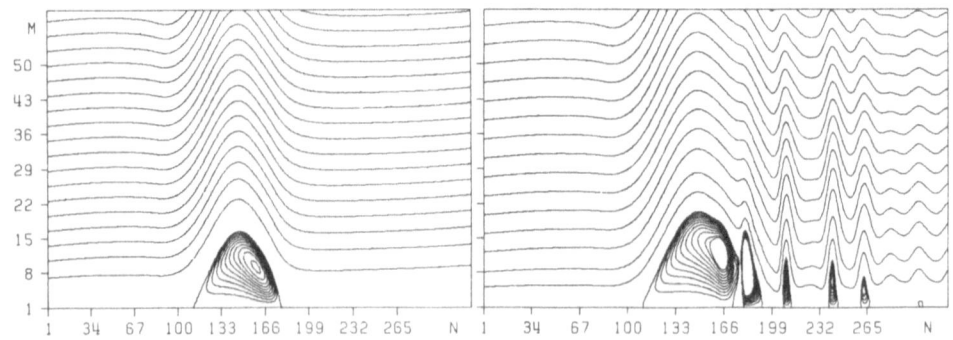

Fig. 3   Instantaneous streamlines at t=300Δt for S=0.05.

Fig. 4   Instantaneous streamlines at t=1150Δt for S=0.052.

76

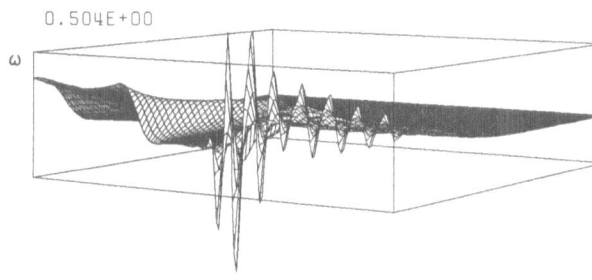

0.504E+00

ω

Fig. 5 Instantaneous values of total vorticity ω at t=1150Δt for S=0.052.

streamline plot show up as traveling regions containing local reversed flow. These wave-like vorticity fluctuations resemble the vorticity disturbances observed in the initial stages of transition in boundary layers.

In order to obtain a better understanding of the hydrodynamic stability characteristics of this type of a laminar separated flow, a calculation was performed for S=0.036, in which the steady separation bubble was perturbed periodically by a small amplitude Tollmien-Schlichting wave introduced at the inflow boundary of the integration domain. For this simulation, the length of the integration domain extends from $Re(x_o)=0.54*10^5$ (A-D) to $Re(x_N)=2.78*10^5$ (B-C). The Reynolds number Re_{δ_1} based on displacement thickness δ_1 of the Blasius boundary layer was $Re_{\delta_1}=400$ at A-D. The global reference Reynolds number was $Re=10^5$. First, a steady laminar separation bubble for S=0.036 was calculated. Then, a Tollmien-Schlichting wave with the dimensionless frequency $F=1.4$ ($F=10^4*\bar{\beta}\nu/U_\infty^2$) and an initial amplitude $a_0=u'_{max}/U_\infty=0.2\%$ was introduced at the inflow boundary A-D, and the reaction of the steady separated flow to the time-periodic disturbances are determined by the numerical solution of the unsteady

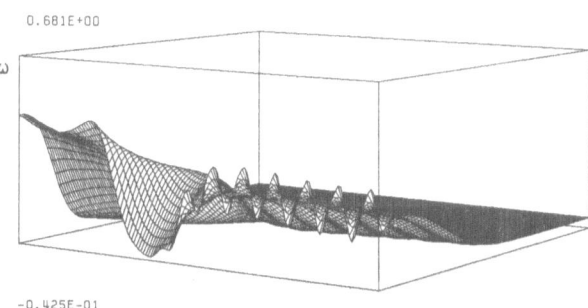

0.681E+00

ω

-0.425E-01

Fig. 6 Instantaneous values of total vorticity ω at t=600Δt for S=0.036 perturbed by a Tollmien-Schlichting wave.

Navier-Stokes equations. In Fig. 6, the instantaneous values of the vorticity ω, at time level t=600Δt after the disturbance input, are plotted in perspective representation. The disturbance waves have spread over the whole integration domain and have grown to large amplitudes downstream of the bubble. Apart from the different amplitudes, the instantaneous total vorticity fields in Figs. 5 and 6 look very similar.

The nonlinear disturbance evolution has been examined in detail by Fourier-analysis of the disturbance signals. The amplitude profiles undergo considerable changes in shape during the downstream evolution of the disturbances. This is shown in Fig. 7, where the profiles for the first harmonic of the u' disturbance at several different downstream locations are plotted versus y/Δy. For these plots, three different scales are used to account for the large differences in magnitudes. From the inflow boundary x=0 up to x=80Δx, the profiles are typical for a Blasius boundary layer, while the amplitude for the first maximum decreases from 0.2% (used for the disturbance input) to 0.03%. At x=100Δx the first indication of a change in the

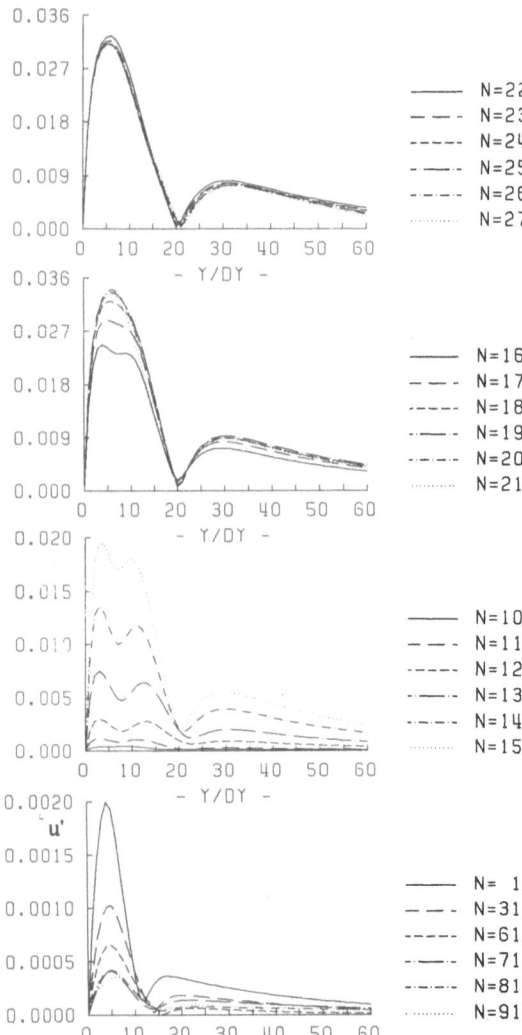

Fig. 7 T-S wave in separated boundary layer. Amplitude profiles for the first harmonic of u' at different downstream locations.

profile shape can be detected with the amplitude starting to grow again. The resulting distortion of the u'-profile leads to a third maximum in between the initial two maxima of the Blasius flow disturbance profile. Profiles of such shape have already been found in experimental measurements [13], as well as in stability calculations for boundary layer flows with or approaching separation. At x=160Δx, a degeneration of the profile shape can be observed while the amplitude is still growing. The third maximum disappears slowly and the resulting u'-profile at positions downstream of x=200Δx resembles that of the Blasius flow, in spite of amplitudes up to 3%. From Fig. 7 it is now obvious that the laminar separation bubble caused disturbance growth from 0.03% up to 3% in amplitude. Higher harmonics of considerable magnitude are also generated during the nonlinear disturbance development, which are not shown here for reasons of brevity.

Attempts were made to compare the results of the numerical simulation with results of linear stability theory calculations. Assuming a locally parallel flow, the spatial stability problem was solved where the velocity profiles of the base flow were taken from the steady Navier-Stokes solution at different downstream locations in the integration domain. For the solution of the Orr-Sommerfeld equation for spatial amplification a finite-difference method was employed. The local amplification rate α_i as well as the wave number α_r for a disturbance with F=1.4, obtained from the linear stability theory at different downstream positions, were compared in Fig. 8 with $\alpha_i(x)$ and $\alpha_r(x)$ determined from the first harmonic of the nonlinear wave (for the maximum of v'). The agreement for both, the amplification rate and the wave number, is surprisingly good.

The highly unsteady flow structures observed in the numerical simulation of a self-excited unsteady laminar separation bubble look very similar to the nonlinear wave structures appearing in a periodically disturbed bubble. Therefore, we conclude that these fluctuations are generated by an amplification process resulting from the hydrodynamic instabilty of the separated flow. Now, the question arises, however, why these fluctuations appear so sudden after the step height S exceeds a certain limit.

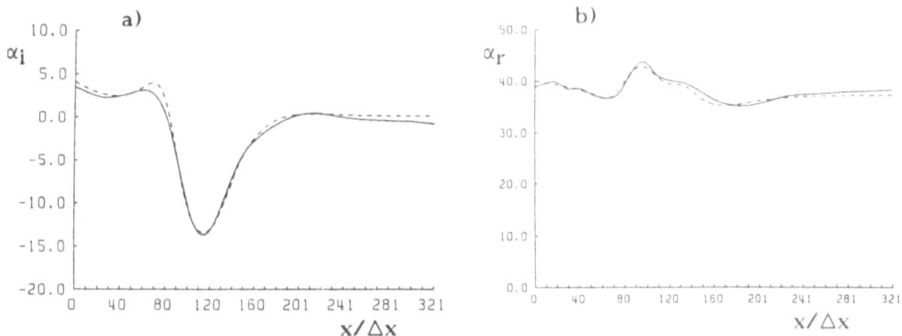

Fig. 8 Comparison of a) spatial amplification rate, b) wave number for S=0.036
(————— numerical simulation; - - - - - linear stability theory)

Based on experience from other numerical investigations we may have an explanation for this phenomenon. Looking at the numerical solution for "steady" laminar separation bubbles (i.e. for S≤0.05) in more detail, we found that a very small disturbance wave is present in the numerical solution. This is shown in Fig. 9, where for the flow variables v and ω the differences between two time steps of the numerical solution for a steady laminar separation bubble for S=0.04 are plotted. This disturbance wave is so small in amplitude that it is not visible when looking at the total flow variables, as plotted in Fig. 2, and therefore the flow appears to be steady. In the numerical model, a kind of "numerical turbulence level" is present due to truncation and round-off errors. Therefore, the very small but finite differences between the grid values feed disturbances into the oncoming Blasius flow. These disturbances are then amplified according to the local stability characteristics of the base flow. With increasing S the flow becomes more and more unstable leading to higher and higher amplification rates. When the amplification rates are large enough the resulting wave motion is visible even in the total flow variables and is then seen as an unsteady behavior of the separation bubble.

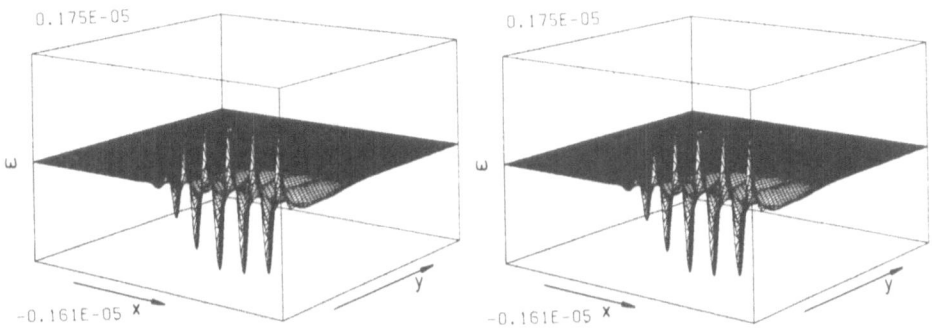

Fig. 9 Difference values for v and ω between time steps t=1449Δt and 1500Δt for a "steady" separation bubble for S=0.04.

ACKNOWLEDGEMENT

The support by the Deutsche Forschungsgemeinschaft, Bonn-Bad Godesberg, FRG, is gratefully acknowledged.

REFERENCES

1. Ward, J.W.: The behavior and effects of laminar separation bubbles on aerofoils in incompressible flow, J. Royal Aero. Soc. 67, 783-790 (1963).

2. Tani, I.: Low-speed flows involving bubble separations, Progress in Aeronautical Sciences, Vol. 5, 70-103 (1964).

3. Mueller, T.J.: Low Reynolds number vehicles, AGARDograph No. 288 (1985).

4. Eppler, R., Somers, D.M.: Airfoil design for Reynolds numbers between 50,000 and 500,000, in: "Proc. Conf. on Low Reynolds Number Airfoil Aerodynamics", Univ. of Notre Dame, June 1985, (T.J. Mueller ed.), UNDAS-CP-77B123, 1-14 (1985).

5. Carter, J.E., Wornom, S.F.: Solutions for incompressible separated boundary layers including viscous-inviscid interaction, NASA SP-347, 125-150 (1975).

6. Veldman, A.E.P.: A numerical method for the calculation of laminar, incompressible boundary layers with strong viscous-inviscid interaction, NLR Report TR 79023 U (1979).

7. Heidsieck, R.D.: Navier—Stokes solutions for laminar incompressible boundary layers with strong viscous-inviscid interaction, Report LR-353, Delft University of Technology, April 1982.

8. Briley, W.R.: A numerical study of laminar separation bubbles using the Navier-Stokes equations, J. Fluid Mech. 47, 713-736 (1971).

9. Vatsa, V.N., Carter, J.E.: Analysis of airfoil leading-edge separation bubbles, AIAA J. 22, 1697-1704 (1984).

10. Briley, W.R., McDonald, H.: Numerical prediction of incompressible separation bubbles, J. Fluid Mech. 69, 631-656 (1975).

11. Crimi, P., Reeves, B.L.: Analysis of leading-edge separation bubbles on airfoils, AIAA J. 14, 1548-1555 (1976).

12. Arena, A.V., Mueller, T.J.: Laminar separation, transition, and turbulent reattachment near the leading edge of airfoils, AIAA J. 18, 747-753 (1980).

13. Dovgal, A.V.: Development of vortical disturbances in flow with laminar separation, in: "Laminar-Turbulent Transition" (V.V. Kozlov ed.), Springer, Berlin, 359-366 (1984).

14. Fasel, H., Bestek, H., Schefenacker, R.: Numerical simulation studies of transition phenomena in incompressible, two-dimensional flows, in: AGARD CP 224 (1977).

15. Gruber, K., Bestek, H., Fasel, H.: Interaction between a Tollmien-Schlichting wave and a laminar separation bubble, AIAA Paper 87-1256 (1987).

16. Gruber, K.: Numerische Untersuchungen zum Problem der Grenzschichtablösung, Dissertation, Universität Stuttgart 1987, Fortschr.-Ber. VDI Reihe 7 Nr. 146, VDI-Verlag, Düsseldorf (1988).

17. Bestek, H., Gruber, K., Fasel, H.: Self-excited unsteadiness of laminar separation bubbles caused by natural transition, in: "Proc. Conf. on the Prediction and Exploitation of Separated Flow", The Royal Aeron. Society, London, April 1989.

18. Fasel, H.: Investigation of the stability of boundary layers by a finite-difference model of the Navier-Stokes equations, J. Fluid Mech. 78, 355-383 (1976).

19. Veldman, A.E.P.: A numerical view on strong viscous-inviscid interaction, NLR Report MP 83049 U (1983).

Global Stability Analysis of 2-D Flows with Closed Separation Bubbles

Marek Morzyński

Hermann Föttinger–Institut

Technische Universität, Berlin

Frank Thiele

Abteilung Turbulenzforschung

DLR, Berlin

Summary

The work presented here describes the investigations of the non-parallel flow stability. The numerical results consist of ones obtained for the circular cylinder, ellipsis and airfoil. For the circular cylinder the critical Reynolds number was sought. The investigation of the eigenvector pattern near the cylinder showed the difference between the flow below and above the critical Reynolds number.

The most important conclusion from this study is that both, the onset of the Karman vortex street and the Tollmien-Schlichting wave are two different aspects of the same phenomenon, represented by non-parallel flow stability theory. The Tollmien-Schlichting wave and the wake instability are two different eigensolutions of the same problem. This conclusion changes significantly the range of possible applications of the non-parallel theory. However, the new demands considering numerical method have to be fulfilled.

1 Introduction

Critical properties of the fluid flow can be concluded numerically on base of parallel flow stability analysis, direct simulation and recently via non-parallel flow linear stability calculations.

From the numerical methods capable to analyze the critical flow conditions only the classical flow stability theory has found the practical application, although with a number of experimental corrections. The direct simulation is still a distant task. It is characteristic for the practical use of the parallel flow theory that the description of some part of the fluid in non-adequate. To generalize the method to different flows and to include the global influences significant changes in its formulation are required. The non-parallel method offers here the potential alternative.

Non-parallel flow stability analysis has been successfully applied to study the wake stability [1, 2, 3, 4, 5]. As the approach is more general than the parallel theory, it can be expected that all the application area of the local analysis should be covered by the global one. Similar results should be obtained for the boundary and shear layer. The advantage of the global approach however should be the fact, that the flow is considered as the whole and the interactions of different flow parts (eg. wake or separation of the boundary layer) are taken into account simultaneously. The first, preliminary results of such an global analysis were already presented [6, 7, 8].

2 Formulation

The problem was solved in the pure (Lagrangian) stream function finite difference formulation. The unsteady incompressible Navier-Stokes equations written in the stream function formulation take the form:

$$\left[\frac{\partial}{\partial t} + (\nabla \times \vec{\psi}) \cdot \nabla - \frac{1}{Re}\Delta\right]\Delta\vec{\psi} = 0 . \tag{1}$$

We assume that the stream function is a sum of a steady part and the unsteady disturbance:

$$\vec{\psi}(x,y,t) = \bar{\psi}(x,y) + \vec{\psi}'(x,y,t) . \tag{2}$$

The disturbance value is assumed to be small compared to the stream function value. In the disturbance equation we separate the time and space dependence:

$$\vec{\psi}'(x,y,t) = \vec{\varphi}(x,y)e^{-i\lambda t} \tag{3}$$

where

$$\lambda = \pi(St + i\sigma) . \tag{4}$$

Introducing the above relationship into (1) and linearization results in the linear partial differential equation:

$$i\lambda\Delta\vec{\varphi} - (\nabla \times \bar{\psi}) \cdot \nabla\Delta\vec{\varphi} - (\nabla \times \vec{\varphi}) \cdot \nabla\Delta\bar{\psi} + \frac{1}{Re}\Delta^2\vec{\varphi} = 0 \tag{5}$$

which discretized, can be written as:

$$(A - \lambda B)\varphi = 0 \tag{6}$$

and represents the generalized eigenvalue problem.

3 Solution

The discretization of the Navier-Stokes equations (1) and disturbance equation (5) is accomplished using the finite difference method. In both cases the thirteen-point stencil was used. For all the calculations the orthogonal O-type mesh obtained by the conformal mapping is applied. The metric coefficients are expressed analytically to assure the maximum accuracy.

For the steady flow calculations the potential boundary conditions (the value of the stream function and its normal derivative) were used on the inflow. On the outflow the convective boundary conditions assured the free flow of vorticity through the boundary. The similar set was applied for the disturbance equation.

The inverse iteration method was used to determine the eigenvalues. Applying the Newton-Raphson method to equation (5) we obtain

$$(A - \lambda^{(n)}B)(\varphi^{(n)} + d\varphi^{(n)}) - d\lambda^{(n)}B\varphi^{(n)} = 0 \tag{7}$$

which can be written as:

$$(A - \lambda^{(n)}B)\eta^{(n+1)} = B\varphi^{(n)} . \tag{8}$$

The iteration process involves the repeated solution of the equation (8), normalization of the eigenvector and correction of the eigenvalue.

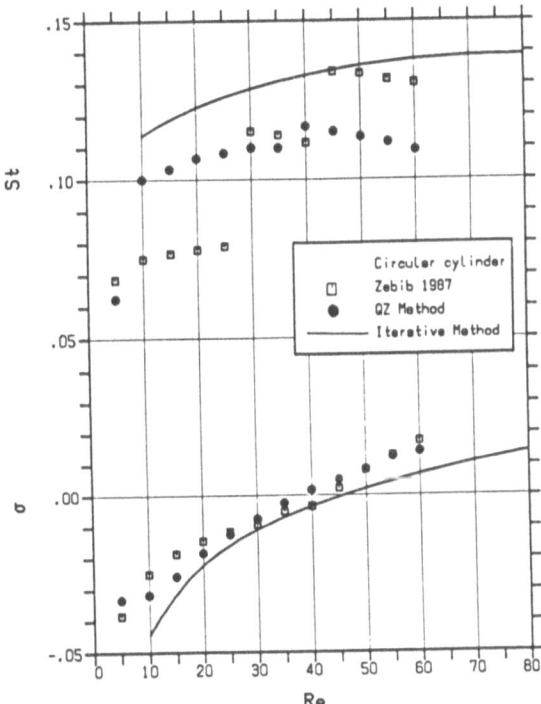

Figure 1: The growth-rate and the Strouhal number for the circular cylinder flow.

4 Numerical results and discussion

The direction of these investigations can be split into two parts. One is the study of the wake instability by means of non-parallel flow stability theory. The another is the generalization of the approach to give insight into the transition phenomenon.

4.1 Circular cylinder results

The result of the calculation consist of the complex eigenvalue for each Reynolds number together with a complex eigenvector. The growth-rate and the corresponding frequency as the function of the Reynolds number is shown in Fig.1. Some results of our previous investigations using the QZ method are also plotted. The results of these calculations are compared with those obtained by Zebib [4]. For the inverse iteration method, used in our computations, the critical values are $Re_c = 46.23$ and $St_c = 0.1345$.

Over a wide range of Reynolds numbers the eigenvector (disturbance) patterns are very similar, showing the physical aspects of the phenomena to be already present in flows of fairly small Reynolds number. The increase in Reynolds number allows these modes to cross the zero-growth-rate line and emerge as instabilities. The problem arises if there is any difference in eigenvector patterns bellow and above the critical Reynolds number. It is known from the parallel flow stability analysis that the wake stability is

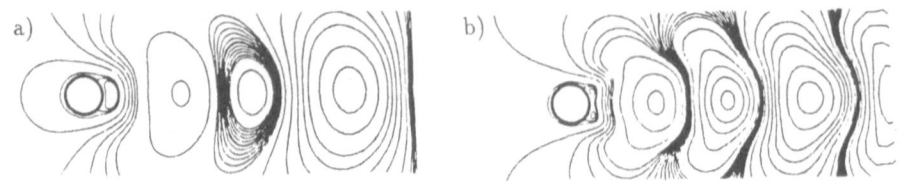

Figure 2: Disturbance streamlines: (a) below Re_c ($Re = 20$) (b) above Re_c ($Re = 60$).

governed by its characteristics in the vicinity of the rear stagnation point. Careful study of the eigenvector values near the cylinder shows (Fig.2) the difference in the disturbance patterns above the Re_c. This enhance the onset of the Karman vortex street.

4.2   Flow around an oblong ellipsis

The steady solution has been obtained for the infinity boundary condition at e^4. Than the results were interpolated on the mesh with e^2 distance for the farfield boundary conditions. For these data the stability of the wake was investigated. The eigenvector patterns similar to those obtained for the circular cylinder flow were obtained for the first bifurcation. The wake becomes unstable for $Re \simeq 130$.

For the same steady solution the higher mode was sought. The result of these calculations is shown in Fig.3.

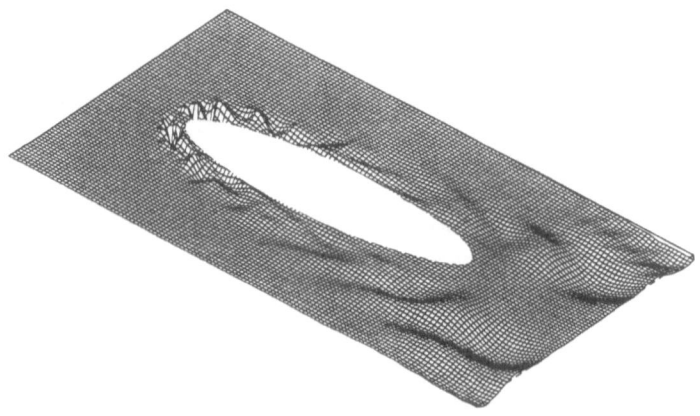

Figure 3: Higher mode eigenvector (real part) for the 1:5 ellipsis flow, $Re = 200$

The characteristic patterns for all higher modes investigated are the family of branches of disturbance streamlines having sequentially positive and negative values. Each branch is ended with a cell located in the vicinity of the maximum velocity gradients in the boundary or shear layer. The temporal evolution of the waves is shown in Fig.4. The amplitude of the wave is rising in the direction of the separation. The waves on the upper and lower surface of the ellipsis are shifted in phase as the result of superposition of the symmetric pattern of disturbances and antisymmetric stream function. The waves are relatively long, and since for $Re = 200$ the boundary layer is thick, such a result could be expected. The maximum Reynolds number, for which the stability analysis has been carried out was $Re = 600$. This maximum value is far below the critical value for the boundary layer. For this reason only the damped Tollmien-Schlichting waves have been found. Higher Reynolds number computations require much finer meshes resulting in significantly larger systems of linear equations.

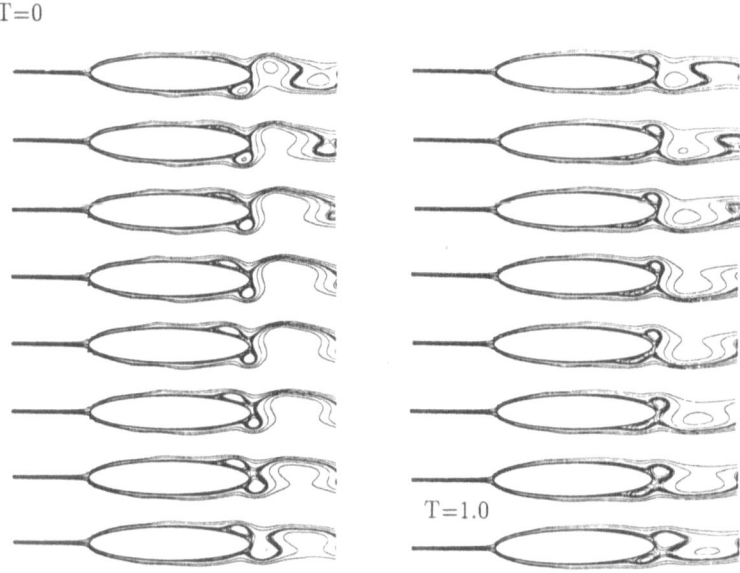

Figure 4: Tollmien-Schlichting waves - temporal evolution for the 1:5 ellipsis flow, $Re = 200$

4.3   Airfoil flow stability

The another cylinder flow which was considered is the airfoil flow. As the example geometry the NACA4412 airfoil is taken. Two different angles of attack were considered. For $\alpha = 15^0$ the stall is evident and the regular Karman vortex street appears for high enough Reynolds number. For $\alpha = 0^0$ dominating phenomena take place in the boundary and shear layer.

First the steady flow solution has been found. The character of the steady flow solution for $\alpha = 15^0$ is different from the circular cylinder one. While for the circular cylinder the wake consist ot two bubles, there is only one for the airfoil flow. The eigenvalue analysis gave the fastest growing mode. For $\alpha = 15^0$ the flow becomes unstable at $Re = 335$. The eigenvector patterns are in this case also very similar to ones for the circular cylinder. In Fig.5 the comparison between the real part of the eigenvector for $Re = 100$ and

$Re = 500$ is shown. The value of the disturbance is growing with the flow direction for both cases. It is normalized, so the disturbance reaches the same maximum, located in the vicinity of the outflow boundary. Because for $Re = 100$ (Fig.5) the growth-rate is negative the disturbance will be damped after a long enough time. The flow for $Re = 500$ is unstable. The disturbance is growing both in time and in the flow direction. The characteristic feature for the higher Reynolds numbers flows is the much larger amplitudes of the disturbance in the wake close to the airfoil.

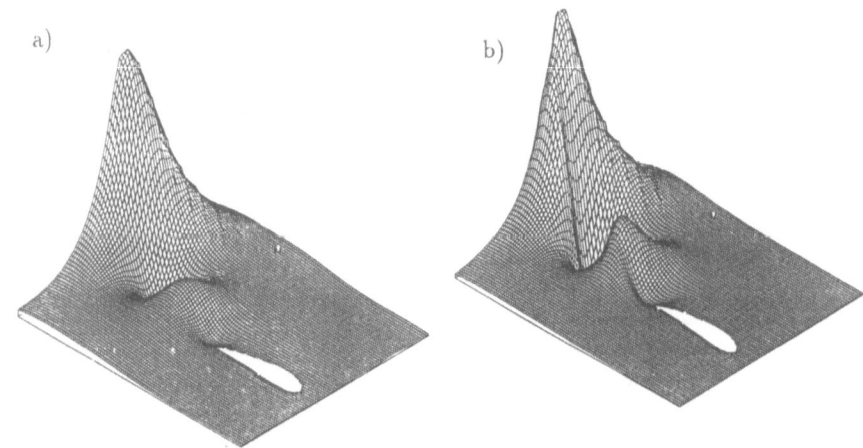

Figure 5: Real part of the eigenvector - airfoil flow, $\alpha = 15^0$ a) $Re = 100$, b) . $Re = 500$

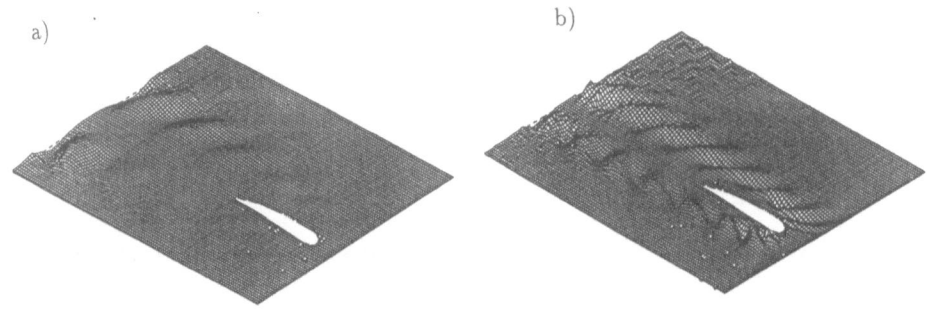

Figure 6: Higher mode solution for the NACA 4412 airfoil $\alpha = 0^0$, a) $Re = 300$, b) $Re = 900$

To compare the obtained eigenvalue analysis results with the real flow patterns the unsteady simulation was used (Fig.7). The simulation was performed for $Re = 1000$. The comparison of the flow patterns for $Re = 600$ (eigenvalue analysis) and $Re = 1000$

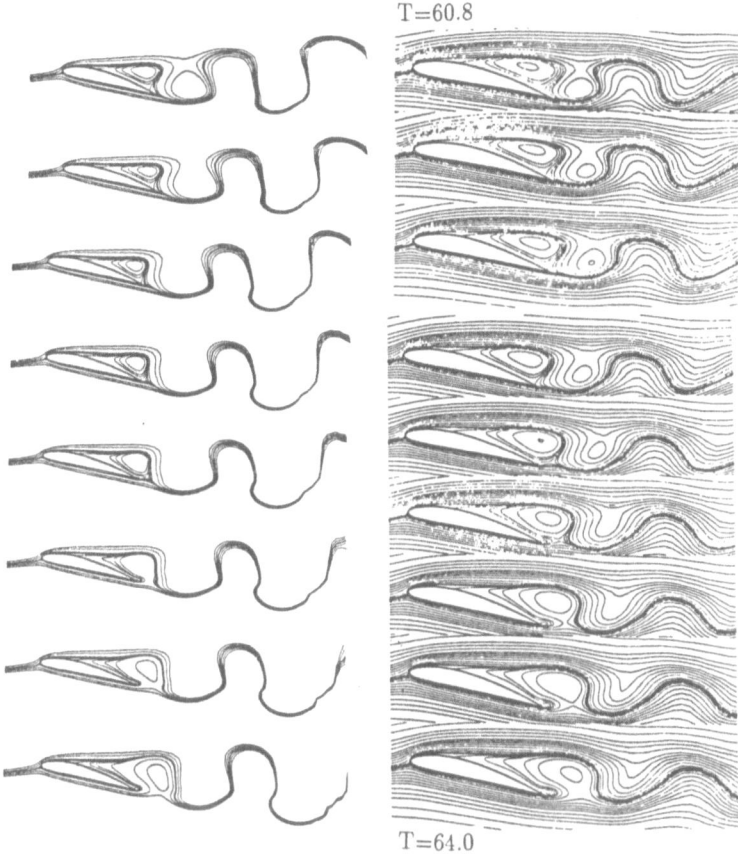

$T=60.8$

$T=64.0$

Figure 7: NACA 4412 airfoil flow: (a) superposition of the steady solution and disturbance fields, $Re = 600$, (b) unsteady simulation, $Re = 1000$

(unsteady simulation) show very good qualitative agreement. All the mechanisms of the vortex shedding are properly reproduced. This fact is one more proof that the Karman vortex street, especially near the body has the linear character.

For the angle of attack equal 0^0 till $Re = 800$ exists no separation on the airfoil. The higher mode solution forms two row of cells (Fig.6) which are close to the airfoil only near the leading edge. For increasing Reynolds numbers the cells are moving closer to the airfoil. The disturbances form now cells attaching the airfoil and forming the "wall" mode. The boundary layer is "modulated" in the way similar to the ellipsis flow. For $\alpha = 0^0$ the Karman vortex street mode also exists, although it is strongly damped for the small Reynolds numbers.

5   Conclusions

It was shown that non-parallel flow stability analysis is a method most suitable for determination of the wake flow instability. Several examples , calculated for different Reynolds numbers and geometries ranging from circular cylinder to the airfoil with the angle of attack, show that the method is a general tool for prediction of the wake instability. It is of advantage of this method, comparing to other numerical approaches, that the critical Reynolds numbers and respective frequencies are determined more precisely. The method is able to handle the unsymmetrical wake flow.

Using the same method higher modes were investigated for ellipsis and airfoil flow. It can be concluded, that the results obtained differ significantly from the first mode solution. The higher mode disturbance patterns added to the steady solution appear to be the Tollmien-Schlichting wave originating in the boundary layer. The wave propagates' further along the mixing layer. Both, the onset of wake instability and the Tollmien-Schlichting waves in the boundary layer are aspects of the same phenomenon. They are different solution of the same eigenvalue problem formulated using the non-parallel flow linear stability analysis.

The method on present stage of its development, with only modest modifications is suitable to analyse the simple wake flows, excluding the extreme difficult geometries. A wide range of interesting applications can be predicted. To increase the accuracy of the method and to step into the analysis of wall (Tollmien-Schlichting) modes, preliminary described here, a great deal of theoretical and numerical development has to be done.

References

[1] D.Wolter, M.Morzynski, H.Schütz, F.Thiele, Numerische Untersuchungen zur Stabilität der Kreiszylinder- strömung, Z. angew. Math. Mech., 69, (1989), 6, 601-604.

[2] M. Morzyński, F.Thiele, Numerical stability analysis of a flow about a cylinder, Z. angew. Math. Mech., 71, (1991), 5, T424-T428.

[3] C.P.Jackson, A Finite-Element study of the onset of vortex shedding in flow past variously shaped bodies, J. Fluid Mech. 182, (1987), 23-45.

[4] A.Zebib, Stability of viscous flow past a circular cylinder, J. of Eng. Math., 21, (1987), 155-165.

[5] I.Kim, A.J.Pearlstein, Stability of the flow past a sphere, J. Fluid Mech. 211, (1990), 73-93.

[6] M. Morzyński, F.Thiele, Non-parallel stability analysis of wake and boundary layer flow, Boundary Layer Control Conference, Cambridge, 9-12 April 1991, 15.1-15.11.

[7] M. Morzyński, F.Thiele, Global non-parallel stability analysis of wake and boundary layer flows, GAMM Jahrestagung, Kraków, 1-5 April 1991.

[8] M. Morzyński, F.Thiele, Stability investigations of airfoil flow by global analysis, Fifth Symposium on Numerical and Physical Aspects of Aerodynamic Flows, 13-15 January 1992, Long Beach, California.

EXPERIMENTAL INVESTIGATION OF LAMINAR SEPARATION BUBBLES

D.Althaus and W.Würz
Institut für Aerodynamik und Gasdynamik,Universität Stuttgart

INTRODUCTION

Laminar separation bubbles play an important role in the flow about airfoils at Reynolds numbers below 3 millions in incompressible flows. Drag as well as stall behaviour can be deteriorated considerably. Designing or analysing airfoils in this flight regime, with some success, requires understanding of the flow phenomena envolved in the formation of these regions with locally separated flow. Detailed experimental studies were carried out on laminar separation bubbles that formed near the midchord of different airfoils at Reynolds numbers from 0.7 to 3 millions. Static pressure, hot wire anemometry, and flow visualisation data were acquired. The data were used to evaluate the applicability of existing separation bubble models. Some details of the materials gathered are presented in the following context.

WIND TUNNEL

The Laminar Wind Tunnel of the Institute (Fig.1) is built as an open return tunnel of the Eiffel design [1]. The high contraction ratio of 100:1 and the screens result in a very low turbulence level of less than $2*10^{-4}$. The rectangular test section measures 0.73m $*$ 2.73 m and is 3.15 m long. The two dimensional airfoil models span the short distance of the test section. The gaps between the model and the tunnel walls are sealed. Blowing air tangential in the corner between the model and the mounting plates is used as a boundary layer control to ensure two-dimensional conditions.

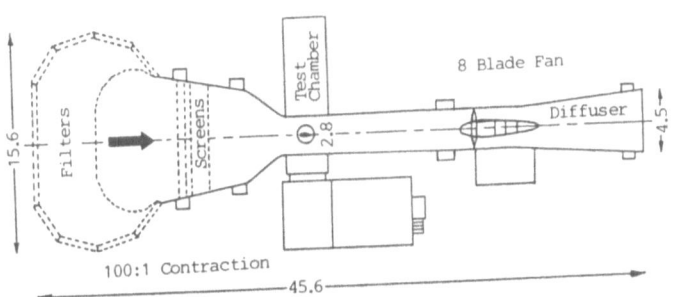

Fig.1: Laminar Wind Tunnel (dimensions in meters)

INSTRUMENTATION AND PROCEDURE

The boundary layers over the airfoils were surveyed by a Disa hot wire anemometry system in constant-temperature mode with a Disa type 55P15 single wire boundary-layer probe. The static

pressure distribution was obtained by using a static pressure
tube of 1mm diameter with its bores at the same streamwise
location as the hot-wire but 20mm apart.
The probes are mounted to a small traversing mechanism
(Fig.2b). A thin support resting on the airfoil surface
defines its position in relation to the wall. By means of a
high precision rack- and pinion drive together with an optical
encoder a resolution of 5μm in wall distance is achieved. To
enable boundary layer traverses vertical to the airfoil
surface this unit can be tiltet about its sting (Fig.2a),
which itself can be positioned at any station on the airfoil
by remotely controled DC-motors. The surface of the airfoils
is coated with a thin layer of graphite-spray which makes it
electrically conductive. To start traversing a boundary layer,
the hotwire probe is moved towards the wall until its prongs
touch the graphite thus closing an electric circuit with high
impedance which stops the motor. The direction of traverse is
then reversed and the probe moves until its contact with the
wall breaks. By this means eventual backlash and bending
effects are removed. This position is taken as the zero wall
distance.

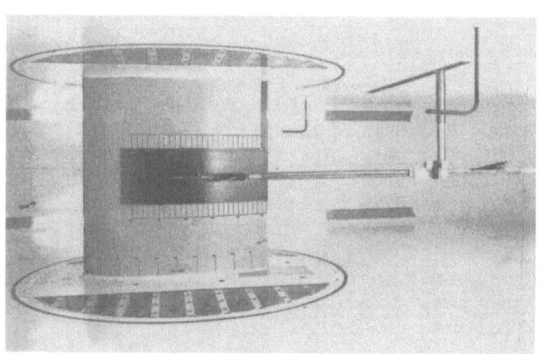 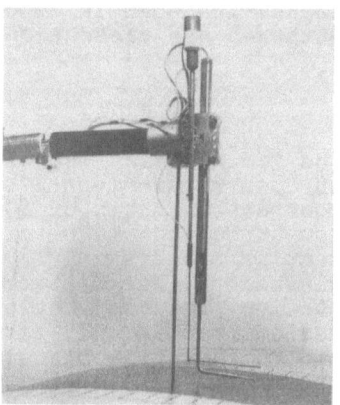

Fig.2: Traversing mechanism for boundary layer measurements

The output of the hotwire anemometer unit is fed to a Disa
55D10 Linearizer. The mean velocity component is integrated by
a low pass filter at 1. or .5 Hz. The fluctuating component is
measured by a Disa 55D35 true-rms-meter. The outputs of these
instruments including the voltages proportional to wall
distance, tunnel speed and static pressure are sampled by a
12-bit analog to digital converter connected to a PDP 11/34
computer. Moving of the probe, delay time for settling of the
instruments and sampling of the data are controlled by the
computer in a cyclic manner. Mean velocity and fluctuating
velocity are immediately plotted in relation to wall distance
as a boundary layer profile. In addition the hot-wire signal
is watched on an oscilloscope. Thirty to sixty points are
acquired in each of the profiles. The locations of these
stations were chosen according to the position of the
separation bubbles.
For FFT-analysis the fluctuating portion of the hot-wire
signal is amplified and low-pass filtered according to the

sampling rate. Normally, the filter is set to 5 KHz at a data rate of 10 KHz. Typical Tollmien-Schlichting frequencies at test conditions are in the order of 1 kHz. At several points of a station spectral data were preferably acquired at the position of maximum turbulence energy. Up to 4096 data points were sampled by a 12-bit a-d converter connected to a PC-80486 computer. The spectral data could be immediately inspected on the monitor, thus enabling to adjust the amplification of the signal.

DISCUSSION OF RESULTS

Length and thickness of separation bubbles depend on the pressure distribution of the outer flow and the conditions in the separation point. With growing Reynolds number and/or growing angle of attack both dimensions are reduced in mid-chord bubbles. The streamline which divides the "dead-air" region and the separated laminar shear layer prevents the exchange of flow between the regions. To find its position above the wall a heating wire with a diameter of 0.1mm and 200mm in length was placed parallel and close to the wall and vertical to flow direction within the "laminar part" of the bubble. This wire was periodically heated by an electric pulse. A temperature probe instead of a hot-wire is attached to the traversing mechanism and moved stepwise away from the wall. As long as this probe moves within the bubble it records the temperature pulses. They are recorded by a flashing light or by the oscilloscope. When the separating streamline is reached the pulses disappear. The height of the dividing streamline ψ is marked (s. Fig.8) in the boundary layer profile (aquired with the hot-wire) together with the position of the line u=0, which is at 2/3 ψ [2]. By repeating this procedure at different stations downstream of the separation, the contour of the streamline can be recorded. As the position

of the line u=0 is close to the height where the hot-wire begins to read u>0 and where the fluctuating velocity grows it is normally found by interactive computer graphics. The evaluation of the streamline data showed that they are nearly straight lines which are slightly bended upward when plotted over a straight axis (Fig.3 u=0 line). Its height, however, was measured from the airfoil surface, which has a curvature. Fig.4 shows the contour of the bubble u=0 line plotted over the airfoil surface. In Figure 5 the tangent of the separation angle of the line u=0 multiplied with the Reynolds number Re_{δ_2} based on conditions at separation, is plotted over m_{sep} which uses the derivative of the velocity distribution and the momentum thickness at separation. The circles represent different test series by their numbers. Dobbinga

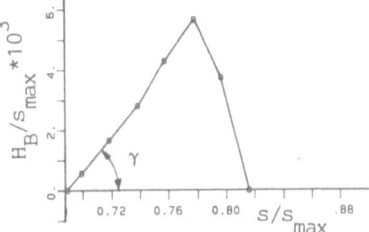

Fig.3: Bubble contour

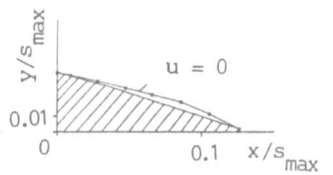

Fig.4: Bubble contour over airfoil shape

et al. [2] found that $B = \tan \gamma * Re_{\delta_{2sep}} = const = 15 - 20$.
The dotted line corresponds to

$$B = \tan \gamma * Re_{\delta_{2sep}} = 2.7 + 416.7 * m_{sep}^2$$

and is a reasonable good approach. Wortmann proposed $B = 64 * P$, where P is based on the gradient $\Delta u / \Delta s$ between separation and recovery. But the point of recovery is not known at the very beginning.

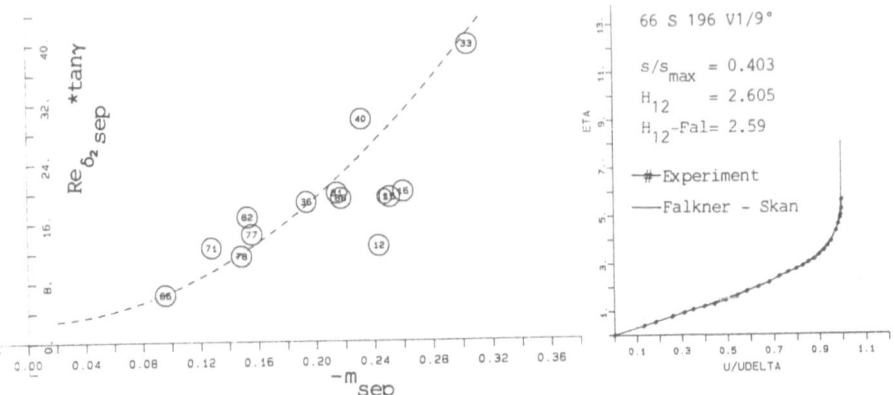

Fig.5: Separation angle γ for different measurements

Fig.6: Laminar boundary layer profile

Boundary layer conditions at separation are essential for modelling the separation bubbles. The velocity at the edge of the boundary layer, however, is modified by the bubble. Therefore it can not be used for a model. With this background the separation bubble is removed by a surface-roughness which causes transition just before the separation point. A new velocity distribution is aquired along the region of the bubble (Fig.10). This is used for boundary layer calculations and in modelling the bubble.
Comparisons are made between experimental and theoretical velocity profiles in the boundary layer. Fig.6 shows experimental data points together with a profile of the similar Falkner-Skan family with equal shape parameter H_{12}. The good correlation is a proof for the quality of the velocity and wall-distance data. When separation is approached however experimental profiles no longer conform to Falkner-Skan profiles (see Fig.7a). Fig. 7b shows that this profile can be matched by a function due to Liu and Sandborn [3]. In Figs.6 and 7a the wall distance is normalized by η.
The experimental profiles within the "dead air" region show zero velocity below the u=0 position (s.Fig.8) down to streamwise stations a short distance before the maximum bubble height is reached where fluctuating and mean velocity begin to grow when transition begins (Fig.9). For computing amplification rates by the Orr-Sommerfeld equation and for modelling the bubble velocity profiles in analytical form are required. In Fig. 8a it is attempted to approximate an experimental boundary-layer profile by a profile of the Falkner-Skan family which affords backflow in the wall region. The hot-wire is unable to distinguish the tangential velocity from the normal one and from reverse flow, but in Fig. 8 it shows no

flow at all. A Green profile [4] which is composed of a part with constant velocity in the wall region and a Coles wake-profile is a very good approximation as Fig. 8b shows. In Fig. 9a the hot wire measures small and constant velocity below the u=0 postion. The Falkner-Skan profile affords higher reverse flow than the Green profile in Fig.9b. An experimental profile in the turbulent part of the bubble can be approximated by a Green profile with or without backflow.

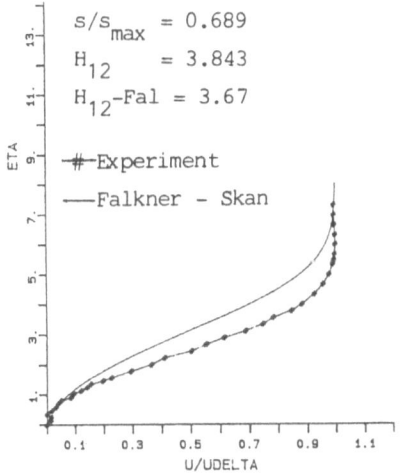 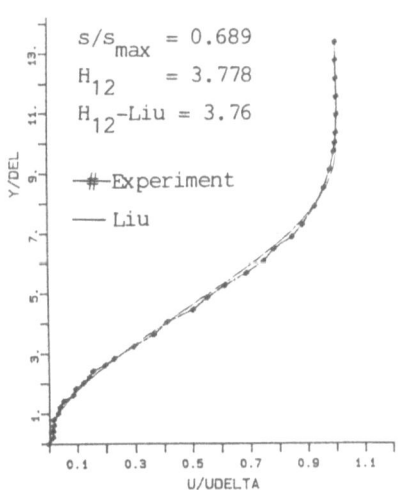

Fig.7: a) **Experimental and Falkner-Skan Profile at Separation**
       b) **Separation Profile from Liu and Sandborn**

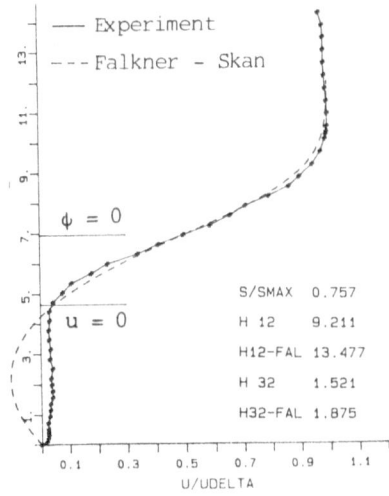 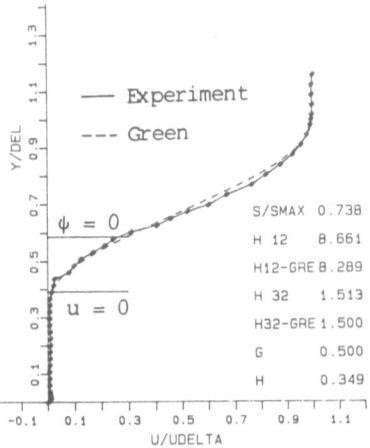

Fig.8: a) **Exp. and Falkner-Skan Profile in separated region**
       b) **Exp. and Green Profile in separated region**

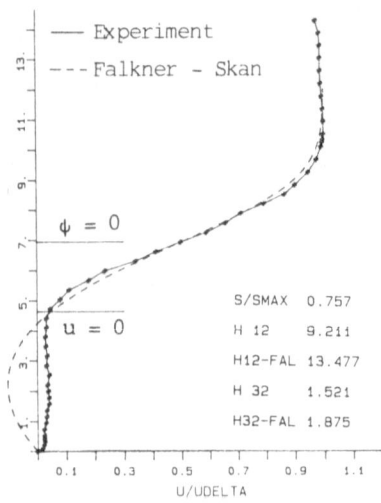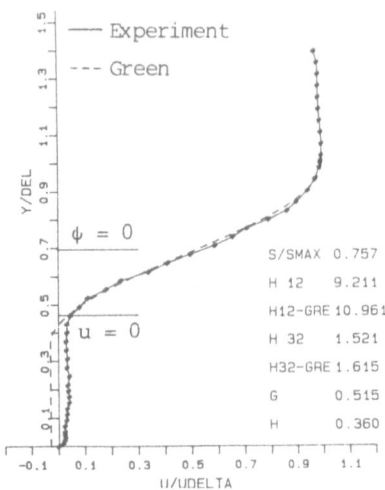

Fig.9: Exp. and analytical profiles in the separated region
       a) Falkner-Skan   b) Green

Hot-wire results are reliable in the laminar part of a laminar
separation bubble. Green profiles yield better approximations
to experimental points than Falkner-Skan profiles. This is
also confirmed by Fitzgerald and Mueller [5] who make
comparisons with Laser Doppler Anemometer measurements at low
Reynolds numbers.
Fig. 10 shows the differences between a boundary layer with a
separation bubble and a boundary layer in which transition is
tripped just before the separation point (dotted lines). The
deceleration of the velocity distribution now begins where
separation occured. The shape factor H_{12} is moderate but grows
towards the trailing edge. Despite the disturbance by the
roughness the momentum thickness is by far smaller than behind
the bubble resulting in less drag.
Detailed flow field observations were taken in the transition
region of a Wortman FX66-196 airfoil at a Reynolds number of
$1.5*10^6$ and zero angle of attack. At this condition a laminar
separation bubble forms on the upper surface between 0.44 and
0.51x/c. Mean and fluctuation velocity profiles as well as
band-pass filtered fluctuation profiles near the most amplified Tollmien-Schlichting frequencies were measured. The
results were compared with linear stability calculations and
show good agreement.
Figure 11 shows the shape of this bubble and the mean velocity
profiles together with the RMS-profiles at four measurement
stations. RMS-data are for a frequency range of 1Hz to 5kHz.
Spectral data were collected at the maximum of turbulence
energy, which belongs to the point of inflection of the mean
velocity profile. Amplification of frequencies in the range of
the TS-waves was first noticed at Point 1. All stations upstream have identical fluctuation spectra.

94

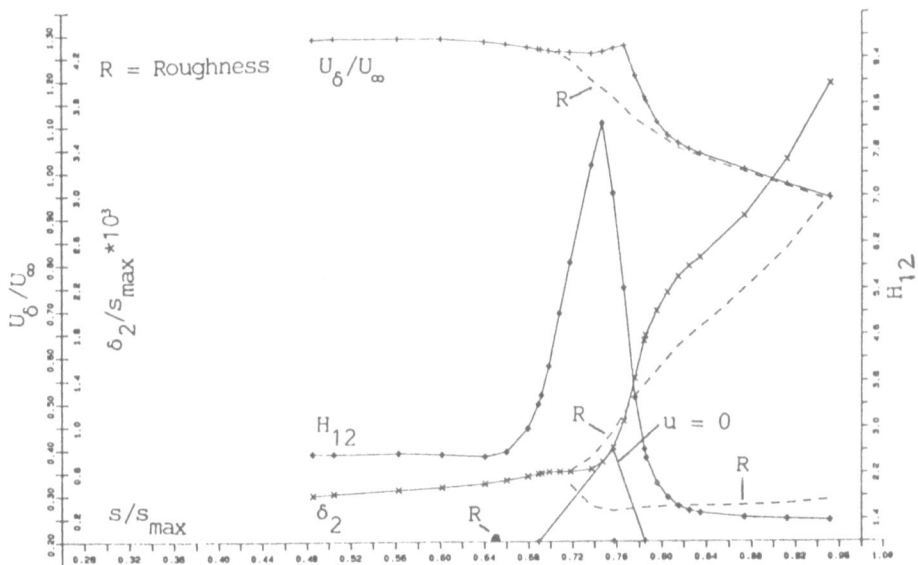

Fig.10: Boundary layer development with and without surface roughness, $Re = 0.7*10^6$

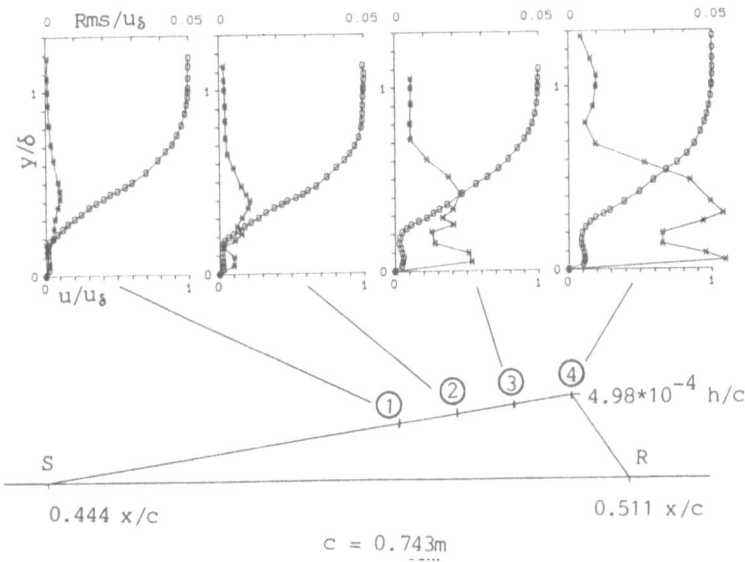

Fig.11: Mean and fluctuation velocity profiles inside a laminar separation bubble with sketch of bubble contour

With a Bruel & Kjaer frequency analyser type 2107 adjusted to an octave selectivity of 40dB band-pass filtered velocity fluctuations were measured through the boundary layer at each station. The filter center frequency was chosen near the most amplified TS-frequency. The result is plotted in Fig.12 to-

95

gether with the mean velocity profile. The measured points are denoted with symbols. Based on the mean velocity profile stability calculations were performed by direct solution of the Orr-Sommerfeld equation. To achieve convergence of the solver, the measured data points must be carefully splined and interpolated to get a minimum of 100 points. The splined mean velocity distribution and the resulting Eigenfunction of the u-fluctuation are shown as a solid line. The maximum of the Eigenfunction and the measured RMS-data are normalized to 1. Outside the bubble the shape of the Eigenfunction is in good agreement with the measured fluctuation velocity but at the upper edge of the bubble the measurements show an additional maximum. The maximum inside the bubble is overpredicted. The calculated amplification ratio $\alpha_i = \alpha_i^* * \delta_1$ for this frequency is -0.22, close to the measured one of -0.21.

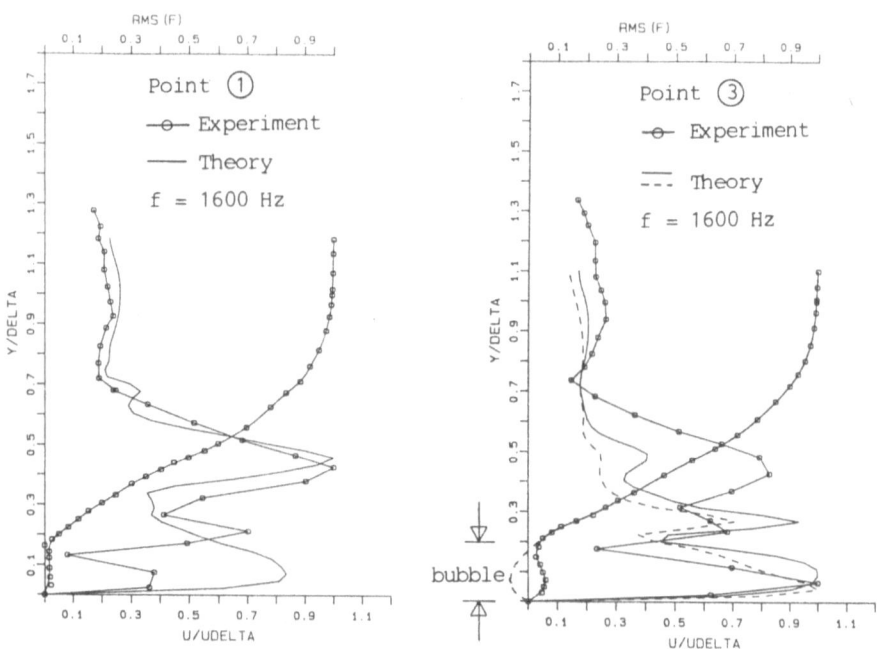

Fig.12: Exp. profile and   Fig.13: Additional calculation
       Eigenfunction of   for reversed velocity
       Orr-Sommerfeld DGL   inside bubble

Figure 13 shows the same comparison for a station (Point 3) only 0.013x/c downstream of the first one. This is nearly identical to one wavelenght of a TS-wave with 1600Hz as calculated for Point 1. The amplitude measured at the inflection point has grown by a faktor of 6.3. Inside the bubble a small mean velocity component is visible. The Eigenfunction and the u-fluctuation profile show only poor agreement outside the bubble but the first maximum inside the bubble and the second one slightly above the bubble is better predicted than for

Point 1. To examine the influence of the mean velocity direction, the velocity data inside the bubble were rectified and the calculations performed again. The result is also plotted in Fig. 13 as a dashed line. The shape of the Eigenfunction shows only astonishing small differences from the first one. The amplification ratio α_i calculated for the original profile is -0.26, with reversed velocity -0.32. The measured α_i begins to decrease and reaches only -0.15.

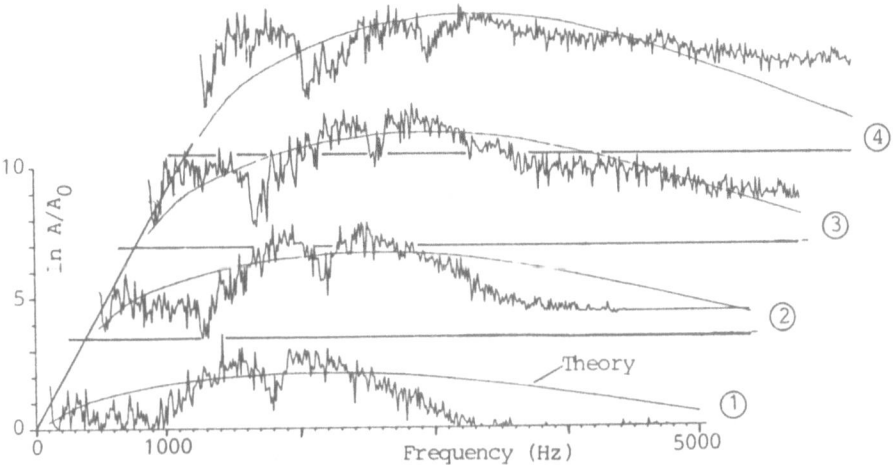

Fig.14: Comparison of amplitudes related to station 0.471 x/c

Fig. 14 shows an overall comparison of amplitude spektra. Measured and calculated amplitudes are related to the amplitudes at the last station upstream of Point 1. For clearness the zero-point is offset at each station and marked by a horizontal line. The growth of amplitudes in this region is well predicted by linear stability theory. It is remarkable that the maximum amplitude ratio ln A/Ao=n, which can be measured reaches only n=6. This problem is also observed by other experimentators [6]. It can be illustrated by plotting the measured spectral energy for a finite frequency range together with the expected spectral energy calculated upstream from transition using theoretical α_i . Fig.15 shows that the measured spectral energy levels out upstream of Point 1, whereas the calculations predict smaller amplitudes. Measurements outside the boundary layer show that the noise level of the hotwire anemometry is about a faktor of 10 smaller than the smallest observed amplitude inside the boundary layer. An explanation for this problem is not found yet.

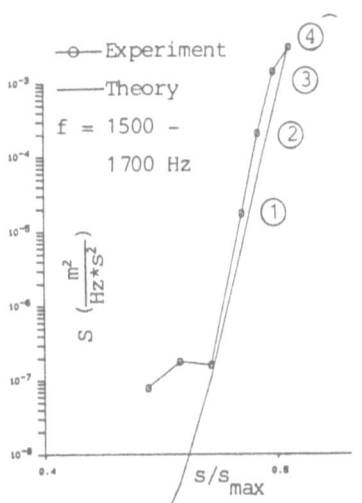

Fig.15: Energy spectrum extrapolated upstream

CONCLUSIONS

In the present paper only a short survey of the experimental work done in the past is conveyed. A lot of data have been gathered and compared with many existing empirical relations and bubble models [7]. It proves difficult to find correlations which conform with the different data sets. Evaluation is going on and additional experiments are needed.

ACKNOWLEDGEMENTS

This research was supported by the Deutsche Forschungsgemeinschaft under Grant AL242/1-3 and by the Institut für Aerodynamik und Gasdynamik der Universität Stuttgart.

REFERENCES

[1] F.X.Wortmann and D.Althaus:Der Laminarwindkanal des Instituts für Aerodynamik und Gasdynamik der Technischen Hochschule Stuttgart.

[2] E.Dobbinga,J.L. van Ingen and J.W.Kooi: Some research on two dimensional laminar separation bubbles. AGARD-CPP-102,1972.

[3] C.Y.Liu and V.A.Sandborn:Evaluation of the sepatation properties of laminar boundary layers.The Aeronautical Quarterly,Aug.1968,Vol.XIX.

[4] Green J.E.:Two dimensional turbulent reattachment as a boundary-layer problem.AGARD CP4,Pt.I,1966.

[5] E.J.Fitzgerald and Th.J.Mueller:Measurements in a separation bubble on an airfoil using Laser Velocimetry. AIAA Journal,Vol.28,Nr.14,1990.

[6] LeBlanc,P.,Blackwelder,R.,Liebeck,R. A Comparison between Boundary Layer Measurements in a Laminar Separation Bubble Flow and Linear Stability Theory Calculations. Department of Aerospace Engineering University of Southern California.

[7] D.Althaus,W.Würz:Schlußbericht für die Deutsche Forschungsgemeinschaft über das Forschungsvorhaben: Experimentelle Untersuchung laminarer Ablöseblasen. Institutsbericht (1988).

Computation of Separated Flow Using the Boundary–Layer Equations

Bernd Schalau

Hermann–Föttinger–Institut

Technische Universität, W–1000 Berlin 12

Frank Thiele

Abteilung Turbulenzforschung

DLR, W–1000 Berlin 12

Summary

This contribution is related to the computation of separated flow regions based on the boundary–layer equations. For laminar and turbulent flow cases the system of boundary–layer equations is solved simultaneously using the Hermitian finite difference approximation. In order to avoid the singularity at the separation point the solution procedure applies the inverse formulation.

Numerical calculations performed for wall boundary layers and wakes show that the inverse procedure is able to predict boundary layer flow with quite large separation regions without any numerical difficulties. With the extension to a viscous–inviscid interaction method the flow around airfoils can be calculated up to high angle of attack where the separated flow strongly affects the aerodynamic properties. The calculations are reported for several turbulence models with different degrees of complexity. From the results it is obvious that higher equation models do not perform better than simpler ones.

1   Introduction

For the calculation of boundary–layer flow with separation the full Navier-Stokes equations are required in general. Here, due to the very fine computational grid which is necessary to resolve the transport mechanism in the boundary layer region significant difficulties in the numerical solution might appear. In addition, the numerical solution of the Navier-Stokes equations can be very time consuming even for two–dimensional flows [1].

Using the boundary–layers equations attached flow can be computed quite successfully at high Reynolds numbers. However these equations become singular at the points of separation and reattachment. This singularity can be avoided by the inverse formulation of the boundary layer problem as the investigation of Catherall and Mangler [2] showed. The inverse formulation uses the displacement thickness as boundary condition rather than the boundary–layer edge velocity.

In comparison to the full Navier-Stokes equations an approximation to the exact solution is achieved by the concept of an interactive coupling between viscous and inviscid flow which is less time consuming but takes into consideration the viscous as well as the inviscid effects. According to the 'defect formulation' of Le Balleur [3] an approximate solution is obtained by superimposing the defect between local inviscid and viscous flow on the inviscid one. The coupling of the two solutions is achieved by an iterative process using an interaction law at the outer boundary.

In connection with the solution of the boundary–layer equations the main difficulty arises from the lack of an appropriate turbulence model for separated flow [4]. Investigations of complex turbulent flows have shown that presently no turbulence model is available which is valid universally. In general sophisticated models do not perform better than simpler ones. The eddy viscosity expressions usually applied run into difficulties when the boundary–layer flow tends to separate. In detached regions lag effects become important, hence the eddy viscosity is predicted lower in connection with the algebraic model. Johnson and King [5] showed that lag effects can be introduced through a transport equation for the maximum shear stress.

The investigation concentrates on the validation of the numerical method with respect to various laminar flow problems where separation occurs in internal as well as external flow configurations. The different turbulence model applied are used to evaluate the behavior of the approach in order to predict separated boundary layers. The main features for flows of engineering interest are demonstrated for the airfoil flow at high angle of attack where separation plays a dominant role with respect to flight characteristics and performance of airplanes.

2   Mathematical Formulation

2.1   Boundary-layer equations

The mathematical formulation of the flow problem will be restricted to incompressible and steady boundary layers in order to represent the main features of the procedure. The two–dimensional boundary layer flow is described according to Prandt's theory by the continuity equation

$$\frac{\partial u}{\partial x} + \frac{\partial v}{\partial y} = 0 \tag{1}$$

and the momentum equation in the main flow direction

$$\gamma u \frac{\partial u}{\partial x} + v \frac{\partial v}{\partial y} = u_e \frac{du_e}{dx} + \frac{\partial}{\partial y}\left[(\nu + \nu_t)\frac{\partial u}{\partial y}\right]. \tag{2}$$

Here, the FLARE approximation neglects the first term ($\gamma = 0$) in the detached flow region and $\gamma = 1$ is applied elsewhere.

2.2   Boundary conditions

For the airfoil flow given in Fig.1 the boundary–layer equations have to be solved according to the conditions for the wall boundary layer

$$y = 0 : u = v = 0; \quad y = y_e : u = u_e(x) \quad direct \tag{3}$$

$$\int_0^\infty \left[1 - \frac{u}{u_e}\right] dy = \delta_1(x) \quad inverse \tag{4}$$

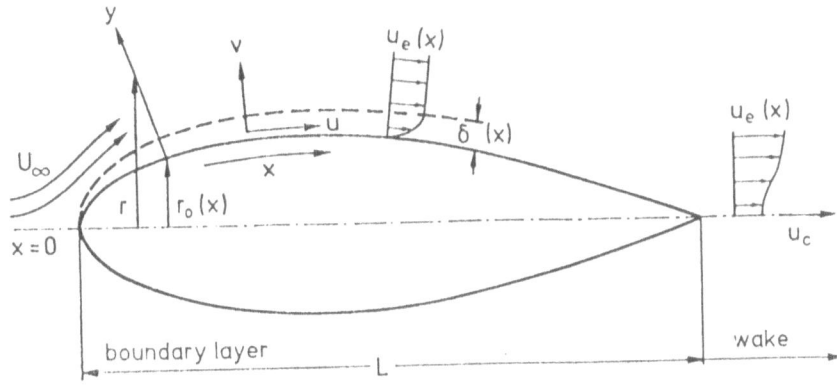

Figure 1: Schematic representation of the boundary layer flow.

and in the wake

$$y = 0 : v = 0; \quad y = \pm y_e : u = u_e(x) \quad direct \tag{5}$$

$$\int_{-y_e}^{+y_e} \left[1 - \frac{u}{u_e}\right] dy = \delta_1(x) \quad inverse . \tag{6}$$

Here, the dividing streamline in the wake is specified at the trailing edge. From the definition of the boundary conditions follows that the wake is considered as a single region where the boundary–layer equations are solved simultaneously for the upper and lower part of the wake.

2.3 Solution procedure

The boundary–layer equations can be cast into a more convenient form by using a coordinate transformation

$$d\eta = \frac{1}{\alpha}\left(\frac{u_r}{\nu x}\right)^{\frac{1}{2}} dy . \tag{7}$$

The grid parameter $\alpha(x)$ takes into account the difference between the growth of the considered boundary layer and that of a corresponding laminar boundary layer on a flat plate. Introducing a streamfunction which satisfies the continuity equation

$$\psi(x,y) = \alpha(\nu u_r x)^{\frac{1}{2}} f(x,\eta) \tag{8}$$

the boundary–layer equations are reduced to the third–order streamfunction equation

$$(bf'')' + \frac{1}{2}ff'' + xf'_e\frac{df'_e}{dx} = x\left[f'\frac{\partial f'}{\partial x} - f''\frac{\partial f}{\partial x}\right]. \tag{9}$$

The parabolic character of eq.(9) allows the finite–difference approximation in each direction x and η, separately. Applying the Newton–Raphson linearisation the marching procedure in the streamwise direction x leads to a linear ordinary differential equation. The advantage of the finite–difference method of Hermitian type used is that the first derivatives at the grid points are unknowns of the finite–difference approximation. Hence complex boundary conditions, such as for the inverse formulation or the wake, can directly be incorporated into the numerical scheme (see Fig. 2). A more detailed representation of the numerical procedure can be found in [6].

Figure 2: System of finite-difference equations for the wake flow and inverse formulation.

Y'_1 Y_2 Y'_2 Y_3 Y'_3 Y_4 Y'_4 Y_5 Y'_5 Y_6 Y'_6	R	Approx.
B_0 $\qquad\qquad\qquad\qquad G_0$	$R_0 = 0$	$Y'_1 = Y'_6$
B_1 C_1 D_1 E_1 F_1 $\qquad\qquad\quad G_1$	R_1 $[-\bar A_1 \bar Y_1]$	DGL j-1
B_2 C_2 D_2 E_2 F_2 $\big\rbrace\, j=2 \quad G_2$	R_2 $I-A_2 Y_1$	DGL j
B_3 C_3 D_3 E_3 F_3 $\qquad\qquad G_3$	R_3 $[-A_3 Y_1]$	DGL j+1
A_4 B_4 C_4 D_4 E_4 F_4 $\big\rbrace\, j=3 \quad G_4$	R_4	$Y''_3 = 0$
A_5 B_5 C_5 D_5 E_5 F_5 $\qquad\qquad G_5$	R_5	DGL j
$j=4 \big\lbrace\, A_6$ B_6 C_6 D_6 E_6 F_6 $\qquad G_6$	R_6	DGL j-1
A_7 B_7 C_7 D_7 E_7 F_7 $\qquad G_7$	R_7	DGL j
A_8 B_8 C_8 D_8 E_8 F_8+G_8	R_8 $[-F_8 Y_6]$	DGL j-1
$j=5 \big\lbrace\, A_9$ B_9 C_9 D_9 E_9 F_9+G_9	R_9 $I-F_9 Y'_6$	DGL j
A_{10} B_{10} C_{10} D_{10} E_{10} $F_{10}+G_{10}$	R_{10} $[-F_{10} Y_6]$	DGL j+1
A_{11} $\qquad\qquad\qquad E_{11}$ F_{11}	$R_{11} = 0$	δ_1

2.4 Interactive boundary–layer procedure

Due to limited space the description of the interactive boundary–layer procedure is not presented here. The basic formulation applied and the detailed presentation of the numerical approach are given in [7].

2.5 Turbulence models

To date no appropriate model is available for separating flows; thus, models developed for attached flows are often extended, by necessity, to predict flow fields including separation. The capability of four such models in predicting separated flows is examined here. The models of Cebeci and Smith (CS) [8] and Baldwin–Lomax (BL) [9] depend solely on the local velocity profile. The one–equation model of Johnson and King (JK) [5] applies an ordinary differential equation for the maximum Reynolds shear stress to account for non-equilibrium conditions; the model requires only a single initial value for the maximum Reynolds shear stress, which can be determined from an equilibrium approximation. The last model is the two –equation k-ε model, using the Low–Reynolds version suggested by Lam and Bremhorst (LB) [10].

3 Results and discussion

3.1 Laminar boundary–layer flow

The boundary-layer procedure developed to calculate separated regions is validated for laminar flow. Briley solved the Navier-Stokes equations for the flat plate flow with a given velocity distribution at the outer boundary-layer edge such that a recirculation region appears. With the displacement thickness prescribed the boundary-layer results [6] are presented in the vicinity of the separation bubble in Fig.3. Compared to Briley's solution the skin friction coefficient is in good agreement with the boundary-layer results.

In steady duct or pipe flows a sudden expansion of the cross section leads to a closed separation bubble downstream of that step. Fig. 4 shows the velocity distribution for

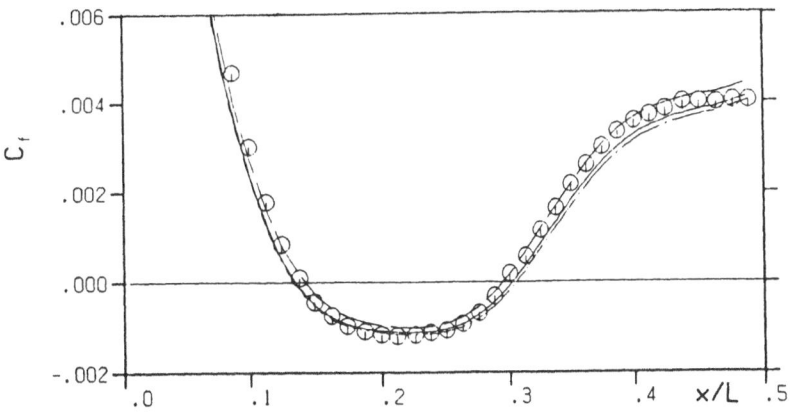

Figure 3: Skin friction coefficient for the Briley's flow (o NS eqs, —— BL eqs).

the flow over a plane symmetric sudden expansion [11]. The calculated velocity profiles agree with the measurements of Durst et. al. which are represented by the symbols. The inverse boundary-layer procedure has been extended to the swirling flow in sudden pipe expansion [12].

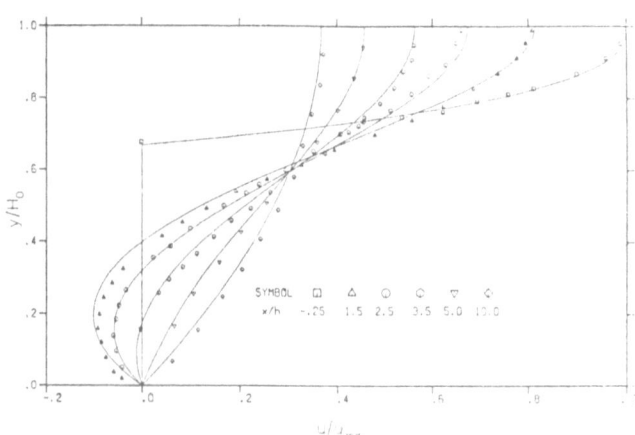

Figure 4: Velocity profiles downstream of the sudden expansion with h/H=1.

3.2   Turbulent boundary–layer flow

Various turbulent flows close to separation or with a recirculation region were investigated in order to evaluate the performance of the turbulence models described in section 2.5. The boundary layer along a flat plate approaches separation due to the prescribed pressure gradient in the experiments of Simpson. In Fig. 5 the skin friction coefficient along the plate is compared with the numerical results obtained with three turbulence models. For this case the JK model results are in good agreement with the measurements of Simpson in the region approaching separation. The CS and LB model lead to larger deviations especially with respect to the point of separation. Fig. 6 shows the turbulence

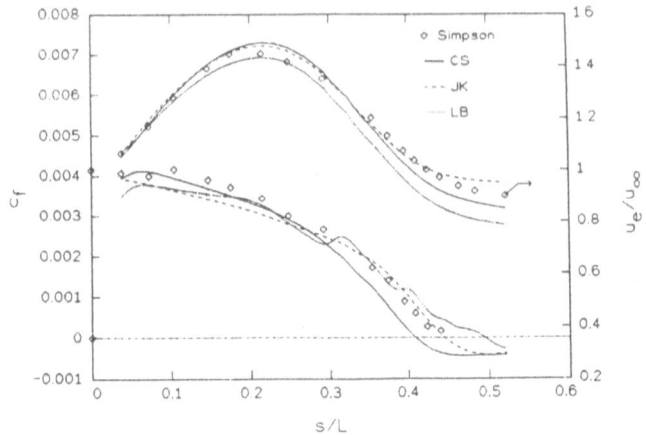

Figure 5: Distribution of skin friction and edge velocity for the Simpson test case.

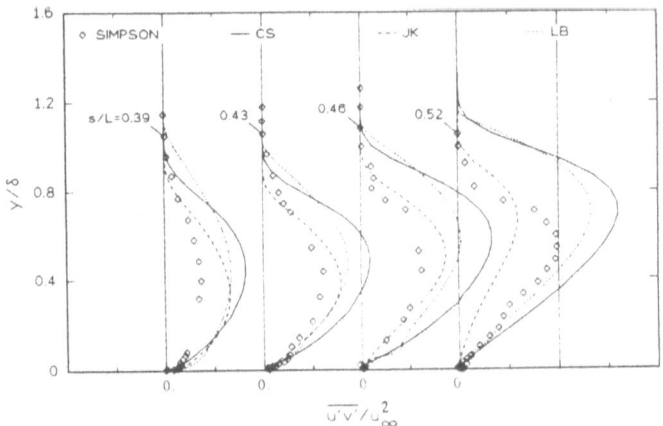

Figure 6: Reynolds shear stress profiles for the Simpson test case.

shear stress profiles for various s–positions. More or less all three turbulence models reproduce the profile shape but do not agree with the measurements. Dengel and Fernholz investigated experimentally a circular cylinder for different streamwise pressure gradients. The influence of the turbulence models (see Fig. 7) is similar to that observed for the Simpson test case. For a more detailed investigation the reader is referred to [13].

3.3 Interactive viscous–inviscid flow

For various angles of attack the flow around the airfoil NACA 4412 has been calculated in [14] by using the interactive boundary-layer procedure. The investigations confirm that the interactive method is able to calculate airfoil flow up to maximum lift where a separated region occurs near the trailing edge and extends up into the wake (Fig.8). The numerical results show a strong influence of the the turbulence models (CS, JK) especially for the separated boundary-layer.

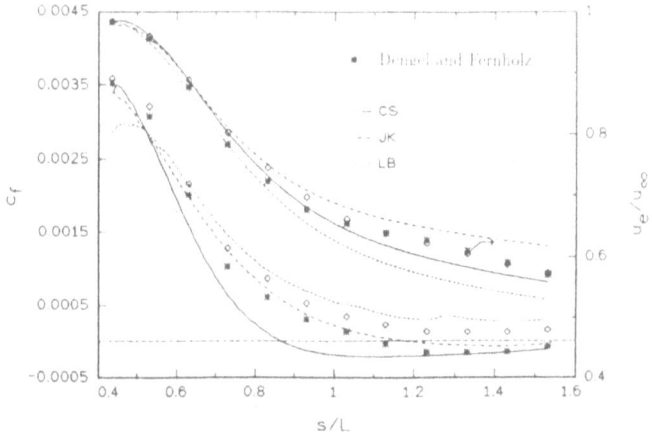

Figure 7: Skin friction distribution for the flow of Dengel and Fernholz. case 3.

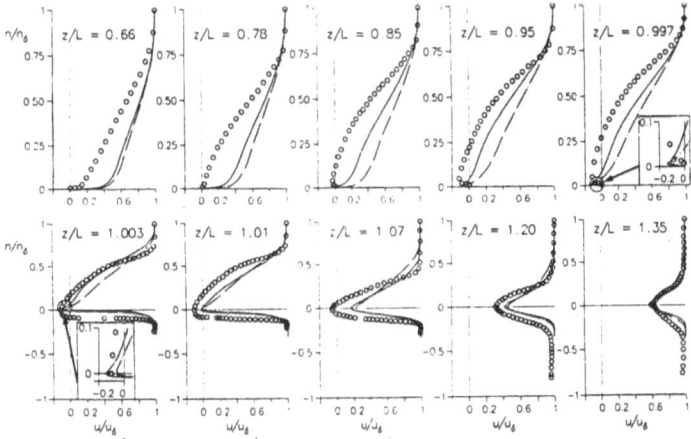

Figure 8: Velocity profiles at the upper surface and in the wake, calculated with CS model (NACA 4412, $\alpha = 12^0$, $Re = 4.2 \cdot 10^6$, o measurements of Hastings and Williams).

4 Conclusions

The investigations of laminar and turbulent boundary–layer flows with separation demonstrate that the boundary–layer equations can be applied to calculate detached regions using the inverse formulation without any difficulty. Even for strong interaction between viscous and inviscid airfoil flow at high angle of attack the interactive boundary–layer method is a reliable procedure to perform flow calculations with separation. Due to the deviations between the numerical and experimental results in the separated region subsequent investigations should be concerned with an improvement of turbulence models.

References

[1] W.J.Mc Croskey, Technical evaluation report on the Fluid Dynamics Panel Symposium on Applications of Computational Fluid Dynamics in Aeronautics, AGARD-AR-240 (1987)

[2] D.Catherall, K.W.Mangler, The integration of the two–dimensional laminar boundary–layer equations past the point at vanishing skin friction, J. Fluid Mech. 26 (1966), 163–182

[3] J.C.Le Balleur, Strong matching method for computing transonic viscous flow including wakes and separations. Lifting Airfoils, Recherche Aerospatiale 1981-3 (1981), 21–45

[4] S.J.Kline, B.Cantwell, G.L.Lilley (Editors), Complex turbulent flows; comparison of computation and experiment, Vols, I, II, III, Stanfort University 1982.

[5] D.A.Johnson, L.S.King, A mathematically simple turbulence closure model for attached and separated turbulent boundary layers, AIAA J. 23 (1985), 1684-1692.

[6] F.Thiele, Ein effektives Differenzenverfahren zur Berechnung komplexer zweidimensionaler Grenzschichtströmungen, IB 01/87 Hermann–Föttinger–Institut, TU Berlin, 1987.

[7] F.Arnold, Ein simultanes Lösungsverfahren mit exaktem Interaktionsgesetz zur Berechnung von inkompressiblen Profilumströmungen, Fortschrittsberichte VDI, Reihe 7, Nr. 188, (1991).

[8] T.Cebeci, R.W.Clark, K.C.Chang, N.D.Halsey, K.Lee, Airfoils with separation and the resulting wakes; Third Symposium on Numerical and Physical Aspects of Aerodynamic Flows. California State University, Long Beach, 1985.

[9] B.Baldwin, H.Lomax, Thin layer approximation and algebraic model for separated turbulent flows; AIAA Paper 78-257, 1978.

[10] C.K.G.Lam, K.Bremhorst, A modified form of the k-ϵ model for predicting wall turbulence; J. Fluids Engineering, 103, (1981), 456-460.

[11] F.Thiele, Berechnung von Kanalströmungen mit plötzlicher, symmetrischer Erweiterung auf der Basis der Grenzschichtgleichungen, ZAMM 67 (987), T333-T335.

[12] W.W.Baumann, F.Thiele, Calculation of separated swirling flows in sudden pipe expansion using boundary–layer equations. Proc. Numerical Methods in Laminar and Turbulent Flow 5/1 (1987), 704-715.

[13] B.Schalau, P.Dengel, F.Thiele, Computation of turbulent boundary–layer flow with separation - a critical evaluation of parameters influencing the numerical results. Advances in Turbulence 2 (1989), 377-382.

[14] F.Arnold, R.Reinelt, B.Schalau, F.Thiele, A viscous–inviscid interaction method for airfoils with separation including the wake flow. Proc. The Prediction and Exploitation of Separated Flow RAeS, London (1989), 61-69.

METHOD FOR CALCULATING THE FLOW AROUND AIRFOILS INCLUDING LOCAL SEPARATION BUBBLES AND MASSIVE SEPARATION

W. H. Isay, J. Marzi

Institut für Schiffbau der Universität Hamburg,
Lämmersieth 90, D–2000 Hamburg 60

ABSTRACT

An integral method for calculating boundary layers has been developed. It can be used in explicit or implicit form and covers attached as well as separated laminar or turbulent boundary layers. Wall curvature effects are included.

The method has been applied for calculating the flow around airfoils when flow separation occurs. Comparisons of the theoretical results with experimental data show, that the integral method can be used in a wide range of technically relevant applications. Further examples elucidate the effects of wall curvature.

REFERENCES

[1]   *Marzi, J.*: Ein Berechnungsverfahren zur Behandlung von Profilströmungen mit lokalen Ablösezonen und offenen Totwassergebieten, Doctoral Thesis of University Hamburg, 1988. See also: Bericht Nr. 481, Institut für Schiffbau der Universität Hamburg
[2]   *Marzi, J.*: Ein Berechnungsverfahren zur Behandlung von Profilströmungen mit lokalen Ablösezonen und offenen Totwassergebieten, Schiffstechnik, Bd. 35, S. 137–152, 1988

Turbulence Modeling for Compressible Separated Flows

J.A. Lieser, B. Ewald
Technische Hochschule Darmstadt,
Fachgebiet Aerodynamik und Messtechnik

Abstract

Because of efficiency, Computational Fluid Dynamics (CFD) in industrial application still makes use of the eddy-viscosity hypothesis. Problems with this hypothesis are well known, but second moment closure needs much more effort for programming and takes more computer time. Most test cases for separated flows use incompressible equations and implicit solvers. There is a lack of research on transonic test cases with separated flow, where density fluctuations are still negligible, but changes in the mean density are important. Turbulence modeling introduces some numerical difficulties which are different in various solvers. In aeronautical sciences often solvers are used that were developed for compressible Euler equations and expanded for Navier-Stokes equations and turbulence modeling. In this paper the implementation of various turbulence models (algebraic and two-equation models with wall functions) into an explicit and implicit finite-volume solver and some results for simple test cases and transonic flow with flow separation are presented. Comparison of algebraic and two-equation models are made and problems with flow separation are discussed.

1   Introduction

For the design of civil aircrafts or missiles and spacecraft the transonic Mach numbers are very important. The aim of this work is to study transonic and supersonic flow behind axisymmetric blunt bodies with or without jets. A solver for the compressible Navier-Stokes equations is used. The influences of turbulence have to be taken into account. The work is restricted on two dimensional axisymmetric flow because of increasing complexity of three dimensional flows.

Some common algebraic and two-equation models are applied and tests will be presented in this paper. Comparisons are made between the explicit and the implicit solver. Algebraic models need about half of the computer time that two-equation models need. Therefore, algebraic models are adapted and tested for special cases.

Reynolds numbers in aeronautics are high, so in the case of large flow separation the periodic part can be neglected and the time dependent equations can be converged to steady state using time stepping scheme togehter with turbulence models which describe the interaction between mean flow and turbulence.

After presentation of mean flow equations and fundamental equations for turbulence modeling comparison of numerical procedure and turbulence modeling will be made. The explicit finite volume method of Jameson [1, 2] has been expanded with the family of time stepping schemes of Beam and Warming [3].

2 Basic Equations

The basic equations are the conservation of mass, momentum and energy in the Reynolds averaged form. In the finite-volume method they are used in integral conservation form. The computer code solves the two dimensional plane or axisymmetric time dependent equations. They are written in cylinder coordinates:

$$\frac{\partial}{\partial t} \iiint_{V(t)} U \, dV + \iint_{S(t)} \vec{F} \cdot \vec{n} \, dS = \iiint_{V(t)} R \, dV \qquad (2\text{-}1)$$

with

$$U = \begin{bmatrix} \rho \\ \rho u_r \\ \rho u_z \\ E \end{bmatrix}, F_r = \begin{bmatrix} \rho u_r \\ \rho u_r u_r - \tau_{rr} \\ \rho u_z u_r - \tau_{zr} \\ E u_r - \tau_{rr} u_r - \tau_{rz} u_z + q_r \end{bmatrix}$$

$$F_z = \begin{bmatrix} \rho u_z \\ \rho u_r u_z - \tau_{rz} \\ \rho u_z u_z - \tau_{zz} \\ E u_z - \tau_{zr} u_r - \tau_{zz} u_z + q_z \end{bmatrix}, R = \frac{1}{r} \begin{bmatrix} 0 \\ -\tau_{\varphi\varphi} \\ 0 \\ 0 \end{bmatrix}.$$

The components of the stress tensor for axisymmetric flow are given by

$$
\begin{aligned}
\tau_{rr} &= \lambda^* \left[\frac{\partial u_r}{\partial r} + \frac{\partial u_z}{\partial z} \right] && + \lambda^* \frac{u_r}{r} && + 2\eta \frac{\partial u_r}{\partial r} && -p \\
\tau_{\varphi\varphi} &= \lambda^* \left[\frac{\partial u_r}{\partial r} + \frac{\partial u_z}{\partial z} \right] && + \lambda^* \frac{u_r}{r} && + 2\eta \frac{u_r}{r} && -p \\
\tau_{zz} &= \lambda^* \left[\frac{\partial u_r}{\partial r} + \frac{\partial u_z}{\partial z} \right] && + \lambda^* \frac{u_r}{r} && + 2\eta \frac{\partial u_z}{\partial z} && -p \\
\tau_{rz} &= \tau_{zr} = \eta \left[\frac{\partial u_r}{\partial z} + \frac{\partial u_z}{\partial r} \right].
\end{aligned}
\qquad (2\text{-}2)
$$

3 Turbulence Models

As mentioned earlier, the eddy viscosity hypothesis is used which is given in general form by

$$\mathbf{T}_t = 2\eta_t \mathbf{E} + \left(-\frac{2}{3}\eta_t \nabla \cdot \vec{u} - \frac{2}{3}\rho k \right) \mathbf{I}, \qquad (3\text{-}1)$$

where $\mathbf{T}_t$ is the Reynolds stress tensor. Thus in the basic equations described in the previous chapter the eddy viscosity has to be added to the laminar viscosity and the same equations as for laminar flow can be used.

To determine the eddy viscosity some more equations are needed, that describe the influence of mean motion on turbulence. In the case of algebraic turbulence models simple algebraic equations are used which describe only local effects. The general form of such an algebraic equation is

$$\eta_t = \rho u_t l_t f(y/\delta). \qquad (3\text{-}2)$$

u_t is a characteristic velocity and l_t a characteristic length. For the inner part of boundary layer we use the mixing length theory of Prandtl together with the damping function of van Driest as most algebraic models do [4, 5].

For the outer part, where the turbulence is not influenced by a wall and eddy viscosity is not a function of wall distance we use the Cebeci Smith model [4]

$$\eta_t = 0.0168 \, \rho \, u_e \delta_1 F_{\text{Kleb}} \qquad (3\text{-}3)$$

or the Baldwin and Lomax model [5]

$$F = y \left| \frac{\partial U}{\partial y} \right| \left[1 - e^{-y^+/A^+} \right] \tag{3-4}$$

$$\eta_{t,o} = 0.0168 \, \rho \, C_{\mathrm{cp}} F_{\mathrm{Wake}} F_{\mathrm{Kleb}}(y) \tag{3-5}$$

$$F_{\mathrm{Wake}} = \min \left\{ y_{\max} F_{\max}; 0.25 y_{\max} u_{\mathrm{diff}}^2 / F_{\max} \right\} \tag{3-6}$$

together with the intermittency function of Klebanoff [6] for $f(y/\delta)$.

For the region between wall bounded shear layer and far wake we have found a simple formulation based on two different physical models, the free shear layer and the far wake which gives good results and has a sounder physical background than commonly used interpolation formulas.

For the far wake the formulation of Rodi and Srinivas is used [7]

$$\eta_t = 0.064 \, \rho \, u_e \delta_1$$

$$\delta_1 = \int_{y(\omega=0)}^{\infty} \left(1 - \frac{\rho u}{\rho_e u_e} \right) dy \, . \tag{3-7}$$

This is a Cebeci Smith like model with modified Clauser constant. For the region directly behind the end of the wall we use an expression similar to the Baldwin and Lomax wake formulation (3-6) with the common expression for the thickness of shear layers, but with a different velocity scale, namely the velocity at the outer edge of the shear layer [8]

$$\begin{aligned} l_t &= u_{\mathrm{diff}} / \left| \omega \right|_{\max} \\ u_t &= u_e \, . \end{aligned} \tag{3-8}$$

In this way the increase of eddy viscosity is obtained in the region between the end of the wall and far wake, as it is expected by the model of free shear layer [9]. The equation for this near wake region becomes

$$\eta_t = 0.0168 \, \rho \, u_e \, 0.25 \frac{u_{\mathrm{diff}}}{\left| \omega_{\max} \right|} f(y/\delta) \, . \tag{3-9}$$

To switch from one model to the other, the eddy viscosity is calculated with both formulas and the lower value is choosen. Results will be presented in chapter 4.

The k-ϵ- and k-τ models with three transport equations are used:

$$\frac{D\rho k}{Dt} = \nabla \cdot (\eta \nabla k) + P_k + D_k - \rho \epsilon \tag{3-10}$$

$$\frac{D\rho \tilde{\epsilon}}{Dt} = \nabla \cdot (\eta \nabla \tilde{\epsilon}) + P_\epsilon + D_\epsilon - \Phi_\epsilon + E \tag{3-11}$$

$$\frac{D\rho \tau}{Dt} = \nabla \cdot (\eta \nabla \tau) + \frac{\tau}{k} P_k - \rho + D_\tau - \frac{\tau^2}{k} P_\epsilon + \frac{\tau^2}{k} \Phi_\epsilon \, .$$

Up to now we don't use the exact viscous terms of Speciales k-τ model [10], because we are not interested in the very near wall behavior of turbulence and numerical problems arise in the exact viscous terms when k and τ approach to zero at the wall.

The production term in cylindrical coordinates is given by

$$P_k = \frac{\eta_t}{\eta} \left[\tau_{zz} \frac{\partial u_z}{\partial z} + \tau_{zr} \left(\frac{\partial u_z}{\partial r} + \frac{\partial u_r}{\partial z} \right) + \tau_{rr} \frac{\partial u_r}{\partial r} \right] - \frac{2}{3} \rho k \left(\frac{\partial u_z}{\partial z} + \frac{\partial u_r}{\partial r} \right) . \tag{3-12}$$

Turbulent diffusion is modeled with the gradient diffusion hypothesis for all equations. The production of dissipation is assumed to be proportional to the production of turbulent kinetic energy. Production and dissipation is modeled as

$$P_\epsilon = f_1 c_1 \frac{\epsilon}{k} P_k, \ \Phi_\epsilon = \rho f_2 c_2 \frac{\epsilon^2}{k} . \tag{3-13}$$

As we want to make turbulence modeling simple for flow fields which include both near wall and free turbulence, we use transport equations with wall functions according to equation (3-13). Wall functions are empirical functions which include wall damping more or less similar to the van Driest function [11] and describe known effects of the wall on the turbulence quantities k, ϵ or τ. They are optimized for non-separated boundary layers, and show similar problems with respect to separation as the van Driest damping.

Some models with different wall functions are implemented and tested, like the models of Jones and Launder [12], Chien[13], Lam and Bremhorst[14], Zhang-So and Speziale[15], Speziale and Abid[10].

4   Results

4.1   Comparison of Numerical Procedures

Couette flow is calculated to compare convergence of the explicit and the implicit time stepping scheme for viscously dominated flow. Figure 1 shows convergence of momentum equation for the explicit scheme with the optimum CFL number and for the implicit scheme with different CFL numbers. It is obvious, that because of the higher possible CFL number, the convergence is reached much faster by the implicit scheme. The main reason for higher possible CFL numbers are the implicit boundary conditions. The same result has been obtained for two-equation models. Decrease of wind tunnel turbulence has been calculated explicitly and implicitly. Convergence is much faster with the implicit scheme as figure 1 shows. All test cases with low Mach numbers have therefore been calculated with the implicit scheme and computer time could be reduced by a factor of ten. For flows with higher Mach numbers and significant nonlinear convective terms the CFL number of the implict scheme has to be reduced because of the linearisations of the used scheme. There is no advantage of the implicit scheme at high Mach and Reynolds numbers.

4.2   Flow behind a Flat Plate

The new algebraic model has been applied for the flow behind a flat plate and compared with the two-equation model. Figure 3 shows the eddy viscosity behind the flat plate calculated with equations (3-7) and (3-9). The lower value is taken and compared with the experimental value which is calculated from the velocity and Reynolds stress profile measured by Pot [16]. Figure 2 shows Reynolds stress profiles behind the plate in the near wake $(x/\delta_2 = 45)$ and in the far wake region $(x/\delta_2 = 900)$ in comparison to experimental values and the two-equation model. Results are as good as the results from the two-equation model and computer time for calculation is about half. No constants were adapted for this special case.

4.3   Seperated Flow about Axisymmetric Nozzle Afterbodies

In figure 5 the geometry and the mesh for the axisymmetric nozzle afterbody is shown. Experimental results are published by Benek [17]. The point of separation is not fixed and this is an critical test case for turbulence models what is calculated by use of the models of Jones and Launder[12], Chien[13], Zhang, So and Speziale[15] and the algebraic model of Cebeci and Smith [4]. Figure 6 shows velocity vectors for the Jones and Launder model. In figure 7 velocity and Reynolds stress profiles at the smallest diameter ($x/D = 0.8$) are compared to the experimental results. Velocity profiles differ appreciably from the experimental values because of two main reasons. First, the calculated eddy viscosity in the recirculating boundary layer is too high. Experimental values and the criterion for relaminarisation given in [18]

$$\frac{u_\tau^3}{\alpha \nu} \approx 30, \quad \alpha = \frac{1}{\rho}\frac{\partial p}{\partial x} \tag{4-1}$$

let us assume that the boundary layer may be laminar. But the described turbulence models are not capable of predicting relaminarisation. The second reason is that the velocity gradient in the outer shear layer is too low. This may be due to the fact that the mesh is not fine enough. With the used stretching functions in the grid generator we can not concentrate mesh points at two positions. So we have to refine the mesh in the whole separation zone and for that a lot of mesh points and computer time is needed. Another problem may be that all used two-equation models are optimized for boundary layers without or with low pressure gradients.

For the calculation all wall functions similar to van Driest damping had to be modified. The exponent in the damping function is calculated by an expression suggested by Cebeci and Smith [4]

$$l_m = \kappa y \left\{ 1 - \exp\left[-\frac{y}{\eta A^+} \left(\rho \left(\tau_w + \frac{dp}{dx} y \right) \right)^{\frac{1}{2}} \right] \right\} . \tag{4-2}$$

The Jones and Launder model calculated the separation point quite good without modifications, because it does not use van Driest like damping. In all other models the separation point without modification was calculated too far in front. Figure 4 shows the pressure contour for all models.

5   Conclusions

Algebraic and two-equation models give good results for compressible and incompressible non-separated flows. Algebraic models and two-equation models with wall functions are adapted to nonseparated flows with moderate pressure gradients. In the case of separated flows and high pressure gradients they have to be modified. Mesh generation is important for turbulence models and critical for separated flows. In the case of flow behind blunt bodies the inability of the eddy viscosity hypothesis to deal with stream line curvature becomes obvious. It is well known that normal Reynolds stresses are predicted inaccurate by linear two-equation models. This will be taken into account by future work.

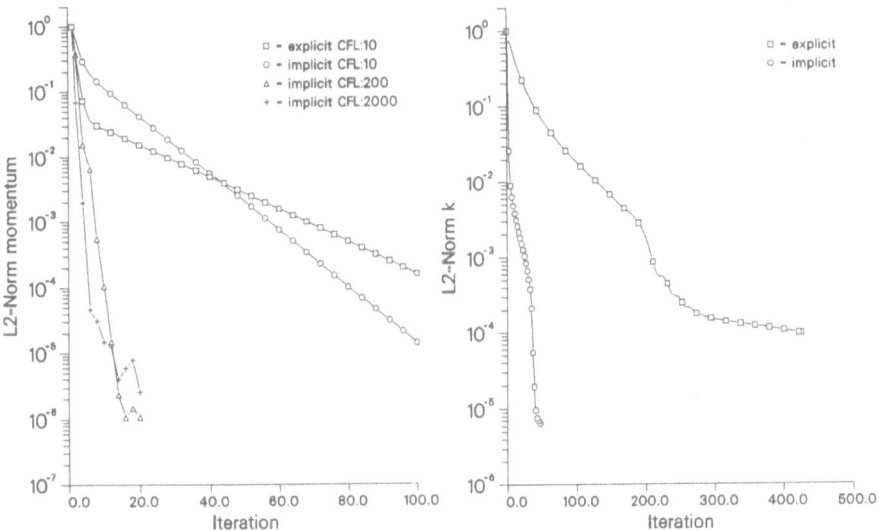

Fig. 1: Comparison of convergence with explicit and implicit time stepping scheme for couette flow and wind tunnel turbulence

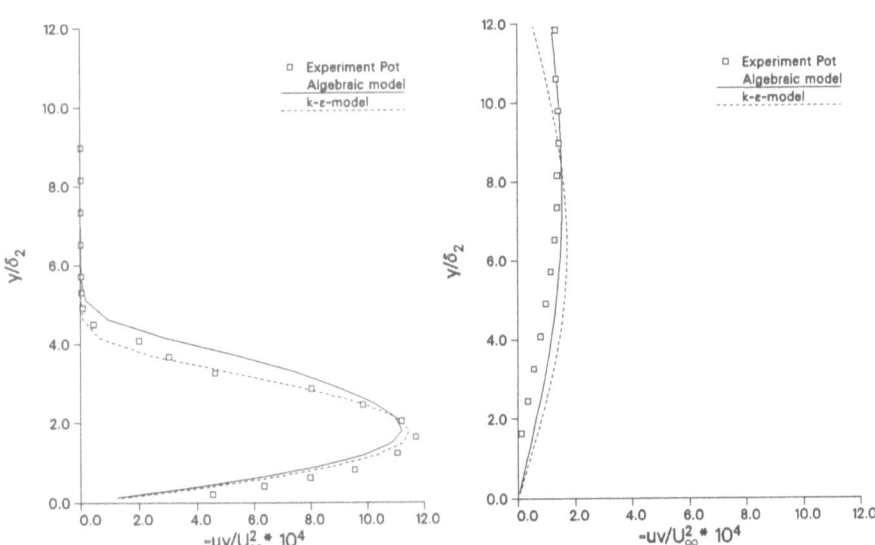

Fig. 2: Comparison of Reynolds stress profiles for the algebraic model and the two-equation model with experimental results in the near wake ($x/\delta_2 = 45$) and the far wake region ($x/\delta_2 = 900$)

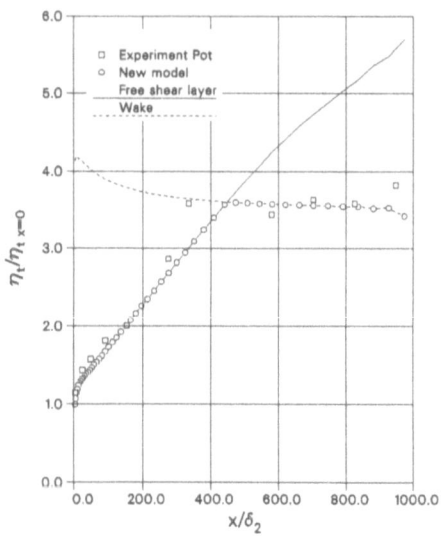

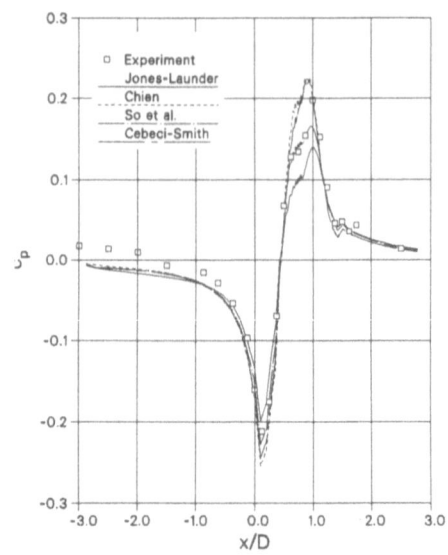

Fig. 3: Maximum eddy viscosity behind a flat plate

Fig. 4: Wall pressure distribution.

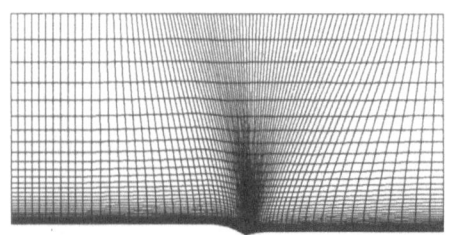

Fig. 5: Axisymmetric nozzle afterbody[17]; Computational mesh.

Fig. 6: Velocity vectors for axisymmetric nozzle afterbody [17]

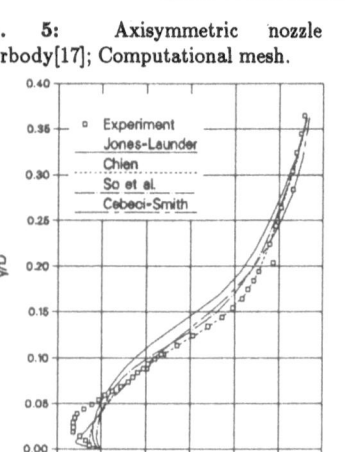

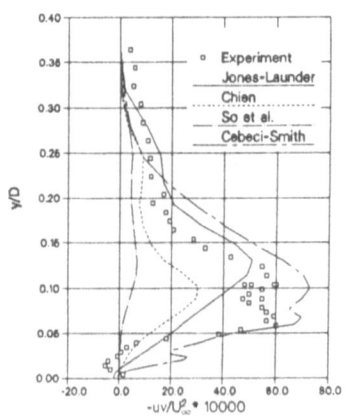

Fig. 7: Velocity and Reynolds stress profiles at the smallest diameter ($x/D = 0.8$). Comparison of different models with experiment.

References

[1] Jameson, A.: *Transonic Flow Calculations*. MAE Report 1651, Princeton University, 1983.

[2] Lieser, J. A.: Explizites Finite-Volumen-Verfahren zur Lösung der Navier-Stokes-Gleichungen. , Bericht, FG Aerodynamik A 81/90, Technische Hochschule Darmstadt, 1990.

[3] Beam, R. M. und Warming, R. F.: An Implicit Factored Scheme for the Compressible Navier-Stokes Equations. *AIAA-Journal April*, Vol. 16, p. 393, 1978.

[4] Cebeci, T. und Smith, A. M. O.: *Analysis of Turbulent Boundary Layers*. Applied Mathematics and Mechanics: Academic Press, New York, 1974.

[5] Baldwin, B. S. und Lomax, H.: Thin Layer Approximation and Algebraic Model for Separated Turbulent Flows. *AIAA-78-257*, 1978.

[6] Klebanoff, P. S.: Characteristics of Turbulence in a Boundary-Layer with Zero Pressure Gradient. , NACA Report 1247, 1955.

[7] Rodi, W. und Srinivas, K.: Computation of Flow and Losses in Transonic Turbine Cascades. *Z. für Flugwissenschaft und Forschung*, Vol. 13, p. 101, 1989.

[8] Lieser, J. A.: Turbulenzmodellierung für zweidimensionale kompressible Heckströmungen. , Bericht, FG Aerodynamik A 85/91, Technische Hochschule Darmstadt, 1991.

[9] Schlichting, H.: *Grenzschicht-Theorie*. Verlag G.Braun, Karlsruhe, 1982.

[10] Speziale, C. G., Abid, R., und Anderson, E. C.: A Critical Evaluation of Two-Equation Models for Near Wall Turbulence. , NASA Contractor Report 182068, 1990.

[11] van Driest, E. R.: On Turbulent Flow Near a Wall. *Journal of Aeronautical Sciences*, pp. 1007–1011, 1956.

[12] Jones, W. P. und Launder, B. E.: The Prediction of Laminarization with a Two-Equation Model of Turbulence. *Int. J. Heat Mass Transfer*, Vol. 15, pp. 301–313, 1972.

[13] Chien, K.-Y.: Predictions of Channel and Boundary-Layer Flows with a Low-Reynolds-Number Turbulence Model. *AIAA-Journal*, Vol. 20, p. 33, 1982.

[14] Lam, C. K. G. und Bremhorst, K.: A Modified Form of the k- Model for Predicting Wall Turbulence. *Journal of Fluids Engeneering Sept*, Vol. 103, p. 456, 1981.

[15] Zhang, H. S., So, R., Speziale, C. G., und Lai, Y. G.: A Near-Wall Two-Equation Model for Compressible Turbulent Flows. *AIAA-92-0442*, 1992.

[16] Pot, P. J.: Measurements in a 2-D Wake and in a 2-D Wake Merging into a Boundary Layer. , NLR TR 79063 I, 1979.

[17] Benek, J. A.: Separated and Nonseparated Turbulent Flows about Axisymmetric Nozzle Afterbodies. , AEDC-TR-79-22, Arnold Eng.Developement Center Tennessee, 1979.

[18] Spurk, J. H.: *Dimensionsanalyse*. Springer-Verlag, Berlin-Heidelberg-NewYork, 1992.

ON TURBULENT SEPARATED FLOWS IN AXISYMMETRIC DIFFUSERS

W. NITSCHE and C. HABERLAND

Institut für Luft- und Raumfahrt der Technischen Universität Berlin
Marchstraße 12, D-1000 Berlin 12, F.R.G.

Summary

This paper describes experimental and numerical investigations concerning the turbulent flow through axisymmetric expansions having different diffuser half-angles ($6 \deg < \alpha < 90 \deg$). Special attention is given to flow separation and reattachment as well as to flow relaxation downstream of reattachment. First, mean flow results (velocity field, pressure and wall shear stress distributions) including their similarity structures are presented. Further investigations refer to measurements concerning turbulent flow quantities (Reynolds stresses, turbulent kinetic energy) in comparison with calculations carried out with a finite-volume method that involved the standard k-ε-turbulence model. Finally, investigations concerning turbulent transport quantities (diffusion, convection, production and dissipation) are discussed by taking as an example the terms contained in the transport equation for the turbulent kinetic energy. In this investigation first experimental results concerning pressure diffusion are also included, which were carried out with a recently developed X-P probe.

1 Introduction

For some time, flows through axisymmetric expansions showing different geometries have served as test cases for investigations concerning separated flows. In most of these studies, the 90 deg expansion (forward or backward facing step flow, see e.g. [1][2]) was the focus of interest, while the transition from this geometry-induced separation to a pressure-induced separation (over-critical diffusers) was examined only exemplarily, e.g. [3]. For this reason, the present study concentrates on systematically investigating the separated flow in an axisymmetric expansion having various diffuser half-angles, Fig.1, in order to contribute thus to the physics of flow separation. The spectrum of these investigations already follows from a simple wall pressure analysis, which is presented in Fig. 2 in the form of the pressure coefficients measured along the normalized axial coordinate x/D_1 for

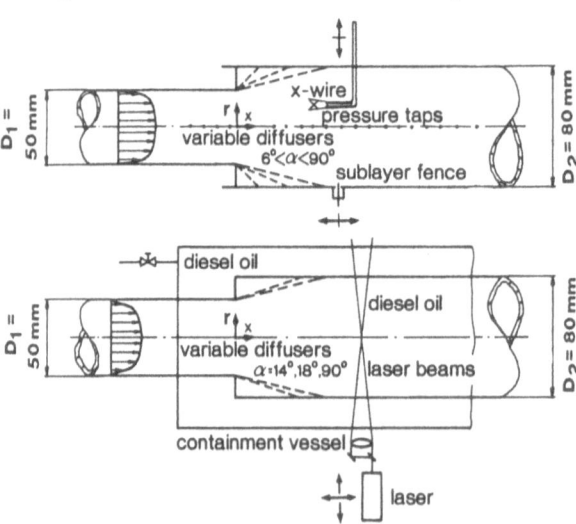

Fig. 1 : Axisymmetric expansion setups

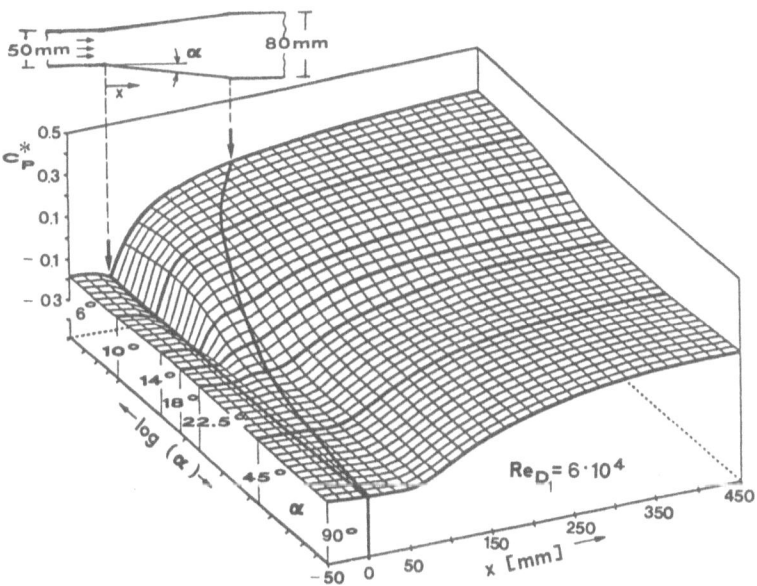

Fig. 2 : Pressure distribution for diffuser half-angles 6 deg $< \alpha <$ 90 deg.

diffuser half-angles between 6 and 90 deg: While in the subcritical 6 deg diffuser the recovery of pressure mainly takes place in the diffuser itself, the c_p- increase in the case of the 90 deg step takes place downstream of the separation point and, due to pressure losses, leads to a significantly smaller pressure recovery. Starting the discussion from the subcritical 6 deg configuration, it becomes evident that a pressure-induced separation starts inside of the diverging section of the diffusers when the half-angle increases. This onset of separation leads to a reorientation of the flow field and, thus, a different pressure distribution characteristic. Significant changes in this characteristic particularly occur in diffusors with half-angles between 14 and 22.5 deg, where the 18 deg configuration constitutes the borderline case between pressure-induced and geometry-induced separation. The present investigation was aimed at analysing the system of these separation phenomena by detailed experiments and attending computation.

2 Test Facilities

The experiments were carried out in two test facilities already shown in Fig.1, which, in principle, are comparable and have both a fully developed turbulent pipe flow as the inlet condition. The interchangeable diffuser sections connecting the inlet pipe of D_1=50mm to the outlet pipe of D_2=80 mm had half-angles of 6, 10, 14, 18, 22.5, 45 and 90 deg. One of the test sections was especially conceived for refractive index matched Laser Doppler measurements using diesel oil as the flow medium. The second test section with air as the flow medium was used for hot-wire measurements and the analysis of wall forces (pressure and wall shear stress distributions). In both cases, the Reynolds number range examined was $10^4 <$ Re$_{D_1} < 6 \cdot 10^4$. The test setups including the measuring techniques employed (LDA, hot-wire probes, sensors / probes for the wall forces) are sketched in Fig.1. A more detailed description of the experimental setups is to be found in [4][5][6]. In addition to these conventional

117

measuring techniques, a recently developed X-P-probe was employed in the last stage of the project, Fig. 3. By means of this probe, first investigations concerning the measurement of pressure diffusion in turbulent flows were carried out. More detailed information about this probe concept is to be found in [7][8][9].

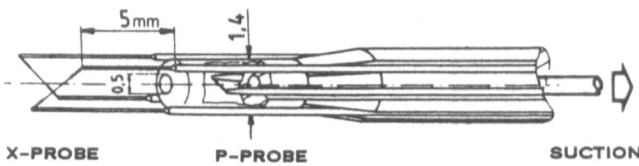

Fig. 3 : X-P-probe for measuring velocity and pressure fluctuations

3 Numerical Approach

For the calculations carried out in parallel with the measurements, the time-averaged continuity and momentum equations for a steady, axisymmetric flow were solved. Neglecting the viscous terms due to the assumption of high Reynolds numbers, these equations read

$$\frac{\partial(\rho U)}{\partial x} + \frac{1}{r}\frac{\partial(r\rho V)}{\partial r} = 0 \tag{1a}$$

$$\frac{\partial(\rho U^2)}{\partial x} + \frac{1}{r}\frac{\partial(r\rho UV)}{\partial r} = -\frac{\partial P}{\partial x} - \frac{\partial(\rho\overline{u^2})}{\partial x} - \frac{1}{r}\frac{\partial(r\rho\overline{uv})}{\partial r} \tag{1b}$$

$$\frac{\partial(\rho UV)}{\partial x} + \frac{1}{r}\frac{\partial(r\rho V^2)}{\partial r} = -\frac{\partial P}{\partial r} - \frac{\partial(\rho\overline{uv})}{\partial x} - \frac{1}{r}\frac{\partial(r\rho\overline{v^2})}{\partial r} + \frac{\rho\overline{w^2}}{r} \ . \tag{1c}$$

Eqs. 1a to 1c were solved by means of a conservative finite-volume method [10] combined with the standard k-ε-turbulence model according to Launder and Spalding. Further details about these numerical calculations are summarized in [5][11][12], for which reason the authors refrain from giving a more detailed description here.

4 Mean Flow Results

As examples of the results concerning the mean flow quantities, the calculated streamlines and the measured U- and V- velocity profiles for three different diffuser half-angles are shown in Fig. 4 . In addition, the measured pressure distributions of six diffusors and the measured shear stress distributions of four different diffusor half-angles are shown in Fig. 5, here again compared with numerical calculations in both cases. Furthermore, the reattachment length x_R was determined by means of the criterion $\tau_w = 0$. It is shown in Figure 6 both in its absolute and normalized values x_R/h for all diffuser half- angles and for different Reynolds numbers. Clearly recognizable, the reattachment length shows a minimum for diffuser half-angles of approximately 18 deg, which means that the transition from pressure-induced to geometry induced separation occurring in this region leads to a significant reorientation of the separation area with strongly increased turbulent transport (see also chapter 5), and hence, to a shorter reattachment length and a shortened relaxation behaviour of the flow [6]. Especially when examining universal flow structures, this reattachment length is of particular importantance as the main standardizing parameter as is shown in Fig.8 for the normalized skin friction coefficients and the presentation of velocity profiles downstream of reattachment in the form of law of the wall coordinates.

118

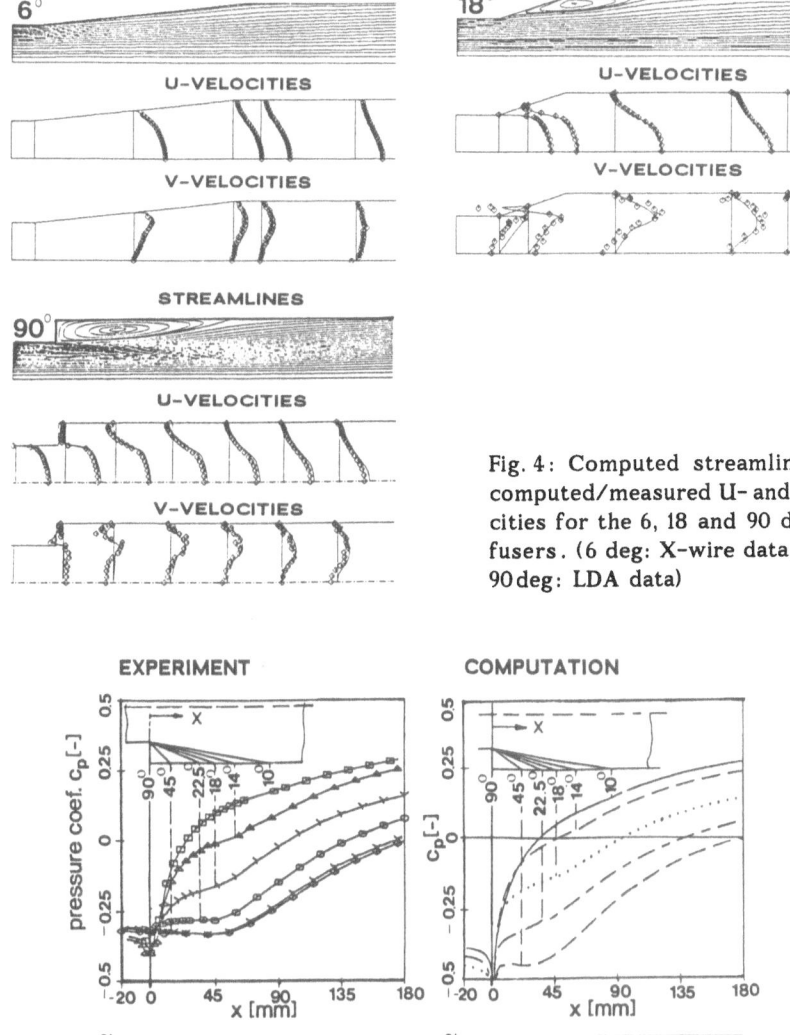

Fig. 4: Computed streamlines and computed/measured U- and V-velocities for the 6, 18 and 90 deg diffusers. (6 deg: X-wire data; 18 and 90 deg: LDA data)

Fig. 5: Pressure and skin friction distribution for different diffuser half-angles. Comparison of experiment and computation ($Re_{D_1} = 5 \cdot 10^4$).

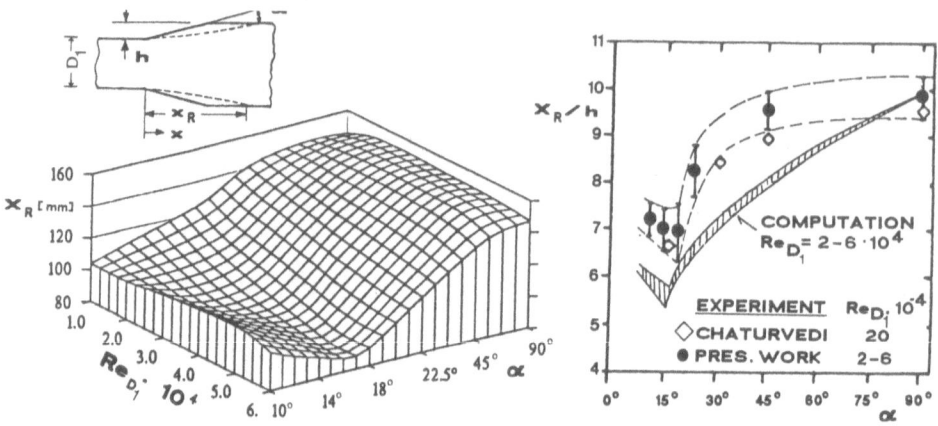

Fig. 6 : Reattachment length depending on Reynolds number and diffuser half-angle.

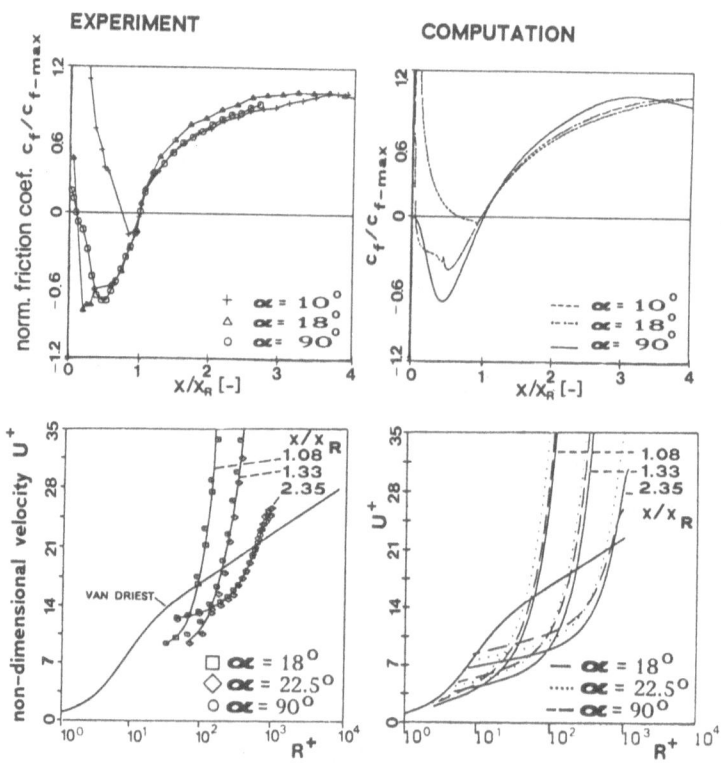

Fig. 7 : Normalized skin friction coefficient and velocity profiles in form of law-of-the-wall coordinates for different diffuser half-angles. Comparison of experiment and computation ($Re_{D_1} = 5 \cdot 10^4$).

5 Turbulent Quantities

To begin with, the investigations concerning the turbulent quantities concentrated on the distributions of the time-averaged Reynolds stresses $\overline{uv}$ and the turbulent kinetic energy k. In the further course of the investigation, measurements were carried out concerning the individual terms of the transport equation of the turbulent kinetic energy. Disregarding the molecular diffusion, this equation reads for axisymmetric flows

$$\underbrace{U \frac{\partial k}{\partial x} + V \frac{\partial k}{\partial r}}_{K} = \underbrace{-\frac{1}{r} \frac{\partial}{\partial r} \left[r \frac{\overline{vu^2 + v^3 + vw^2}}{2} + \frac{\overline{pv}}{\rho} \right] - \frac{\partial}{\partial x} \left[r \frac{\overline{u^3 + uv^2 + vw^2}}{2} + \frac{\overline{pu}}{\rho} \right]}_{D_t}$$

$$\underbrace{- \overline{uv} \frac{\partial U}{\partial r} - (\overline{u^2} - \overline{v^2}) \frac{\partial U}{\partial x} - (\overline{w^2} - \overline{v^2}) \frac{V}{r}}_{P} \quad \underbrace{- \quad \varepsilon}_{DI} \quad . \qquad (2)$$

For this balance equation, first measurements were carried out by means of the X-P-probe shown in Fig.3. They concerned the convective transport of k (K), the turbulent diffusion (D_t) including the pressure diffusion, the production (P) and — in the form of a residual term determination - the dissipation (DI). Since the velocity fluctuations w could not be measured by means of the X-wire probe, the assumptions

$$\overline{w^2} = 0.5 (\overline{u^2} + \overline{v^2}) \qquad (3)$$

and

$$k^2 = 0.75 (\overline{u^2} + \overline{v^2}) \qquad (4)$$

were made here. In addition, the differences of the quadratic terms of the velocity fluctuations in the production term were neglected.

Fig. 8 shows the Reynolds stresses measured in the 6, 18 and 90 deg diffusers for $Re_{D_1} = 5 \cdot 10^4$, while Fig.9 depicts the distributions of the turbulent kinetic energy, both compared with numerical simulations carried out in parallel. All in all, agreement is found to be thoroughly satisfactory. Measurement and calculation deviate, however, more distinctly with regard to the relaxation behaviour of the diffusor flows in general as well as with regard to the position of the maxima and, to some extent, the level of the k-distributions and the Reynolds stresses. As a further evaluation, some data concerning the individual terms of the k-equation (2) are compared with the computation in Fig. 10. To begin with, only the production term and the radial term of convection are depicted, which are normalized here respectively by U_{∞}^3 / L with L=1m. Being

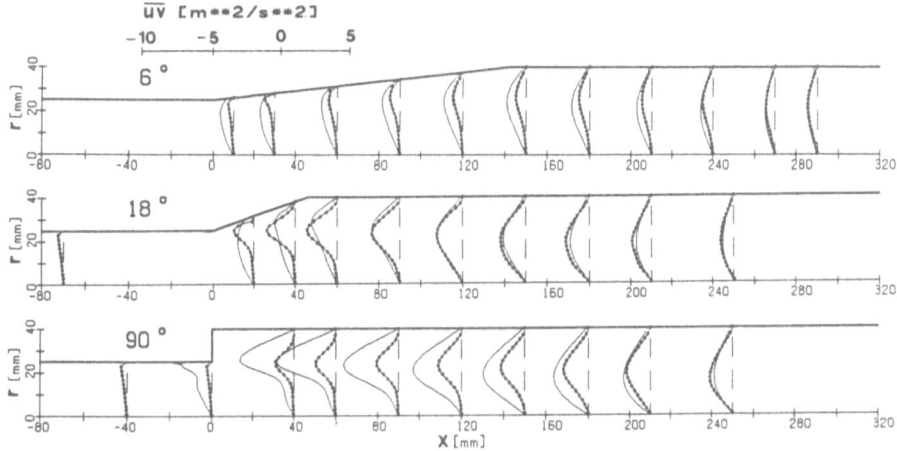

Fig. 8 : Reynolds stresses in the 6, 18 and 90 deg diffusers ($Re_{D_1} = 5 \cdot 10^4$).

121

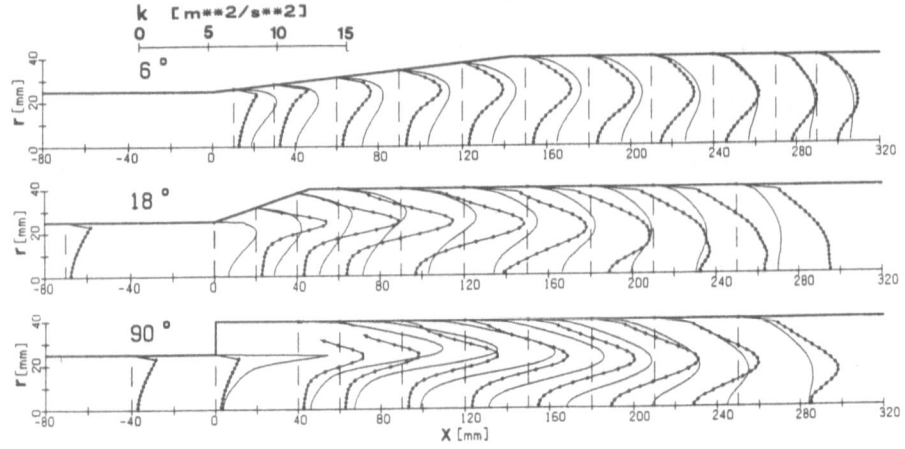

Fig. 9: Turbulent kinetic energy in the 6,18 and 90 deg diffusers ($Re_{D_1}=5 \cdot 10^4$).

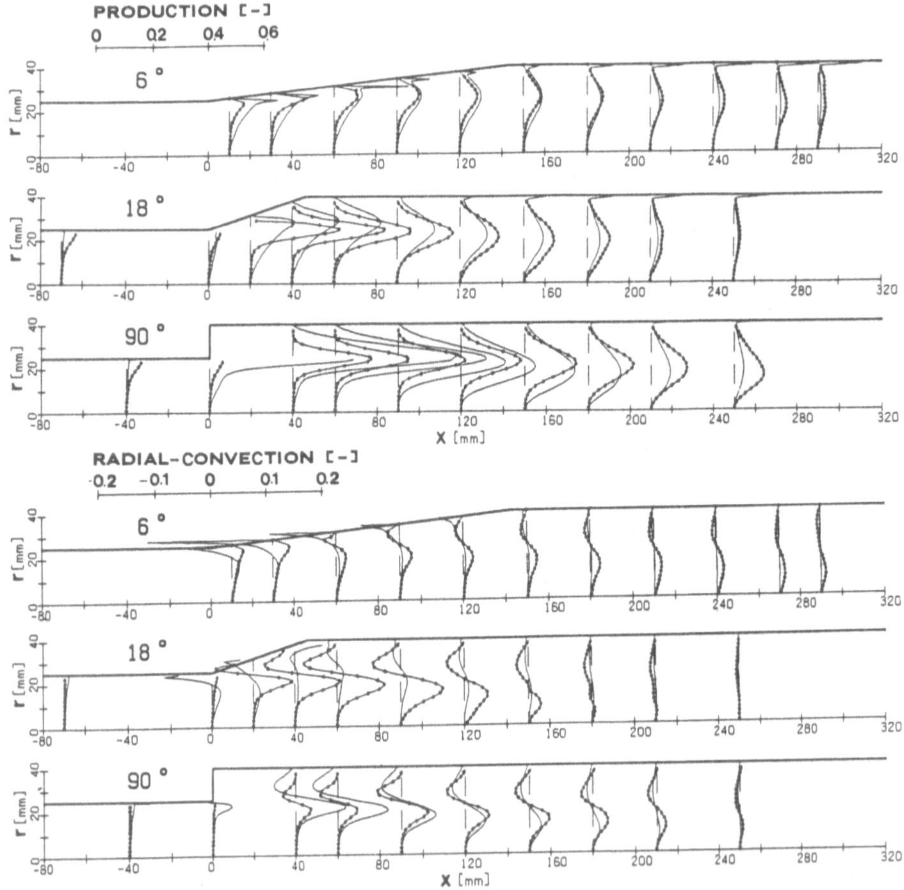

Fig.10 : Turbulent production and turbulent radial-convection in the 6,18 and 90 deg diffusers ($Re_{D_1}=5 \cdot 10^4$).

exclusively obtained from the X-wire signals of the X-P-probe, these measured results also show a satisfactory agreement with the numerical simulations; only the radial convection in the 18 deg diffuser, which the calculations predict to be much smaller, is an exception.

As a provisional conclusion of these investigations, first results of the X-P-probe including the pressure diffusion are shown in Figs. 11 and 12. Fig.11 shows the production term as well as the dissipation term measured at the 6 deg diffuser. Here, the dissipation followed as a residual term from equation (2), i.e. all balance terms of this equation including the pressure diffusion were taken into consideration. Since the flow in question is nearly an equilibrium flow, production and dissipation balance out because diffusion and convection do not make any considerable balance contributions.

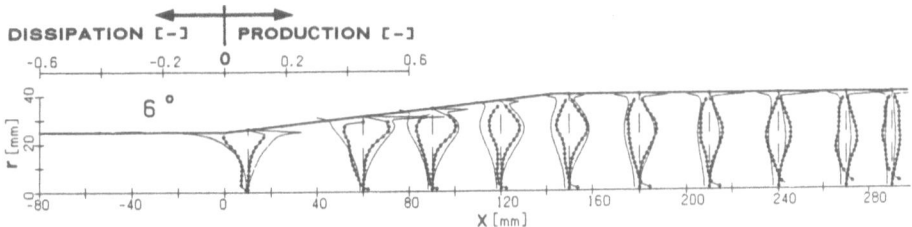

Fig. 11: Turbulent dissipation and production in the 6 deg diffuser (Re$_{D_1}$=5·10^4).

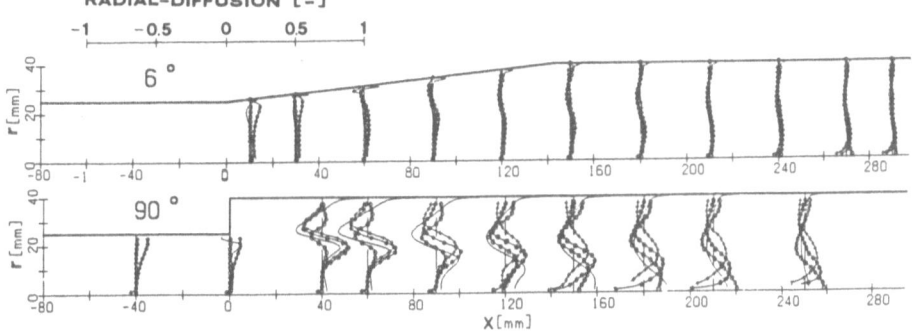

Fig.12 : Turbulent radial-diffusion and diffusion shares in the 6 and 90 deg diffusers
Total diffusion : -•-•- Exp. ——— Comp. (Re$_{D_1}$=5·10^4)
Pressure diffusion: -+-+- ; Diffusion due to u,v,w : -o-o-

The evaluations summarized in Fig. 12 are related to first measurements concerning the complete diffusion term D_t in equation (2), including pressure diffusion. For the radial share of the diffusion, here again normalized by U_{∞}^3 /L with L=1m, this figure shows the total radial diffusion measured by means of the X-P-probe as well as both the separated radial diffusion shares resulting from the correlation of velocity fluctuations and resulting from the correlation of pressure-velocity fluctuations, respectively. To begin with, it follows from these results that - as has already been discussed in connection with Fig. 11 - the turbulent diffusion is generally rather small in the 6 deg diffuser, whereas in the 90 deg diffuser the diffusion increases in toto and the pressure diffusion in particular. Here, the increase for the 90 deg diffuser is described satisfactorily in its tendency by the numerical simulation. Moreover, it becomes evident that, in particular, the diffusion shares resulting from the velocity correlations are relatively well modelled by the k-ε-turbulence model. However, this result is plausible inasmuch as the pressure diffusion is not modelled explicitly in the the standard k-ε-turbulence model.

6 Summary

The investigations conducted in axisymmetric expansion flows having various diffuser half-angles showed that the transition from pressure-induced separation (overcritical diffuser half-angles) to geometry-induced separation (step-like flow configurations) leads to a systematic reorientation of the flow field. The mean flow quantities show a satisfactory agreement between experiment and computation carried out with the standard k-ε- turbulence model. Moreover, these investigations revealed that the reattachment length of the separation region is the main normalizing parameter for the mean flow field. Apart from deviations occurring especially in the borderline case of pressure-induced to geometry-induced separation in the range of diffuser half-angles of 20 deg, experiment and numerical calculation show, in principle, a good agreement in the case of the time-averaged turbulent quantities, too. Preliminary investigations concerning the individual balance terms in the transport equation for the turbulent kinetic energy involving a recently developed X-P- probe have shown that, with increasing non-equilibrium of the flow, the pressure diffusion may constitute a significant part of the diffusion term and, therefore, has to be taken into consideration when modelling turbulent transport quantities in a more detailed way.

Acknowledgements

It is gratefully acknowledged that the Deutsche Forschungsgemeinschaft funded the present investigations within the scope of the DFG -research programme "Physik abgelöster Strömungen". Furthermore, the authors are indebted to N.Weiser, M.Nasseri and P.Bartsch for their valuable ideas and contributions to this project.

References

[1]   Ha Ming, H.,Chassaing,D.(1982) "Pertubations of Turbulent Pipe Flow" *Turbulent Shear Flows I, Springer Verlag, pp.178-197.*

[2]   Khezzar, L. et al.(1985) "An Experimental Study of Round Sudden Expansion Flows" *Proc. 5th. Symp. Turbulent Shear Flows, pp. 5-25 (1985).*

[3]   Chaturvedi, M. (1963) "Flow Characteristics of Axisymmetric Expansions" *Journ. Hydr. Div./Proc. of the ASCE , pp. 61-92.*

[4]   Stieglmeier, M. ,Tropea,C. Weiser,N., Nitsche, W.(1989) "Experimental Investigation of the Flow Through Axisymmetric Expansions" *Journ. of Fluids Eng. Vol. 111 , pp. 464 -471.*

[5]   Weiser, N., Bartsch, P., Nitsche,W. (1990) "On Turbulent Flow Separation in Axisymmetric Diffusers. *Eng. Turb. Mod. and Exp. , Elsevier Publ. , pp. 227-236.*

[6]   Weiser, N. (1992) "Untersuchungen zur Systematik abgelöster Diffusorströmungen" *Diss. TU-Berlin.*

[7]   Nasseri, M. (1992) "Entwicklung eines Sondenkonzepts zur simultanen Erfassung von Druck- und Geschwindigkeitsschwankungen in turbulenten Strömungen" *Diss. TU-Berlin.*

[8]   Nasseri, M.,Nitsche, W. (1991) "Development of a Probe for Measuring Pressure Diffusion" *Proc. 8th Symp. Turbulent Shear Flows, pp.441-446.*

[9]   Nasseri, M., Nitsche,W. (1991) "A Probe for Measuring Pressure Fluctuations in Flows" *ICIASF '91 Rec., pp.284-294.*

[10] Perić, M. (1985) "A Finite-Volume Method for the Prediction of Three-Dimensional Fluid Flow in Complex Ducts" *PhD-Thesis, University of London.*

[11] Bartsch, P. ,Weiser, N.,Nitsche,W. (1989) "Numerical Calculations of Variable-Angle Diffuser Flows " *Comp. and Exp. in Fluid Flow, Springer, pp. 251- 263.*

[12] Bartsch,P.,Weiser, N.,Nitsche,W.(1991) "Experimental and Numerical Investigations on Turbulent Separated Diffuser Flows" *Separated Flows and Jets, Springer Verlag , pp. 327-332.*

ASYMPTOTIC ANALYSIS OF
TWO—DIMENSIONAL TURBULENT SEPARATING FLOWS

K. Gersten, J. Klauer, D. Vieth

Institut für Thermo— und Fluiddynamik, Ruhr—Universität Bochum,
Universitätsstr. 150, D—4630 Bochum 1

SUMMARY

A universal law of the wall is developed by applying asymptotic theory for turbulent flows at high Reynolds numbers. This law is valid for attached as well as for separated flows and hence describes correctly the change of the law of the wall from attached to separated flows. This change is demonstrated for two examples: Couette—Poiseuille flows and equilibrium boundary layers.

The asymptotically correct flow resistance formulae for the Couette—Poiseuille flows are developed by applying an indirect turbulence model in addition to the universal law of the wall. The results for the equilibrium boundary layers are used to develop an integral method for calculating turbulent boundary layers including those with separation. The theoretical results are compared with experiments.

INTRODUCTION

Since turbulent flows are characterized by high Reynolds numbers, it is appropriate to use asymptotic methods to describe those flows. In the asymptotic analysis the turbulent flows are considered as perturbations of the flows at infinite Reynolds numbers, i.e. as perturbations of the inviscid flows. The perturbation parameter is a properly chosen characteristic wall—shear—stress value, which approaches zero for the limit of infinite Reynolds number. According to the asymptotic analysis the turbulent flows at high Reynolds numbers exhibit a multi—layer structure typical for singular perturbation problems. For attached flows the flow equations of the various layers have been developed by *Mellor* [10], see *Gersten* [5]. It was shown that the flow in the layer adjacent to the wall, the so—called viscous wall layer, is characterized by a universal velocity distribution.

The purpose of this investigation is the extension of the law of the wall to flows with separation. This generalized law of the wall will then be used for calculating flows where the wall shear stress goes to zero or even changes the sign. Examples will be the Couette—Poiseuille flows and the equilibrium boundary layers.

GENERALIZED LAW OF THE WALL

The asymptotic analysis shows that the inertia forces in the wall layer are negligible and the viscous forces are of the same order of magnitude as the turbulent shear forces. The relationship for the velocity gradient in the wall layer is given by:

$$\frac{\partial u}{\partial y} = f[y, \nu, \tau_t/\rho, (dp_W/dx)/\rho], \tag{1}$$

where $\tau_t = -\rho \overline{u'v'}$ is the turbulent shear stress. The shear stress distribution in the very thin wall layer can be represented by the first two terms of a Taylor series expansion with respect to the wall distance y :

$$\tau(y) = \tau_t + \rho \nu \frac{\partial u}{\partial y} = \tau_W + \frac{dp_W}{dx} y . \tag{2}$$

Introducing the dimensionless variables $\tau^* = \tau/\tau_w$, $\tau_t^* = \tau_t/\tau_w$, $u^* = u/u_\tau$,

$y^* = y\, u_\tau/\nu$, $u_\tau = \sqrt{|\tau_w|/\rho}$ yields

$$\tau^*(y^*) = \tau_t^* + \frac{|\tau_W|}{\tau_W} \frac{\partial u^*}{\partial y^*} = 1 + K\, y^* , \tag{3}$$

with

$$K = \frac{\nu}{\tau_W\, u_\tau} \frac{dp_W}{dx} = \frac{\nu\,\sqrt{\rho}}{\tau_W\sqrt{|\tau_W|}} \frac{dp_W}{dx} \tag{4}$$

being a very important dimensionless parameter. From dimensional analysis of (1) and (2) the generalized law of the wall follows:

$$\frac{du^*}{dy^*} = F'(y^*,K) , \qquad u^* = F(y^*,K) . \tag{5}$$

Asymptotically flow separation is characterized by a double limiting process, where $\tau_w \to 0$ as well as $\nu \to 0$ is required such that the combination parameter K is kept constant. Cases $\tau_w \neq 0$ are represented by $K \to 0$, whereas $K \to \infty$ corresponds to zero wall shear stress.

The overlap region between the viscous wall layer and the neighbouring fully turbulent layer is characterized by vanishing viscosity effects at constant K . Hence in the overlap layer a reduced form of (1) must be valid:

$$\frac{\partial u}{\partial y} = f[y, \tau_t/\rho, K] . \tag{6}$$

By using dimensional analysis this relation reduces to

$$\frac{y}{\sqrt{|\tau_t/\rho|}} \frac{du}{dy} = \lim_{y^* \to \infty} \left[\frac{y^*}{\sqrt{1 + Ky^*}} \frac{du^*}{dy^*} \right] = \frac{1}{\kappa(K)} . \tag{7}$$

This is the universally valid generalized matching condition, which has been formulated without applying any turbulence modelling. Integration of du^*/dy^* over the entire wall layer finally leads to the velocity distribution in the overlap layer:

$$\lim_{y^* \to \infty} u^*(y^*,K) = \frac{1}{\kappa(K)} \left[\ln y^* + 2(\sqrt{1 + Ky^*} - 1) + 2\ln\left(\frac{2}{\sqrt{1 + Ky^*} + 1} \right) \right] + C(K), \tag{8}$$

with the two universal functions $\kappa(K)$ and $C(K)$. The following limiting cases exist:

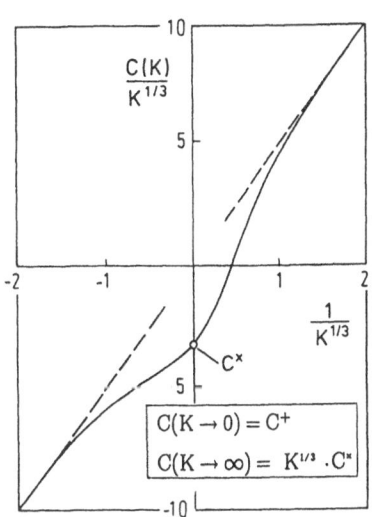

Figure 1: Function $C(K)$ (——) and its asymptotes for $K \to 0$ (— — —)

a) Attached flow: $K \to 0$,
$C(0) = C^+$,
(smooth wall: $C^+ = 5$),
$\kappa(0) = 0{,}41$

$$\lim_{y^+ \to \infty} u^+(y^+) = \frac{1}{\kappa} \ln y^+ + C^+ \qquad (9)$$

b) Separation: $K \to \infty$,
$C(\infty) = K^{1/3}\, C^\times$,
$\kappa(\infty) = \kappa_0$

$$\lim_{y^\times \to \infty} u^\times(y^\times) = \frac{2}{\kappa_0} \sqrt{y^\times} + C^\times \qquad (10)$$

where the dimensionless variables $y^\times = yu_s/\nu$ and $u^\times = u/u_s$ with

$$u_s = [\nu\,(dp_w/dx)/\rho]^{1/3}$$

have been introduced, see *Gersten* [4]. Experimental data for the universal constants κ_0 and $C^\times$ are rare and fall in the ranges $0{,}41 \leq \kappa_0 \leq 0{,}8$, $-3{,}2 \leq C^\times \leq 2$. In what follows the values $\kappa_0 = 0.41$ and $C^\times = -3.2$ will be used. Figure 1 shows a possible form of the universal function $C(K)$ for $C^\times = -3{,}2$.

COUETTE–POISEUILLE FLOWS

Couette–Poiseuille flows occur between two parallel plane walls, whereby the upper one is moving in the direction of a positive pressure gradient, see Figure 2. The flows are considered to be two–dimensional, incompressible and fully developed. The parameters of these flows are: half channel height H, wall velocity u_w, and the shear stresses at the upper wall τ_{Wu} and the lower wall τ_{Wl}, respectively. The characteristic dimensionless parameter is $\gamma = \tau_{Wl}/\tau_{Wu}$. Figure 2 shows the distributions of the mean velocity and the shear stress for four values of γ, see *Gersten* [3].

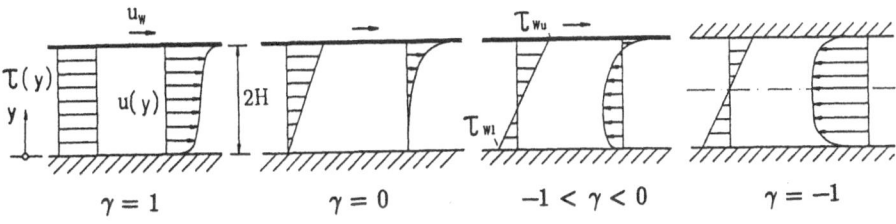

Figure 2: Couette–Poiseuille flows, fully developed

According to asymptotic analysis, see *Gersten* and *Herwig* [8], the flows consist of three layers, i.e. the core region and two wall layers. Since the wall layer behaviour is universal, it is sufficient to calculate the core region only. The generalized matching condition (7) is now a boundary condition for the core region. Using (7) as a boundary condition in the calculation of the core region is called the "method of wall functions" in the literature. It is a consequence of the asymptotic analysis.

None of the existing turbulence models is able to satisfy the boundary conditions (7) and (8) for all values of K . Therefore, a so—called "indirect turbulence model" will be used here. The following equation for the velocity gradient is chosen:

$$\frac{du^+}{d\eta} = \frac{G(\eta)}{\kappa(K)\eta} + \frac{1}{\kappa(2-\eta)} - \frac{2-\eta}{4\kappa} - \frac{\eta}{4\kappa(K)} + \frac{1}{\kappa} F(\eta,\gamma_R) , \qquad (11)$$

$$G(\eta) = \sqrt{\gamma + \frac{1 - \gamma}{2} \eta}$$

$$G(\eta) = \gamma_R \sqrt{1 + \frac{1 - \gamma}{2\gamma} \eta}$$

$$\left\{ \begin{array}{ll} 0 \leq \gamma \leq 1: & 0 \leq \eta \leq 2 \\ -1 \leq \gamma \leq 0: & -2\gamma/(1-\gamma) \leq \eta \leq 2 \\ -1 \leq \gamma \leq 0: & 0 \leq \eta \leq -2\gamma/(1-\gamma) \end{array} \right.$$

with $\eta = y/H$, $K = (1-\gamma)/(2\gamma| \gamma_R | Re_{\tau u})$, $\gamma_R = sign(\gamma) \sqrt{|\gamma|}$, $Re_{\tau u} = u_{\tau u} H/\nu$, $u_{\tau u} = \sqrt{\tau_{Wu}/\rho}$.

It should be noted that the reference friction velocity has been constructed with the non—zero shear stress at the upper wall. This ensures that the logarithmic boundary condition is valid at the upper wall and the generalized boundary condition at the lower wall, respectively. The function $F(\eta,\gamma_R)$ is regular for $\eta \rightarrow 0$ and $\eta \rightarrow 2$, respectively. Integrating $du^+/d\eta$ over the entire core region and satisfying the matching condition (8) at the lower wall and condition (9) at the upper wall finally leads to the following results, see *Gersten* and *Herwig* [8]:

a) $\quad u_c^{*}(\gamma,K) = \gamma_R C(K) + \frac{1}{\kappa(K)}[W(\gamma) + \gamma_R ln(\frac{4}{|K|})] + \hat{C}_1(\gamma),$

b) $\quad u_W^{*}(\gamma,K) = u_c^{*} + \frac{1}{\kappa}lnRe_{\tau u} + C^* + \bar{C}_2(\gamma),$

c) $\quad u_m^{*}(\gamma,K) = u_c^{*} + \bar{\bar{C}}(\gamma),$

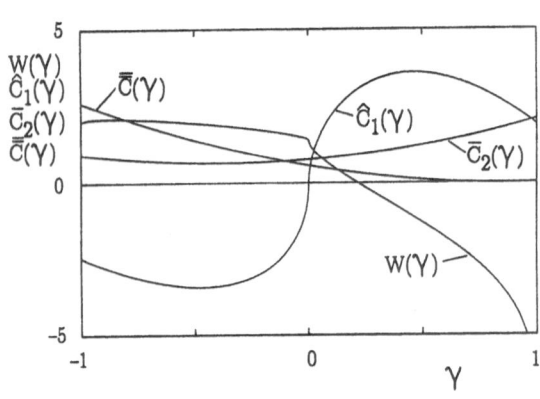

with $u_c^{*}(\gamma,K)$, $u_W^{*}(\gamma,K)$ and $u_m^{*}(\gamma,K)$ being the center line velocity, moving wall velocity and the average velocity, respectively. The functions $W(\gamma)$, $\hat{C}_1$, $\bar{C}_2(\gamma)$ and $\bar{\bar{C}}(\gamma)$ are given in Figure 3. They are regular in the neighbourhood of $\gamma = 0$. The last three functions depend on the choice of the function $F(\eta, \gamma_R)$ in (11) and have been deter— mined by use of experi— mental data of *El Telbany* and *Reynolds* [2]. The function $\kappa(K)$ has been

Figure 3: Functions $\hat{C}_1(\gamma), \bar{C}_2(\gamma), \bar{\bar{C}}(\gamma), W(\gamma)$

128

approximated by $\kappa(K) = 0{,}41$. Figure 4 shows velocity profiles for four values of γ.

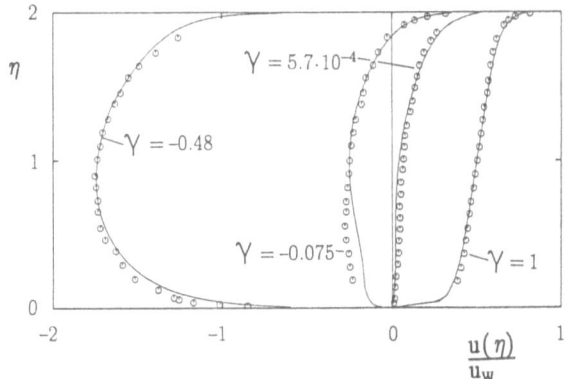

Figure 4:

Velocity profiles according to eq. (11) in comparison with measurements of *El Telbany* and *Reynolds* [2]

EQUILIBRIUM BOUNDARY LAYERS

Similar to channel flows, wall bounded flows have a multi–layer structure at high Reynolds numbers. One result of the asymptotic analysis of attached boundary layers is that the flow field exhibits three regions. One distinguishes between the inviscid outer flow and the boundary layer, whereby the latter consists of the viscous wall layer and an outer part where the flow is essentially governed by the turbulent motion.
Since the law of the wall (8) is universally valid, only the fully turbulent part of the boundary layer has to be calculated. The boundary conditions can be found with the help of the law of the wall and by matching the outer part of the boundary layer with the inviscid outer flow.
In the following the results of the analysis of so–called equilibrium boundary layers will be presented. These kinds of flows have a self–similar structure in the outer part of the boundary layer. Again, we distinguish between attached flows and the special flow case where $\tau_W = 0$ holds (the so–called *Stratford* flow [13]).

Clauser flows ($\tau_W \neq 0$)

Self–similar solutions can be found with the following defect approach:

$$\frac{u(x,\eta)}{U(x)} = 1 - \gamma\,f'(\eta), \quad \gamma(x) = \frac{u_\tau(x)}{U(x)}, \quad t = \frac{\tau_t(x,\eta)}{\tau_W(x)}, \quad \eta = \frac{y}{\Delta(x)}. \tag{12}$$

This approach leads to an ordinary differential equation:

$$(1 + 2\,\beta)\,\eta\,f'' + 2\,\beta\,f' = t', \quad \beta = \delta_1\frac{dp_W}{dx}\,/\,\tau_W. \tag{13}$$

Matching the outer layer with the wall layer leads to

$$\lim_{\eta\to 0} f'' = -\frac{1}{\kappa\eta}.$$

A further result of this matching process is the following implicit formula for the friction velocity:

$$\frac{1}{\gamma} = \frac{U(x)}{u_\tau(x)} = \frac{1}{\kappa} \ln[\, u_\tau(x)\, \Delta(x)\, /\nu] + C^+ + \overline{C}(\beta) , \qquad (14)$$

with

$$\overline{C}(\beta) = \lim_{\eta \to 0} \left[f'(\eta) + \frac{1}{\kappa} \ln \eta \right] .$$

Again, this result has been achieved without applying any turbulence modelling. A turbulence model influences eq. (14) only via $\overline{C}(\beta)$.
The defect approach has proven to be a proper way to calculate turbulent boundary layers, provided that the Reynolds number is high enough and $\tau_W \neq 0$.

Stratford flow ($\tau_W = 0$)

The boundary layer flow where $\tau_W = 0$ holds has at first been investigated by *Stratford* [13]. The asymptotic analysis leads to a three–layer structure, see *Klauer* [9].

The outer region of the boundary layer is again governed by the turbulent motion. In contrast to attached flows the velocity defect is not small enough to be treated as a perturbation of a homogeneous velocity profile. The expansion parameter for the outer layer can be related to a turbulent Reynolds number with the turbulent viscosity replacing the molecular viscosity. The turbulent Reynolds number can be related to the model constants of a turbulence model. For the Cebeci–Smith model the perturbation parameter can be shown to be $\alpha = \nu_{to} /(U(x)\, \delta_1(x))$. With $\nu_{to} = $ const. the solution approaches a smooth analytic behaviour near the wall with a slip velocity at the wall. On the other hand, the square root law holds at the outer edge of the viscous wall layer. Thus it is obvious that the outer part cannot be matched to the wall layer without introducing a third layer, the so–called intermediate layer. In this layer the streamwise velocity is represented as a perturbation of the slip velocity. The shear stress there is a linear function of the wall distance. The equations for the intermediate layer can be solved if the outer flow is given and a proper turbulence model is chosen. Matching the intermediate layer to the outer layer yields the boundary conditions for the latter. Again, self–similar solutions can be found for the outer region of the boundary layer with $u(x,\eta)/U(x) = f'(\eta)$, $\eta = y/\Delta(x)$. The ordinary differential equation

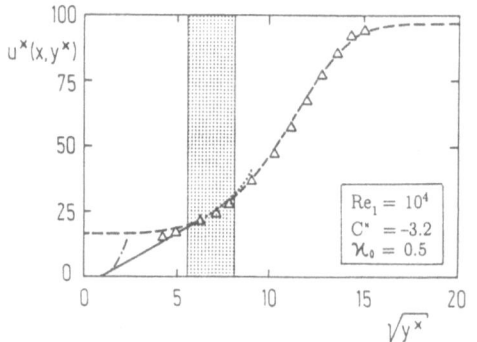

Figure 5: The three–layer structure of a separating boundary layer in comparison with measurements of *Dengel* and *Fernholz* [1]

130

can be solved by using the matching conditions between the outer region and the intermediate layer. Figure 5 shows a result using the Cebeci–Smith model and illustrates the three–layer stucture.

Due to the matching to the viscous wall layer a Reynolds–number influence for the results in the middle and outer layer exists. Thus the shape parameter H_{21} for $\tau_W = 0$ increases with increasing values of Re.

Development from attached to separated flow

An asymptotic correct approach for separating boundary layers was given by *Melnik* [11]. *Melnik's* theory is consistent with the aforementioned theories of equilibrium boundary layers and of the flow case with $\tau_W = 0$.

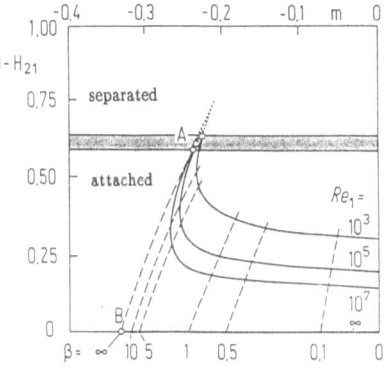

The equilibrium boundary layers as well as the Stratford boundary layers can be characterized by two independent parameters, for example β and $Re_1 = \delta_1 U / \nu$ or $H_{21} = \delta_2 / \delta_1$ and $m = \Delta U' / (\Delta' U)$ or any two out of these four global parameters.

Figure 6 shows the states of all possible equilibrium boundary layers, see also *Gersten* [6, 7, 8]. The values for the Reynolds number dependent separation points A are calculated by using the three layer analysis with $\kappa_0 = 0.41$ and $C^x = -3.2$.

Figure 6:   Relation between the shape parameter and the parameters of equilibrium boundary layers

Integral method for calculating turbulent boundary layers

An integral method for calculating turbulent boundary layers with separation has been developed, see *Klauer* [9], by using the integrated momentum equation and the integrated mean flow kinetic energy equation. For attached boundary layers the outer flow velocity distribution U(x) is given and the parameters $H_{21}(x)$ and $Re_1(x)$ are determined. For boundary layers with separation an inverse form of the method was applied, where $\delta_1(x)$ is given and U(x) and $H_{21}(x)$ are calculated. The two ordinary differential equations of the integral method contain additional "closure" relationships, which in this case have been taken from the results of equilibrium boundary layers. Hence, the integral method gives "exact" results for equilibrium boundary layers.

Unfortunately the integral method could not be intensively tested for separated flows due to the lack of suitable experimental results. Figure 7 allows a comparison of this integral method with experimental results of *Simpson* et al. [12]. As long as the flow is attached, the measured points are well predicted. The discrepancy for the separated area (x > 3.45m) in Figure 7 may be explained by the massive separation that occurred in the measurements.

131

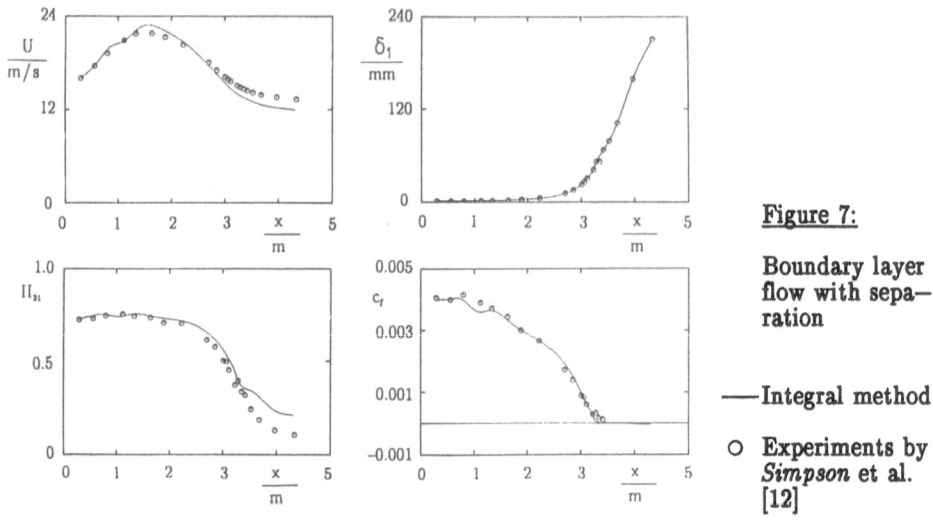

Figure 7:

Boundary layer
flow with sepa—
ration

——Integral method

o Experiments by
Simpson et al.
[12]

REFERENCES:

[1] *Dengel, P.; Fernholz, H.H.* : An experimental investigation of an incom—
pressible turbulent boundary layer in the vicinity of separation, J. Fluid Mech.,
vol. 212, pp. 615—636, 1990

[2] *El Telbany, M.M.M; Reynolds, A.J.* : Velocity distribution in plane turbulent
channel flows, J. Fluid Mech., vol. 100, pp. 1—29, 1980

[3] *Gersten, K.* : Asymptotische Theorie für turbulente Strömungen bei großen
Reynolds—Zahlen, Ernst—Becker—Gedächtnis—Kolloqium, TH Darmstadt,
Schriftenreihe Wissenschaft und Technik, Nr. 28, S. 67—94, 1985

[4] *Gersten, K.* : Some contributions to asymptotic theory for turbulent flows, in:
Proc. of 2nd Int. Symp. on Transport Phenomena in "Turbulent Flows",
Tokyo, pp. 201—214, 1987

[5] *Gersten, K.* : Introduction to asymptotic theory for turbulent flows, ZAMM,
Bd. 69, S. T 555—T 558, 1988

[6] *Gersten, K.* : Die Bedeutung der Prandtlschen Grenzschichttheorie nach 85
Jahren, Z. Flugwiss. Weltraumforsch., Bd. 13, S. 209—218, 1989

[7] *Gersten, K.* : Some open questions in turbulence modelling from viewpoint of
asymptotic theory, Proc. of Tenth Australian Fluid Mechanics Conference,
Melbourne, vol. II, pp. 12.1—12.4, 1989

[8] *Gersten, K.; Herwig, H.* : Strömungsmechanik. Grundlagen der Impuls—,
Wärme— und Stoffübertragung aus asymptotischer Sicht, Vieweg—Verlag,
Braunschweig, 1992

[9] *Klauer, J.* : Berechnung ebener turbulenter Scherschichten mit Ablösung und
Rückströmung bei hohen Reynoldszahlen, VDI—Fortschrittbericht, Reihe 7:
Strömungstechnik, Nr. 155, VDI—Verlag, Düsseldorf, 1989

[10] *Mellor, G.L.* : The large Reynolds number asymptotic theory of turbulent
boundary layers, Int. J. Engng. Sci., vol. 10, pp. 851—873, 1972

[11] *Melnik, R.E.* : An asymptotic theory of turbulent separation, Computers &
Fluids, vol. 17, no. 1, pp.165—184, 1989

[12] *Simpson, R.L.; Chew, Y.T.; Shivaprasad, B.G.* : The structure of a separa—
ting turbulent boundary layer, J. Fluid Mech., vol. 113, pp. 23—51, 1981

[13] *Stratford, B.S.* : An experimental flow with zero skin friction throughout its
region of pressure rise, J. Fluid Mech., vol. 5, pp. 17—35, 1959

TURBULENT SEPARATING FLOW OVER HUMPS IN AN OPEN CHANNEL

P. Larsen and U. Kertzscher

Institut f. Wasserbau und Kulturtechnik, Univ. Karlsruhe,
Kaiserstr. 12, 7500 Karlsruhe 1, Germany

SUMMARY

Two dimensional flow past various obstacles placed in a rectangular cross section flume was studied. Position and extent of
separation and the distribution of turbulence produced in the
shear zone were key elements of the study. Parameters varied
were geometry of the obstacles (vertical wall, wave shaped
sills), depth of flow and discharge. For comparison measurements
were carried out in a density stratified flow with some of the
obstacles.

1. INTRODUCTION

Flow over humps, hills and similar wave shaped obstacles has
been the topic of several experimental, theoretical and numerical studies [2]. These studies mainly focussed on the effect of
the obstacle on a well defined approach flow. Loy [2], however,
in addition considered the separation on the lee side of the
obstacle. He investigated a sub-critical, turbulent flow over a
hump placed in a flume with a rectangular cross section.

Fig. 1 shows the zones of the flow studied:

Zone a) undisturbed approach flow

The time averaged velocity profile and the turbulence characteristics correspond with those of a fully developed, turbulent
flow.

Zone b) flow acceleration on the upstream side of the obstacle

The time averaged velocity profile becomes bulkier, the turbulence of the outer flow is essentially unchanged, whereas it
becomes intenser close to the bottom due to the steepened velocity gradient. The depth is reduced in accordance with the
energy equation prediction.

Zone c) flow deceleration and separation at downstream side of
        the obstacle

Depending on flow and obstacle parameters the flow separates and
a recirculating sub-zone exists near the boundary. A free shear
zone exhibits strong turbulence production. The depth of flow
increases due to the deceleration.

Zone d) readjustment

Downstream from the re-attachment a new boundary layer develops. Characteristic of the velocity profiles in this zone is a point of inflexion. The turbulence produced in the shear layer diffuses over the depth of flow.

2. STUDY PROGRAM, GEOMETRIES AND MEASURING METHODS

The geometries of the obstacles, also used by Lawrence [1], are shown in Fig. 2. Additionally a vertical plane wall of the same height, 20 mm, was used. The averaged steepness of the obstacles was 5°, 7.5°, 10° and 15° for obstacles D, C, B, A, respectively. Reynolds number were 5000 and 8000, based on approach flow depth and volumetric velocity. These correspond with discharges of 1.0 and 1.6 l/s. The Froude numbers varied between 0.1 and 1 corresponding with the depths of flow between 100 and 40 mm. The flume was kept horizontal and the bottom, including the obstacles, was hydraulically smooth.

Measurements were performed with a resistance thermometer, a point gauge and a one component laser doppler anemometer.

3. RESULTS

3.1 Approach flow

Velocity profiles of the approach flow are shown in Fig. 3 for various discharges and depths of flow. The wall coordinates used were normalized with shear velocities obtained from LDA measurements of velocity distribution in the viscous sublayer. Fig. 3 shows that the measured velocities comply with the theoretical predictions in the viscous sublayer, with that of van Driest in the transition zone and with the logarithmic law in the outer flow. The corresponding curves are shown in the figure. Thus the approach flow can be considered as a fully developed turbulent boundary layer flow.

Comparison of turbulence velocities based on measured turbulence fluctuations with the semi-theoretical results of Nezu and Rodi [3] shown in Fig. 4, also confirm that the approach flow can be considered as a fully developed turbulent flow.

As a comparison to the measurements performed in homogeneous flow Fig. 5 and 6 show measurements in density stratified flows. The volume flow rate of 1 l/s was composed of 0.4 l/s 1% by weight salt water flow and 0.6 l/s fresh water flow. The stable interface divided the flow into layers of approximately equal depths. Fig. 5 shows that the velocity distribution near the bottom, essentially up to the interface, approximates that of a turbulent, homogeneous flow. In the upper, fresh water layer the velocity distribution, however, deviates markedly from a devel-

oped turbulent velocity distribution. This is because a distribution develops for which the interface can be considered as a
co-moving boundary.

The turbulence velocities of the two layers, plotted in Fig. 6,
show a similarly good agreement with the semi-theoretical distribution as for the homogeneous flow, (Fig. 4). The turbulence
velocities of the upper layer are lower than that of the homogeneous flow case due to the damping effect of the interface.

3.2. Integral dimension of the separation zone

In the following the dependance of the position of the point of
separation and that of re-attachment on geometry and flow depth
over the sill is described. For the latter a parameter defined
as the ratio of downstream flow depth to depth over the obstacle
is applied, i.e. an expansion ratio. The point of separation
could be well determined (± 1 mm) by simply observing the flow
and checking with LDA measurements for confirmation. The reattachment point, however, cannot be equally well determined
because of the instationarity of the flow. A considerably more
laborious determination was adopted: in the re-attachment area
LDA measurements were done near the bottom at various points
along the flume. The re-attachment point was determined as that
point where the mean velocity was equal to zero. In some instan-
ces this was checked by extrapolation of the separation streamline down to the bottom.

Of the two parameters, geometry and expansion ratio, the geometry has the stronger influence on the positions of separation
and re-attachment points, Fig. 7. The horizontal distance of the
separation point from the obstacle crest and the length of the
separation zone were normalized with the height of the obstacle.
The observations show that the separation depends on a local,
"critical" angle. The steeper the obstacle the closer is the
point of the critical angle to the crest. The critical angle was
found to lie between 4.5° and 8° with the geometries of this
study. This finding agrees, particular for the steeper obstacles, with the observations pertaining to straight diffusers,
for which the angle of separation is approximately 7°.

The length of the separation zone with the vertical wall is, as
can be expected, considerably longer than those of the humps.
With the wall the separation streamline is upward directed at
the separation point in contrast to the case with the humps. An
effect of the expansion ratio on the position of the separation
point was observed only with the flatter sills, Fig. 7a. The
length of the separation zone increases slightly with separation
ratio for the steeper obstacles whereas it decreases slightly
with the flatter ones. The former observation is in agreement
with the findings of Tropea [4], who studied the flow over an
adverse step.

The results of measurements with density stratified flow are
shown in Fig. 8. Two different fresh water - salt water combinations were selected in order to better show the effect of

stratification. It is noted, that the behaviour of the point of separation is similar to that for homogeneous flows in all cases studied. In contrast, the length of the separation zone is markedly influenced by the stratification, Fig. 8b.

3.3. Streamline pattern

The normalized stream function can be determined from the measured velocity by integration over the depth and normalized with the discharge. The streamlines are shown in Fig. 9 for all obstacles studied. For the vertical wall and sill A the length of the separation zone is independent of discharge. However, the length and the height of the separation zone reduces the flatter the sills and the greater the discharge are. It is probable that with sill D and still higher discharge the separation would disappear.

3.4. Isolines of turbulence velocities

The distribution of the turbulence, produced in the shear zone, is shown in Fig. 10. The turbulence has been normalized using the shear velocity of the approach flow, computed from measurements 300 mm upstream of the obstacles. The isoline indicating a turbulence velocity two times the shear velocity is drawn with a thicker line. This value corresponds approximately with the maximum value of this quantity in the undisturbed flow, for comparison see Fig. 4. Fig. 10 shows isolines obtained with all obstacles for one discharge and one depth of flow.

As may be seen in Fig. 10 the marked isoline reaches the surface closer to the obstacle the steeper the obstacle is. The vertical diffusion of turbulence becomes less intense with the flatter sills. With the flattest humps, C and D, the turbulence produced in the shear zone is dissipated in the vicinity of the bottom as it is transported downstream. This is in contrast to the case with steep obstacles.

REFERENCES

[1] Lawrence, G.A., 1985: "The Hydraulics and Mixing of Two-Layer Flow over an Obstacle", Report No. UCB/HEL-85/02, Berkeley, California, USA.

[2] Loy, T., 1990: "Turbulente abgeloeste Stroemung ueber Grundschwellen in einem offenen Gerinne", Dissertation, Heft 39, Inst. f. Hydrologie und Wasserwirtschaft, Univ. Karlsruhe.

[3] Nezu. I., Rodi, W., 1986: "Open-Channel Flow Measurements with a Laser Doppler Anemometer", J.of Hydr. Res., Vol. 112, No.5.

[4] Tropea C., 1982: "Die turbulente Stufenstroemung in Flachkanaelen und offenen Gerinnen", Dissertation, Univ. Karlsruhe.

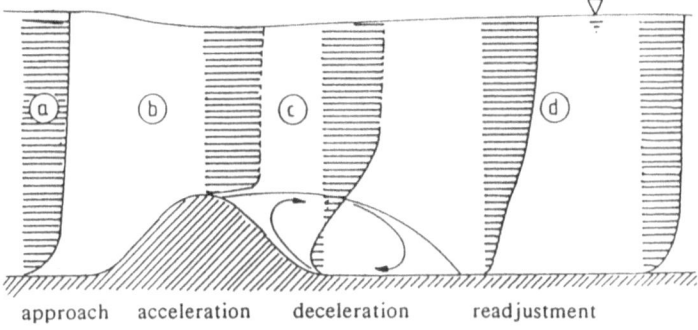

Fig. 1: Zones of the flows studied

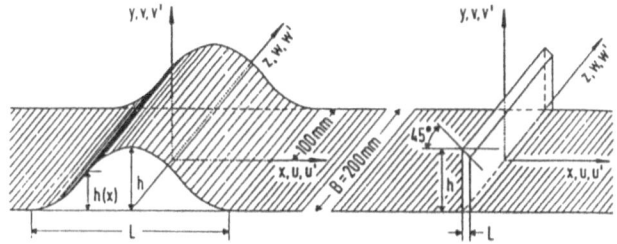

Fig. 2: Obstacle geometries

$$h(x) = h \cdot \cos^2\left(\frac{x}{h \cdot a}\right) \qquad -\frac{\pi}{2} \cdot h \cdot a \le x \le \frac{\pi}{2} \cdot h \cdot a$$

137

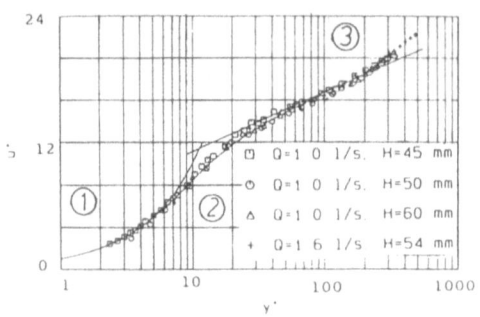

Fig. 3: Velocity distribution of the approach flow (homogeneous flow)

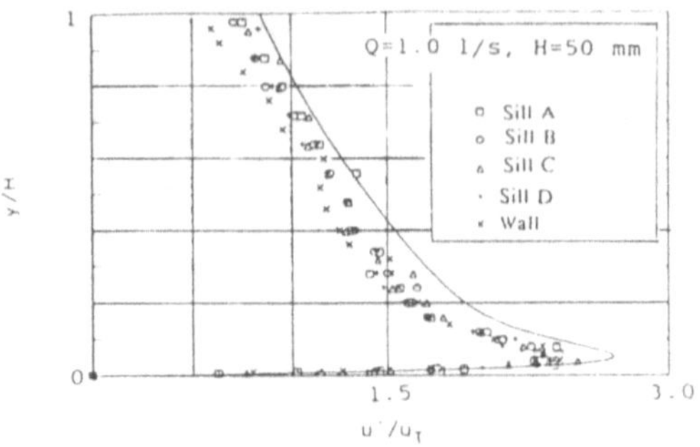

Fig. 4: Turbulence velocity of the approach flow (homogeneous flow)

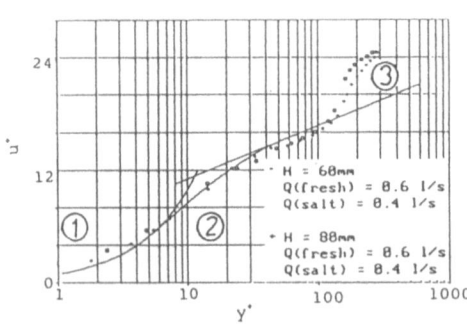

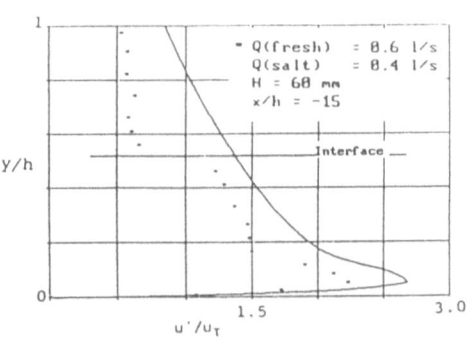

Fig. 5: Velocity distribution of the approach flow (stratified flow)

Fig. 6: Turbulence velocity of the approach flow (stratified flow)

$$u^+ = y^+$$

①

$$u^+ = \frac{1}{\kappa} \ln y^+ + C$$

③

$$du^+ = \int_0^{y^+} \frac{2 dy^+}{1 + [1 + 4(\kappa \cdot y^+ \cdot \Gamma)^2]^{0.5}}$$

②

$$\kappa = 0.41 \quad C = 5.49$$

$$\frac{u'}{u_\tau} = D_u \cdot \exp\left(-\lambda_u \cdot \frac{y}{H}\right) \cdot \Gamma + 0.3 \cdot y^+ \cdot (1 \cdot \Gamma)$$

$$\Gamma = 1 \cdot \exp\left(-\frac{y^+}{B_2}\right)$$

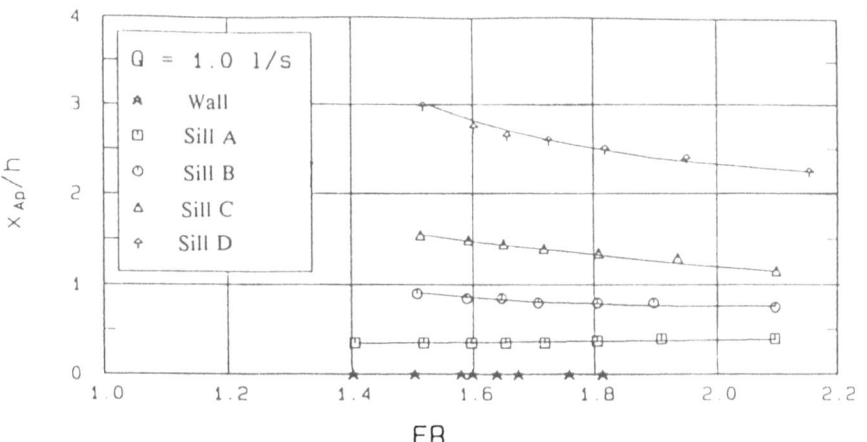

Fig. 7a: Point of seperation for different geometries
(homogeneous flow)

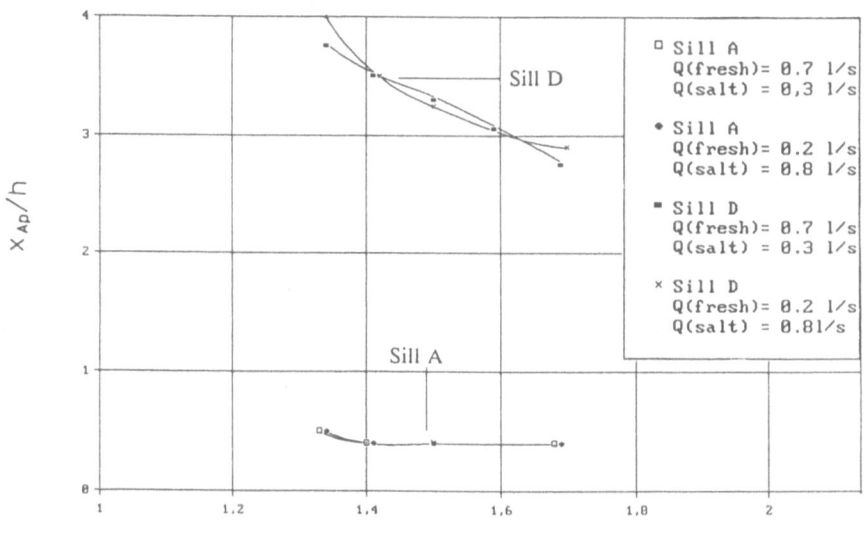

ER

Fig. 8a: Point of seperation for different geometries
(two-layer stratified flow)

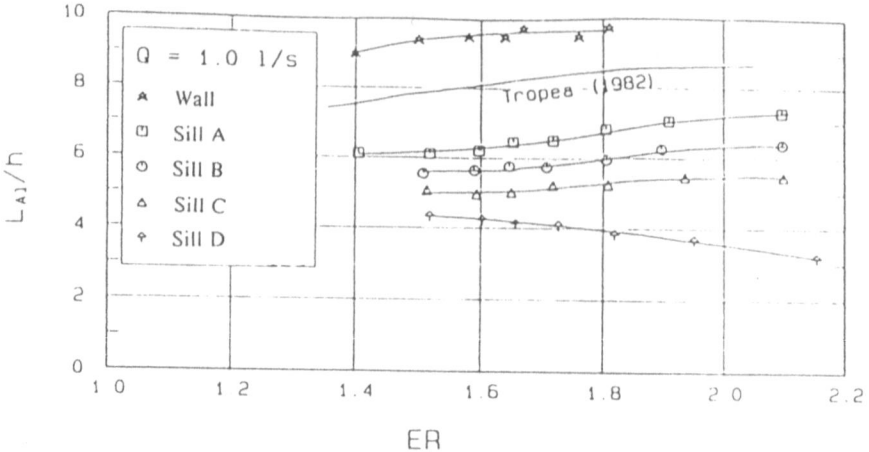

Fig. 7b: Length of seperation zone with various geometries
(homogeneous flow)

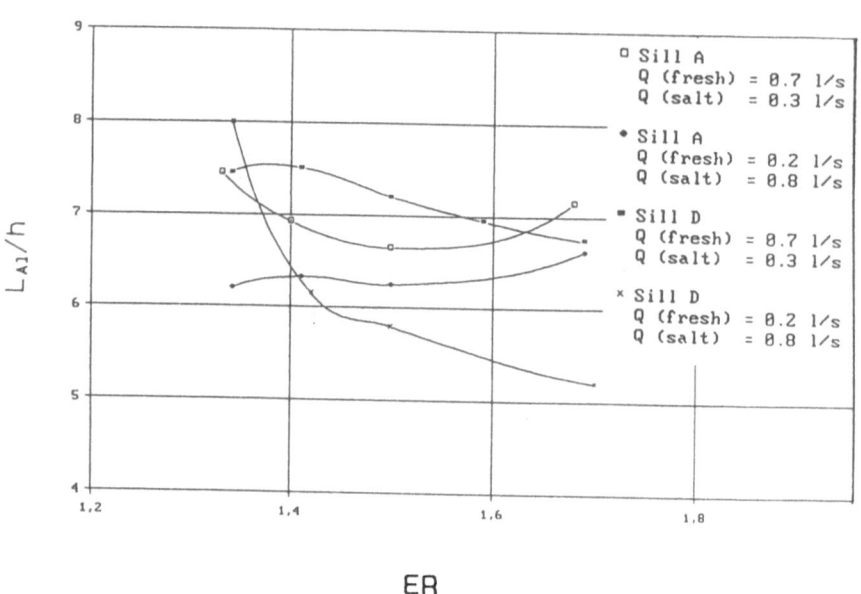

Fig. 8b: Length of seperation zone with various geometries
(two-layer stratified flow)

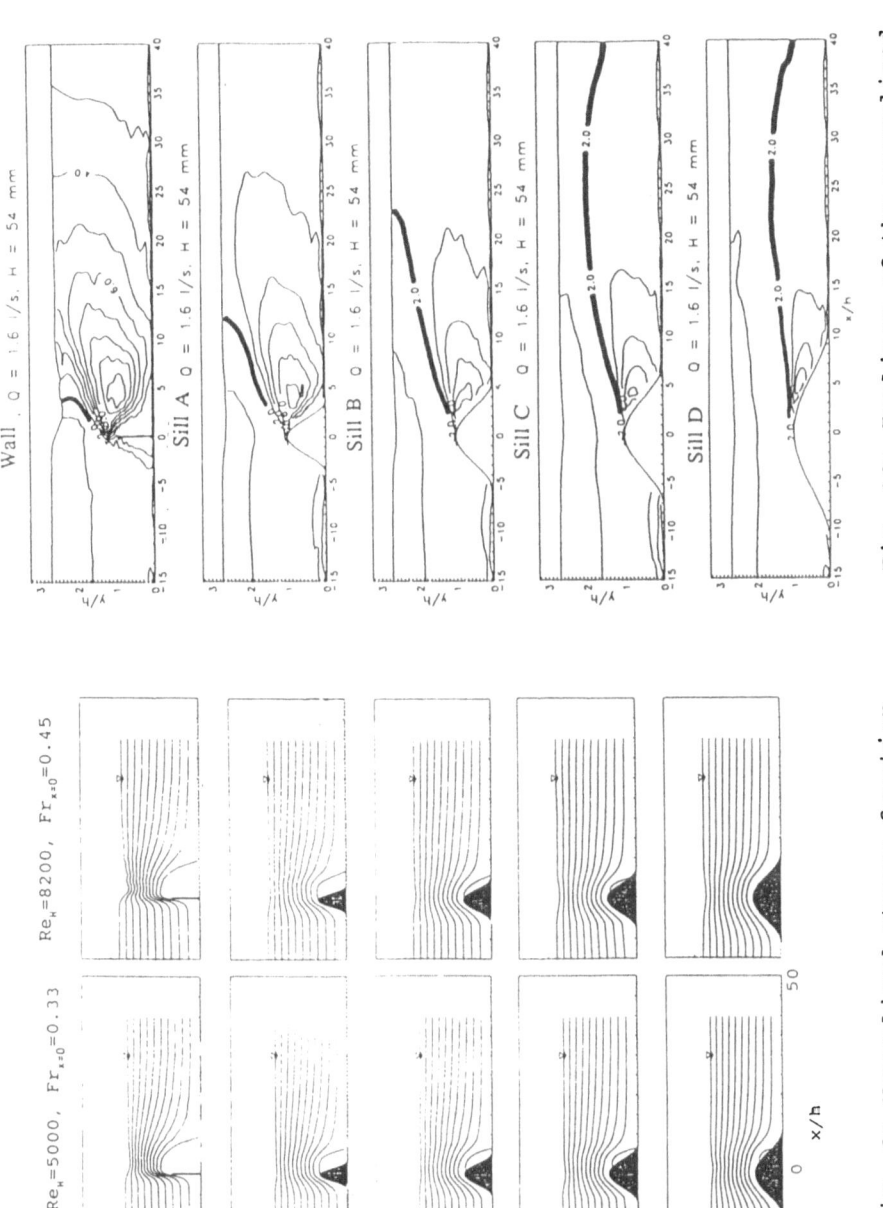

Fig. 10: Isolines of the normalized turbulence fluctuations

Fig. 9: Normalized stream function

141

The Flow Structure in the Wake and Separated Region of Plane and Axisymmetric Bodies

D. Geropp, A. Leder

Institut für Fluid- und Thermodynamik
Universität – GH – Siegen, Paul – Bonatz – Straße 9-11
D – 5900 Siegen, Germany

SUMMARY

In this project the physical structure of separated flows and wakes behind plane and axisymmetric bodies in the incompressible flow regime was investigated. To detect the instationary behaviours of the flowfields the velocity measurements were performed with a specifically developped laser-Doppler anemometer, the phase-locked LDA. The experimental results reveal characteristic distributions in the velocity field with the occurence of saddle points as well as characteristic distributions in the terms of the Reynolds stress tensor. Saddles are quasiperiodically formed by interactions of free shear layers. The distributions of the Reynolds stresses indicate that maximal amplitudes in the normal stresses and the shear stress term are provoked by positive pressure gradients in the shear layer flows upstream of saddle points. Regions with enhanced Reynolds stresses and saddles are convectively transported within the flowfield. Beside the database for the validation and improvement of numerical schemes, the results of these studies deliver a deep insight into the physical mechanisms of vortex sheddings in the lee of plane and axisymmetric bodies.

1 INTRODUCTION

Numerical procedures, e.g. Navier-Stokes calculations including k-ϵ turbulence modeling, are still failing to predict real properties of separated flows [1], [2]. In many cases the accuracy of drag calculations for example to design airfoils or to reduce the aerodynamic drag of vehicles are far from being applied in the technical practise. Improvements of numerical schemes can only be expected on the basis of more detailed data concerning the turbulence structures of these complex flows. For this task the experimental data should have the highest degree of accuracy. The precision of experimental data depend essentially on the measuring technique used. Because of the noninvasive character and directional sensitivity, optical techniques like laser-Doppler anemometry are well suited to analyse turbulent flow structures past bluff bodies.

The properties of separated regions like its extensions or distributions of the turbulent terms are mainly affected by the behaviour of the free shear layers, emanating from the separation lines of the models. The characteristics of the free shear layers depend, e.g., on its geometry, the Reynolds number Re, the turbulence intensity, and the pressure gradient in the flow. The individual influences of these parameters on the flow structure were discussed in [3].

The main intentions of our studies were the improvement of the knowledges about the physical processes and interactions in separated flows as well as the description of characteristics in the flow and turbulence structures. We focused the experiments on plane and axisymmetric model geometries. This paper gives an overview over the results of these studies.

2 DESCRIPTION OF THE EXPERIMENTS

The LDA system used to investigate the turbulent flow structures consists of a two-channel version operating in backscatter mode and a specifically developed phase detector to extract means at constant phase from recorded measurement ensembles. Fig. 1 shows the hardware configuration of the measuring system.

The oscillations of the flow due to vortex sheddings were detected by a hot-wire sensor, integrated in the phase detector circuit, see fig. 1. The sensor was positioned in the external flow just outside the separated region, within the coherence length of the first developing vortex structure. Thus phase jitter in the periodic velocity components could nearly be eliminated. The principles of the measuring technique are discussed in detail in [4] and [5].

The experiments were performed in the closed-circuit wind tunnel at University of Siegen. The open working section has a 1 m · 1 m cross section and is 1.8 m long. The thyristor driven blower allowed the adjustment of the test section velocity between 1 m/s – 50 m/s. The freestream turbulence level is about 0.5 %. To obtain a nominally two-dimensional flow, the plane models were mounted with endplates horizontally into the test section. The extension of the endplates was 8D downstream the models (D is the model dimension vertical to the direction of free stream velocity). To investigate turbulent near wake flow characteristics of axisymmetric bluff bodies like cones, spheres, and discs an active model support with integrated boundary layer suction was designed. The special advantages of this construction like the minimizing of disturbances in the external flow were discussed in [6].

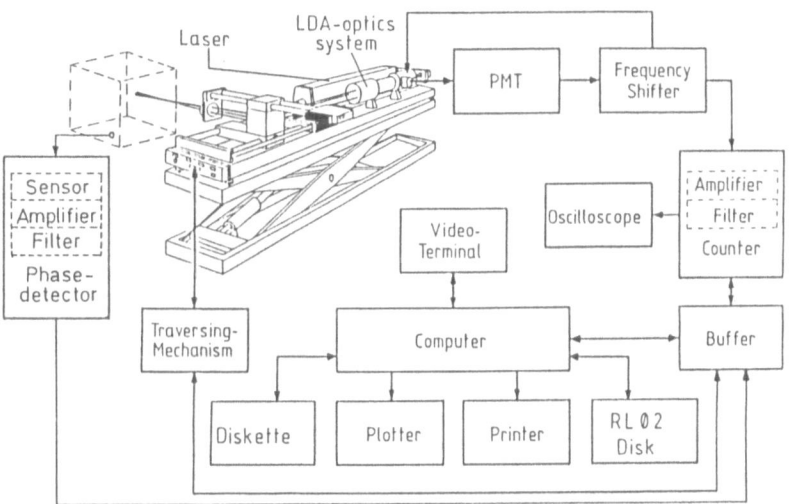

Fig. 1 Hardware configuration for phase-locked LDA measurements.

3 RESULTS

The experimental results confirmed the prognostics that the flowfields past plane and axisymmetric bodies show fundamental differences, especially in its dynamic behaviours. The amplitudes of the quasiperiodic velocity components and the Reynolds stress terms in the separated region of plane models are substantial higher than in the case of an axisymmetric flow. In the discussion of the results, according to fig. 2, we distinguish between the development of plane and axisymmetric free shear layers. A vertical flat plate was choosen as the generic example for the flow class with two interacting plane shear layers, the sphere

Class I : development of two interacting plane free shear layers

circular cylinder   vertical flat plate   airfoil, $\alpha > \alpha_{crit}$

Class II : development of an axisymmetric free shear layer

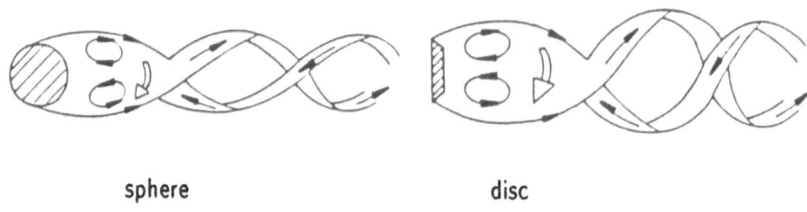

sphere   disc

Fig. 2 Classification of separated flows.

model is a representative of the axisymmetric case. Within the same class the flow structures, the amplitudes of the quasiperiodic velocity components, and the Reynolds stress terms show similar behaviours [7], [8], [9], [10].

3.1 Structure of the time averaged flow

The results of the time averaged flowfield, measured with the LDA system, are plotted in fig. 3 for the case of the vertical flat plate and in fig. 4 for the sphere model. In both cases the separated region is formed by a vortex configuration. The external flow at first accelerates along the deadwater-contour, defined by the zero-streamline Ψ_0, see fig. 4. Ψ_0 can be calculated by an integration of the velocity profiles [7]. In reaching the position of the maximal width of the separated region, the flow decelerates until reaching the end of the recirculating region. The flow than accelerates again compensating the defect velocity in the wake profiles.

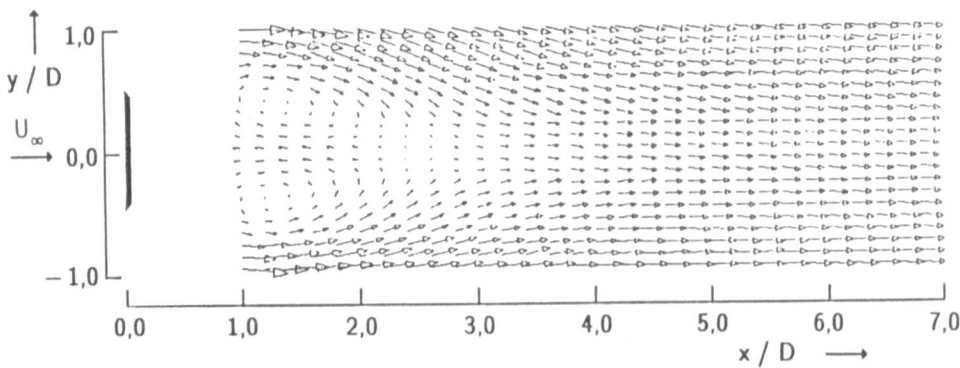

Fig. 3 Velocity-vector plot of the time averaged flow, flat plate, Re = $2.8 \cdot 10^4$.

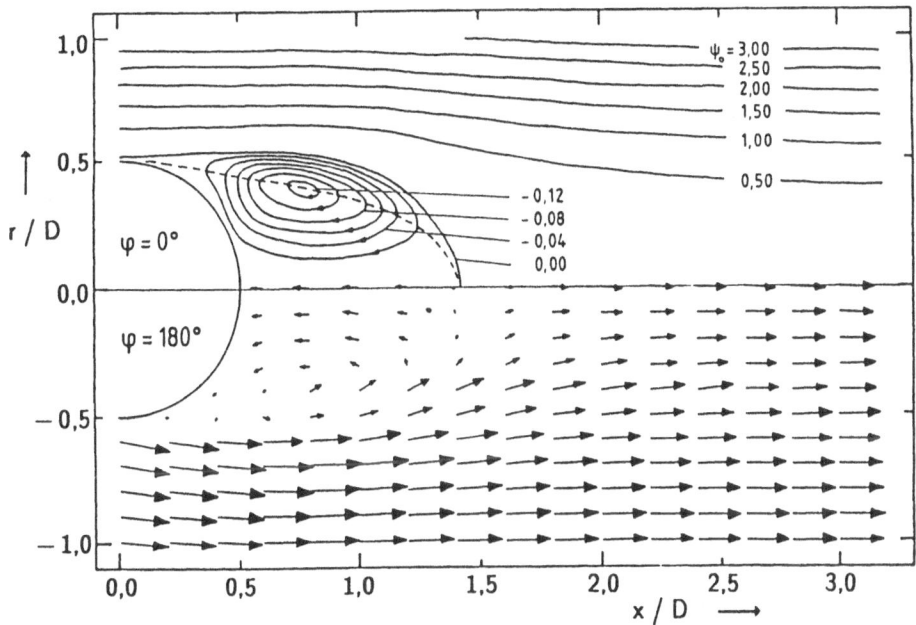

Fig. 4 Velocity-vector plot and mean streamlines of the time averaged flow past a sphere model, $Re = 5 \cdot 10^4$.

The acceleration followed by a deceleration of the external flow along the zero streamline indicates that the separated region can be understood as a displacement contour. This was already taken into account by Walz [11].

The characteristic distributions of the Reynolds stresses are displayed in fig. 5 — fig. 7. The terms are nondimensionalized with the squared free-stream velocity U_0. Because these distributions are similar in both flow classes they are discussed below only for the example of the sphere.

Fig. 5 illustrates the behaviour of the longitudinal Reynolds normal stress component u'. This term reaches a maximum after a steep gradient shortly behind the flow separation. The extremum indicates the transition from the laminar into the turbulent state of the free shear layer flow.

In fig. 6 the results for the transversal Reynolds normal stress component v' are plotted. The isoline-diagram indicates that v' reaches its maximum on the axis of symmetry around the position of the free stagnation point at $x/D = 1.5$.

The distribution of the shear stress u'v' is represented in fig. 7. This term shows a similar behaviour like u' with the exceptions that the gradient in u'v' is not as steep as in the longitudinal component and that the location of maximal shear stress appears further downstream, between the positions of the extremum in u' and v'.

As already mentioned the amplitudes of the Reynolds stress terms of both flow classes (see fig. 2) differ significantly. Fig. 5 — fig. 7 indicate the extremum values for the case of an axisymmetric developping free shear layer: the longitudinal and the transversal terms in flow class II reaches an amplitude of 0.09, while in flow class I the corresponding terms have an amplitude of 0.23 respectively 0.35. A similar amplification can be observed for the shear stress: the term increases from 0.04 in axisymmetric separated flows to approximately 0.11 in flow class I.

The reason for the reduced amplitudes in flow class II is the minimized interaction-strength of the free shear layers resulting in smaller quasiperiodic oscillations of the flow. The instationary behaviour of the flow was investigated with the phase-locked LDA technique.

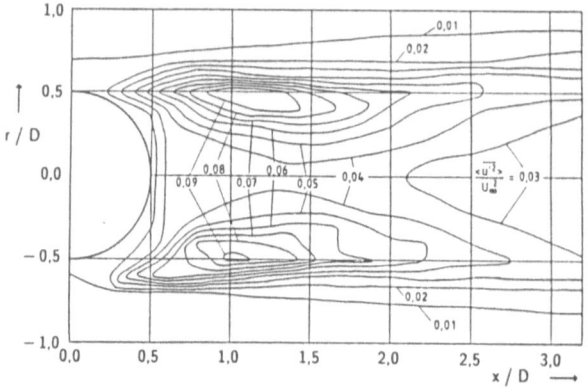

Fig. 5 Distribution of the Reynolds normal stress $\overline{u'^2}/U_\infty^2$, Re $= 5 \cdot 10^4$.

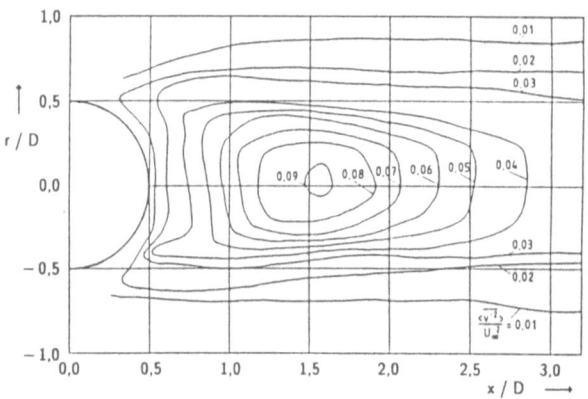

Fig. 6 Distribution of the Reynolds normal stress $\overline{v'^2}/U_\infty^2$, Re $= 5 \cdot 10^4$.

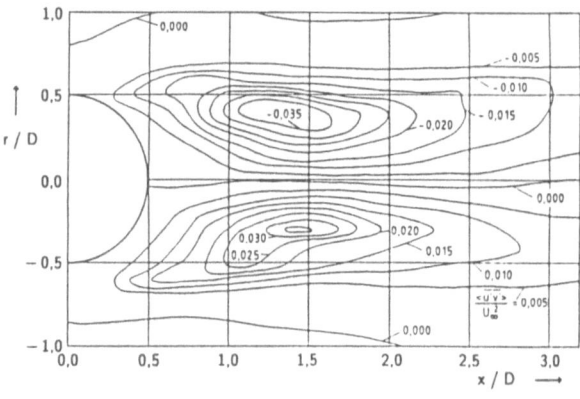

Fig. 7 Distribution of the Reynolds shear stress $\overline{u'v'}/U_\infty^2$, Re $= 5 \cdot 10^4$.

146

3.2 Dynamics in separated flows

The wake flow of axisymmetric and plane bodies exhibits a quasiperiodic component at the Strouhal frequency. The frequency depends on the width of the separated region and the spreading velocity of disturbances to cross the separated region [3]. In the investigated Reynolds number range from 10^3 - 10^5 the Strouhal number remained constant.

With the help of the instationary measurement technique the developing vortex structures could be resolved. In flow class II, the axisymmetric case, a helical vortex, emanating from the rear end of the separated region, could be detected. The vortex is formed by the rotation of a saddle point at the rear end of the recirculating region around the axis of symmetry [3].

In flow class I the dynamics of the saddle points are initiated by intense interactions between the two free shear layers [12], [13], [14]. Fig. 8 represents the dynamics of the flow past the vertical flat plate. In this case the interactions between the two free shear layers occur near the free stagnation point, which is located in the time averaged graph of fig. 3 at $x/D = 2.5$. The presentation of the results in fig. 8 therefore concentrate on the flow region enclosing this area.

Fig. 8 illustrates the developing flow structure in the section from $x/D = 1.0$ to 4.5 and $y/D = -1.0$ to 1.0. The periodic process of vortex shedding is resolved into six phases. The time difference between consecutive phases is $\Delta t = 0.02$ sec. Each displayed velocity vector represents the mean at constant phase calculated from the data of more than 300 vortex shedding cycles.

In phase 1 a vortex forms in the upper shear layer with its center at $x/D = 1.0$. Its formation was initiated in the preceding phases (see phases 6, 5, ..) by the development of a saddle point positioned at this time near $x/D = 2.4$. Saddle points are emphasized in fig. 8 by short arrows indicating instantaneous streamlines. Saddles originate from interactions between the lower and upper free shear layer flows.

The decreasing lengths of the velocity vectors in the shear layer flow approaching the saddle point indicate the local increase in the pressure. At phase 2 the center of the developing vortex shifts to $x/D = 1.3$, and 20 ms later, at phase 3, the center occures at $x/D = 1.7$. From phase 4 to phases 5 and 6 the acceleration of the clockwise rotating structure enhances. Supported by an intense entrainment process, originating at this time from the lower shear layer, the vortex structure leaves the formation region and enters the near wake. The equivalent sequence for the formation of a vortex structure with counter-clockwise rotation can be observed with a time shift of $T/2$. In phase 4 the center of the counter—clockwise rotating vortex enters the plotted flow field. The development of the vortex continues in phases 5 and 6. After the acceleration in phases 1 and 2, induced by the entrainment process from the upper shear layer, the vortex structure detaches at phase 3 the recirculating flow region.

When the saddles leave the formation region and enter the near wake, see the sequences from phase 1 to phase 2 and phase 4 to phase 5 in fig. 8, the saddle—structures are accelerated. In these cases, the centers of the saddles are no longer directly visible. In transforming the velocity field by subtracting the convective velocities of these flow structures, the saddles would reappear for the stationary observer.

The triangle, square, and star symbols in fig. 8 represent the locations of extremum values in $<u'^2>$, $<v'^2>$, and $<u'v'>$ respectively. According to the notation of Reynolds and Hussain [15] the brackets characterize means at constant phase. Obviously maximal amplitudes of the normal stresses $<u'^2>$ and $<v'^2>$ as well as of the shear stress $<u'v'>$ occur in the decelerating shear layer flows upstream the saddle points. The dynamics of of the terms of the Reynolds stress tensor are discussed in detail in [3] and [14].

The production of turbulent energy in twodimensional flow is the sum of three terms, representing the product of Reynolds stresses and velocity gradients. Analyses described in [16] clarified that the main contribution to the production of turbulent energy is traced back to the shear stress component. These results indicate, that in order to minimize the aerodynamic drag of bluff bodies it is essential to reduce the amplitude of the shear stress term.

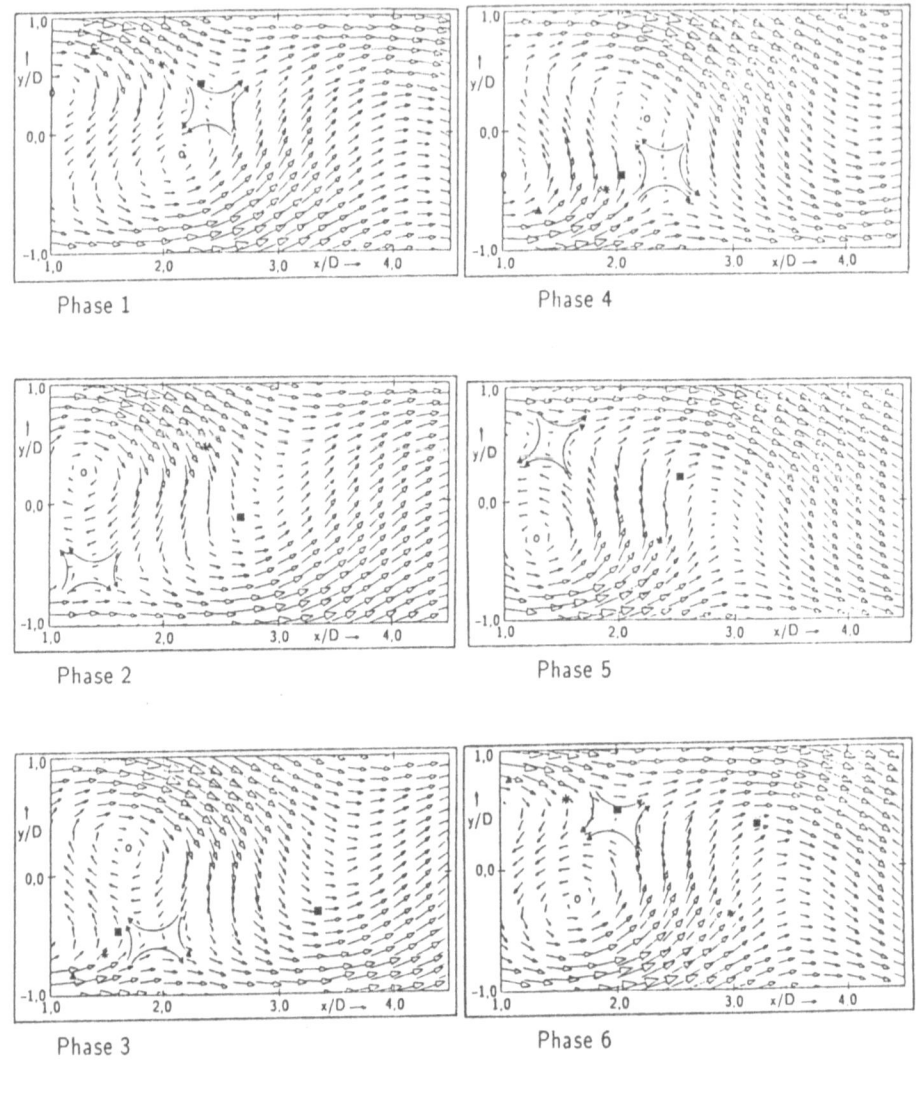

Fig. 8 Dynamics of vortex shedding past a vertical flat plate, Re = 2.8·10⁴.
Saddle points are emphasized by short arrows, indicating instantenous streamlines.

▲ : $<u'^2>/U_\infty^2 \geq 0.24$; • : $<v'^2>/U_\infty^2 \geq 0.29$; ; ⋆ : $<u'v'>/U_\infty^2 \geq 0.12$;

o : vortex center .

ACKNOWLEDGMENT

This project was partially financed by the DFG. The authors gratefully acknowledge this support.

148

REFERENCES

[1]   Majumdar, S.; Rodi, W.: Numerical calculations of turbulent flow past circular cylinders. Third Symp. on Numerical and Physical Aspects of Aerodynamic Flows, Long-Beach, CA, USA, 21-24 January 1985.

[2]   Leder, A.: Physical properties of separated flows behind two-dimensional bluff bodies in uniform flows. ASME Winter Annual Meeting, Miami Beach, Florida, USA, 17-22 November 1985.

[3]   Leder, A.: Abgelöste Strömungen - Physikalische Grundlagen. Vieweg-Verlag, Braunschweig, Wiesbaden (1992).

[4]   Leder, A.; Geropp, D.: Phase-averaged LDA measurements in turbulent separated flows. In: Durão, D.F.G. (Editor): Proceedings of the fourth International Symposium on Applications of Laser-Anemometry to Fluid Mechanics, Lisbon, Portugal, paper 3.3 (1988).

[5]   Leder, A.; Geropp, D.: Laser-Doppler measurements in quasiperiodic flows. In: Durão, D.F.G. (Editor): Proceedings of the sixth International Symposium on Applications of Laser-Anemometry to Fluid Mechanics, Lisbon, Portugal, (1992).

[6]   Leder, A.; Geropp, D.: An active model support to investigate turbulent near wake flow characteristics of axisymmetric bluff bodies. In: 12th International Congress on Instrumentation in Aerospace Simulation Facilities (ICIASF), pp. 33-40. IEEE Service Center, Piscataway, New York, USA, (1987).

[7]   Leder, A.: Laser-Doppler Untersuchungen und einige theoretische Überlegungen zur Struktur von Totwasserströmungen. Fortschritt-Berichte der VDI-Zeitschriften, Reihe 7, Nr.: 78 (1983).

[8]   Geropp, D.; Leder, A.: Turbulent separated flow structures behind bodies with various shapes. In: Turner, J.; Stanbury, J. (Editors): Proceedings of the International Conference on Laser Anemometry - Advances and Application, Manchester, England, paper 12, pp. 219-231. BHRA, Fluid Engineering Center Cranfield, Bedford, England, (1985).

[9]   Geropp, D.; Leder, A.: A study of turbulent near wake flow characteristics behind the 50⁰-cone using LDA and visualization techniques. In: Turner, J.; Frazer, S. (Editors): Proceedings of the second International Conference on Laser Anemometry-Advances and Applications, pp. 165-174. BHRA, Fluid Engineering Center Cranfield, Bedford, England, (1987).

[10]   Geropp, D.; Leder, A.: Abgelöste Strömungen hinter axialsymmetrischen Körpern. In: Felsch, K.O.; Marcinowski, H.; Zierep, J. (Hrsg.): Mitteilungen des "Instituts für Strömungslehre und Strömungsmaschinen" der Universität Karlsruhe. Bd. 39/88, S.51-59. Verlag G. Braun, Karlsruhe, (1988).

[11]   Walz, A.: Berechnung der Druckverteilung an Klappenprofilen mit Totwasser, Jahrbuch der deutschen Luftfahrtforschung 1, S. 265-277, (1940).

[12]   Leder, A.; Geropp, D.: Ermittlung von kohärenten Strömungsstrukturen in hochturbulenten Ablösegebieten mit dem Laser-Doppler-Anemometer. In: Deutsche Gesellschaft für Luft- und Raumfahrt-DGLR (Hrsg.): Jahrbuch 1988, Bd. I, S. 345-353, (1988).

[13]   Leder, A.; Geropp, D.: Experimentelle Untersuchung instationärer Ablösungen. In: Deutsche Gesellschaft für Luft- und Raumfahrt-DGLR (Hrsg.): 7. DGLR-STAB Symposium vom 7.-9. November 1990 in Aachen. DGLR-Bericht 90-06, S. 214-218, (ISBN 3-922010-55-5), (1990).

[14]   Leder, A.: Dynamics of fluid mixing in separated flows; Physics of Fluids A, vol. 3, Issue 7, pp. 1741-1748, (1991).

[15]   Reynolds, W.; Hussain, A.: The mechanics of an organized wave in turbulent shear flow. Part 3. Theoretical models and comparisons with experiments. Journal of Fluid Mechanics, vol. 54, part 2, pp. 263-288, (1972).

[16]   Leder, A; Geropp, D.: Dynamics of turbulent energy production in separated flows. In: Rodi, W., Ganic, E.N. (Editors): Engineering Turbulence Modelling and Measurements, pp. 399-408. Elsevier Science Publishing Co., New York, (1990).

A new model of turbulence for separated flow behind bluff bodies

D. Geropp, G. Schumann

Institut für Fluid- und Thermodynamik
Universität Siegen, Germany

Summary

Periodical pressure and velocity fluctuations influence decisively the separated flow field behind bluff bodies next to the statistical turbulence. These phenomena are investigated theoretically and represented in a new model of turbulence. The coherent large scale vortex structures are analysed with the help of foreign experiments and own calculations simulating the unsteady laminar flow over a bluff body, e.g. over a vertical plate. Results of this analysis are used to model a deduced integral relation for the pressure velocity correlation. This characterises the part of large scale vortex motions and is combined with the "second order closure" model from Launder, Reece and Rodi [1]. This extended model permits to describe correctly the physics of separated flow behind a bluff body, e.g. a vertical plate.

1 Introduction

The flow about bluff bodies at subcritical Reynolds-numbers contains separated flow regions, in which nearly periodical large scale vortex motions are observed besides statistical turbulence fluctuations. Figure 1 shows a separated flow about a vertical plate.

The flow is visualised by white and yellow smoke. High turbulent eddy regions (yellow smoke) are apparent side by side with low turbulent entrainment motions (white smoke). A chronological sequence of such pictures would illustrate the nearly periodical large scale motions. A time averaged process of velocity fluctuations indicates that the part of large scale Reynolds-stresses has about 25% to 40% of total stresses. In addition, the stochastical part of turbulence is increased by the shear action in the large scale vortex motions.

A stationary calculation of such separated flows is very important, e.g. for flows about buildings or cars. It requires a turbulence model, which can describe the reflections of large scale turbulent motions on the time averaged flow field.

The stationary calculation of large scale separated flows on the basis of classic statistical turbulence models shows incorrect results particulary for distributions of Reynolds-stresses, for the length of the recirculating flow in wake and for the drag coefficient. Also extensive turbulence models, such as the "second order model" from Launder e.a. [1] do not supply substantially better results than more simple models, such as the standard $k - \varepsilon$ model from Launder and Spalding [2] (see figures 5 and 6).

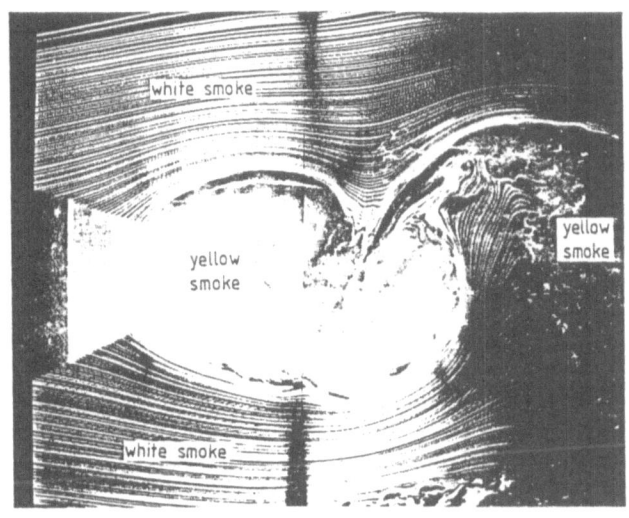

Fig. 1   Visualisation of a separated flow behind vertical plate

2   Effect of periodical fluctuation

A time dependent variable $u(x_i, t)$ in a nearly periodical turbulent flow can be generally divided into three parts: a time average $\bar{u}$, a periodical fluctuation $\tilde{u}$ and a superposed stochastic part u'.

$$u(x_i, t) = \bar{u}(x_i) + \tilde{u}(x_i, t) + u'(x_i, t) . \tag{1}$$

The periodical and the stochastic part can be combined to total fluctuation.

$$u^*(x_i, t) = \tilde{u}(x_i, t) + u'(x_i, t) . \tag{2}$$

Analogous division and combination can be made for the pressure p. With the help of this triple decomposition and after a time averaged process binary correlations can be splitted up into a stochatic part $< \overline{u'_i \cdot u'_j} >$ and a pure periodical part $\overline{\tilde{u}_i \cdot \tilde{u}_j}$. Both these parts can also be combined to a total correlation.

$$\overline{u^*_i \cdot u^*_j} = \overline{\tilde{u}_i \cdot \tilde{u}_j} + \overline{u'_i \cdot u'_j} . \tag{3}$$

At this the sign $<\quad>$ marks an ensemble average of correlation $u'_i \cdot u'_j$. Such averaged process concludes also mixed correlations, especially correlations of velocity and pressure fluctuations, e.g. $\overline{\tilde{u}_i \cdot \partial \tilde{p}/\partial x_i}$. Moreover, triple correlations can contain relations between stochastic and periodical fluctuations. Cantwell and Coles [3] as well as Leder and Geropp [4],[9] have measured suitable velocity fluctuations and correlations in different separated flows. In figure 2 (left side) the time averaged periodical stress correlations $\overline{\tilde{u} \cdot \tilde{u}}, \overline{\tilde{v} \cdot \tilde{v}}, \overline{\tilde{u} \cdot \tilde{v}}$ measured by [3] in wake behind a circular cylinder are represented exemplary.

The isolines of correlations show characteristical maxima and minima, which can not be calculated by classic turbulence models, such as used in [1] or [2].

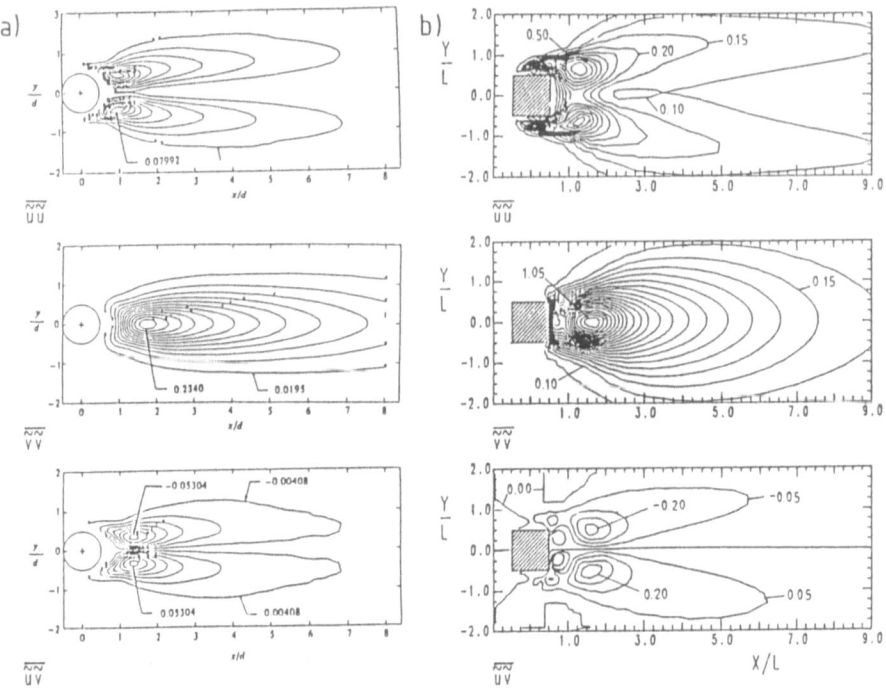

Fig. 2   Periodical correlations: a) measurements [3] b) simulating calculations [5]

Unfortunately, suitable correlations of velocity and pressure fluctuations, symbolizing the pressure diffusion in turbulent flow, can not be measured because a measuring method does not exist to this day. But the experiments from [3],[4],[9] combined with calculations indicate that pressure fluctuations decisively play a great part in the large scale vortex motions in wake. Therefore, the pressure diffusion is an important element of turbulent modelling for separated flows close to the body. In consequence of missing measurements information about pressure fluctuations could be obtained only by simulating calculations in an unsteady wake flow.

3   Simulation of unsteady wake flow

Since wake flows with periodically separating vortices already exist for low Reynolds-number, a wake is calculated by the unsteady laminar Navier-Stokes equations for $Re = 1000$. As body is used a twodimensional cylinder with quadratic cross-section, which has fixed separation lines. The chosen Reynolds-number is high enough that viscous effects damping on the vortex motions are avoided and the unsteady wake is preserved. Finally, correlations of the velocity and pressure fluctuations are determined on the basis of the results of the unsteady calculation by Schumann [5], [6].

First, the Reynolds-stresses for this case are put opposite to the measurements from [3] in figure 2 (right side). The plots demonstrate good qualitative reproduction of low-frequency fluctuations, respectively periodical correlations by the simulating calculations. Further, these calculations show that the interaction between pressure and velocity fluctuations influence essentially the effect of vortex separation. High $\tilde{u}$ values are assigned to high negative pressure gradients $\partial\tilde{p}/\partial x$ in x-direction and by analogy high $\tilde{v}$ components to high gradients $\partial\tilde{p}/\partial y$. It follows also, that the correlation $\overline{\tilde{v}\cdot\partial\tilde{p}/\partial y}$ is a significant quantity which influences the large scale fluctuation in the wake region close to the body. In figure 3 these correlations are compared with results for model assumptions used by [1]. The comparison shows clearly that the distributions both quantitatively and qualitatively do not correspond to those from [5], [6].

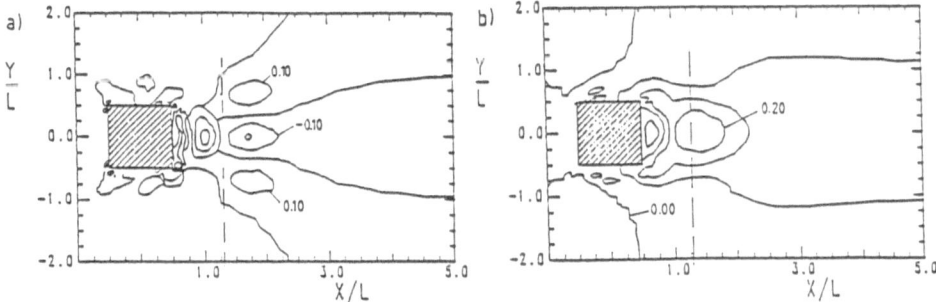

Fig. 3   Correlation $\overline{\tilde{v}\cdot\partial\tilde{p}/\partial y}$: a) simulating calculation unsteady laminar flow [5] b) calculation [1]

Thus, very close to the body the wake flow is accelerated by pressure fluctuations and so anisotropy between the velocity fluctuations is given. This stands in contrast to Rotta's assumption [7] that energy of the biggest velocity component passes over directly to that of smaller velocity components. But this isotropy is observed only about 2.5 body dimensions downstream in wake. Therefore, stationary turbulent calculation for the separated flow region close to the body needs a new modelling of pressure-velocity correlation. Downstream the new model must be adapted to relations of classic turbulence models.

4   The new model

The decisive correlation $\overline{\tilde{v}\cdot\partial\tilde{p}/\partial y}$ was represented by integral expressions developed from a Poisson equation for the periodical pressure fluctuations in [6]. First the superposed stochastic fluctuations are neglected.

In these integral expressions there are e.g. triple correlations of the velocity fluctuations referred to spatially different points. For those modelling information is obtained by analysis of the mentioned measurements in [3], [4] and of results of the simulating calculation for unsteady laminar separated flows. First of all the information relates to the different positions of maximal fluctuations and to the increments of velocity fluctuations. In [6] some principal relations for these phenomena are deduced by the available calculation and measurement results.

Finally, from these the following decisive equation of the modelling can be derived:

$$\frac{\overline{u_i^* \, \partial p^*}}{\varrho \, \partial x_j} + \frac{\overline{u_j^* \, \partial p^*}}{\varrho \, \partial x_i} = \left. \left(\frac{\overline{u_i^* \, \partial p^*}}{\varrho \, \partial x_j} + \frac{\overline{u_j^* \, \partial p^*}}{\varrho \, \partial x_i} \right) \right|_{\mathrm{LRR}}$$

$$+ \; C_3 \cdot (f_x)^{C_2} \sqrt{\overline{v^* \cdot v^*}} \, \frac{\partial \overline{u^* \cdot v^*}}{\partial y} \delta_{i2} \delta_{j2} \; . \tag{4}$$

The first term on the right represents the classic modelling from [1] and the second, framed term is an extension of the turbulence model. It describes the special above mentioned phenomena of separated flows.

In it f_x is a wall function, which restricts the investigated fluctuation effects to the wake region close to the body (about 2-3 body dimensions) and simulates a downstream reduction of these effects. In other respects, f_x is the same function, which is used also in the wall damping model from Gibson e.a. [8].

The new expression contains two freely eligible constants C_2 and C_3. C_3 determines fundamentially the level of the $\overline{v^* \cdot v^*}$ maximum and C_2 the position of the x-coordinate. The equation (4) is valid on the assumption that the back of the body is vertical to the x-coordinate and the incident flow is parallel to x. Altogether, the new approach represents an important supplementation to known and tested turbulence models.

5 Numerical results

The calculated results base on the new turbulence model, the wall function from [8] and a set of fundamental equations, namely the Navier-Stokes and Reynolds-stress equations and in addition the transport equation for ε. Herewith the quantities $\bar{u}, \bar{v}, \bar{p}$, $\overline{u^* \cdot u^*}$, $\overline{u^* \cdot v^*}$, $\overline{v^* \cdot v^*}$, $\overline{w^* \cdot w^*}$ and ε are calculated for twodimensional flow fields. In figure 4 the results for the pressure correlation behind a cylinder of square section are in good accord with the simulated flow.

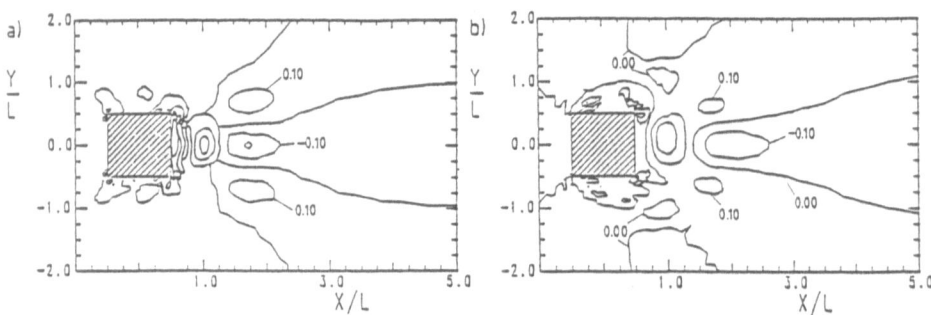

Fig. 4   Correlation $\bar{v} \cdot \partial \bar{p} / \partial y$: a) simulating calculation unsteady
laminar flow [5]   b) calculation with the new model [6]

The position of maximal and minimal values, the relative levels and the isolines of zero levels are reproduced very well by the calculation, respectively by the model assumptions.

The great importance of the new model is becoming evident when you look at the separated flow behind a vertical twodimensional plate in the figures 5 and 6. The calculated time averaged streamlines in figure 5, especially the length of the recirculating region and the drag coefficient c_W are confirmed by experiments from [2], [9]. Contrary to this, the results of the $k - \varepsilon$ method from [2] or the Reynolds-stress method from [1] are very different from the measurements.

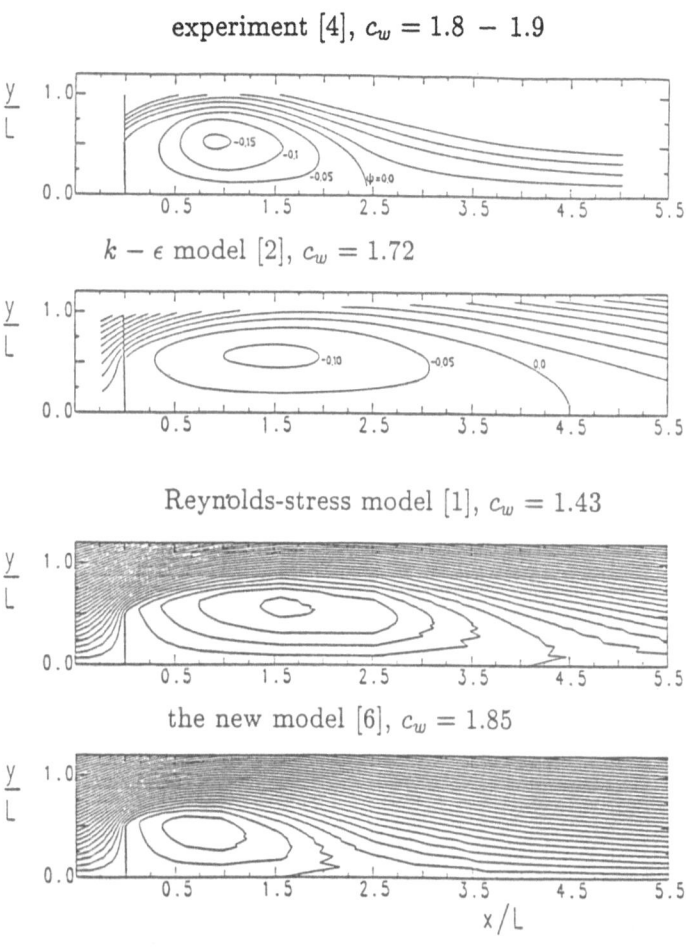

Fig. 5 Time averaged streamlines of the experiment [4] and of different turbulence models

But the high quality of the new model is evident above all in figure 6, where the distribution of Reynolds-stress $\overline{v^* \cdot v^*}$ is presented exemplary. The position of maximum of $\overline{v^* \cdot v^*}$ and its value agree with the experiment from [9] while the results obtained by the turbulence model from [1] show noticeable differences.

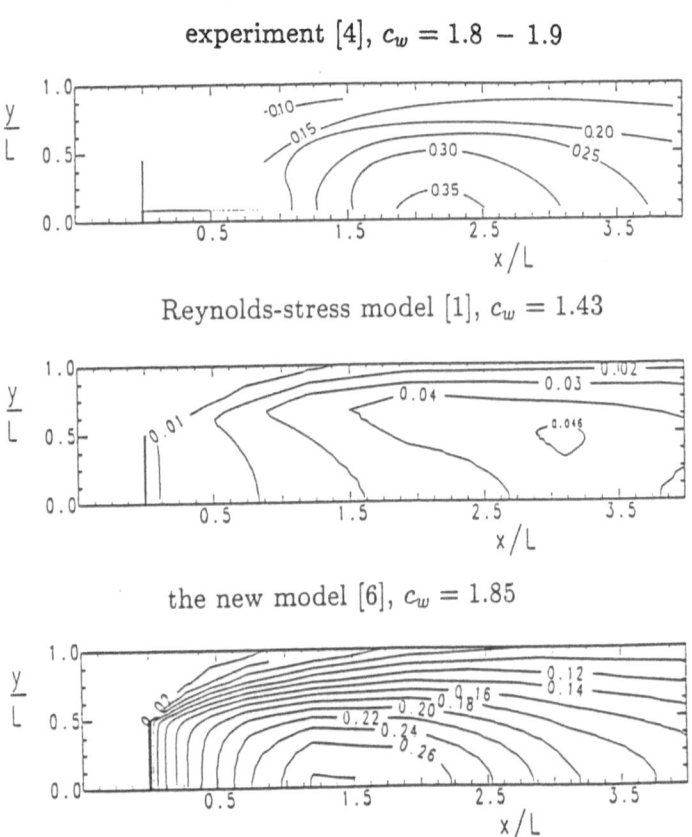

Fig. 6   Correlation $\overline{v^* \cdot v^*}/u_\infty^2$ of the experiment [4] and of different turbulence models

6   Conclusion

The calculation of time averaged flows behind bluff bodies is very disappointing on the basis of the present turbulence modellings. Therefore, a new model was developed recently, which takes into account particulary the large periodical vortex motions in separated flow regions. The required physical information for this model could be found by analysis of known measurements and in addition by special simulations on basis of calculation for unsteady laminar flows. Results calculated by the model are corresponding very well with experiments.

References

[1]   Launder, B.E.; Reece, G.J. , Rodi, W.: "Progress in the development of a Reynolds-stress turbulence closure." Journal of Fluid Mechanics, Vol. 68, Nr.3, pp. 537 – 566, (1975).

[2]   Launder, B.E.; Spalding, D.B.: "The numerical computation of turbulent flows" Computer Methods in Applied Mechanics and Engineering, Vol. 3, pp. 269 – 289, (1974).

[3]   Cantwell, B.; Coles, D.: "An experimental study of entrainment and transport in the turbulent near wake of a circular cylinder." Journal of Fluid Mechanics, Vol. 136, S. 321 – 374, (1983).

[4]   Leder, A.; Geropp, D.: "Dynamics of turbulent energy production in separated flows." In: Rodi, W.; Ganic, E.N. (Editors): Engineering Turbulence Modelling and Measurements, pp. 399 – 408. Elsevier Science Publishing Co., New York, (1990).

[5]   Schumann, G.: "Eine neue Turbulenzmodellierung in ebenen Strömungen mit großräumigen Ablösegebieten." Workshop "Turbulenzmodellierung abgelöster Strömungen" des DFG Schwerpunktprogrammms PAS an der Universität Siegen, pp. 64 – 84, (1991).

[6]   Schumann, G.: "Turbulenzmodellierung in ebenen Strömungen mit großräumigen Ablösegebieten." Dissertation der Universität Siegen, erscheint Ende 1992.

[7]   Rotta, J.: "Statistische Theorie nichthomogener Turbulenz." Zeitschrift für Physik, Bd. 129, pp. 547 – 572, (1951).

[8]   Gibson, M.M.; Jones, W.P.; Younis, B.A.: "Calculation of turbulent boundary layers on curved surfaces." The Physics of Fluids, Vol. 24, Nr.3, pp. 386 – 395, (1981).

[9]   Leder, A.: "Abgelöste Strömungen – Physikalische Grundlagen." Habilitationsschrift der Universität Siegen, Vieweg-Verlag, Braunschweig, (1992).

157

AN INVESTIGATION OF THE SEPARATED FLOW PAST AN

OSCILLATING CIRCULAR CYLINDER

I. Teipel

Institute of Mechanics

University of Hannover

3000 Hannover, Germany

Summary

An investigation was carried out concerning the separated
flow past an oscillating circular cylinder. Experimental as well
as theoretical work have been performed. First experiments were
made in a low speed wind tunnel to discuss in detail the position of the separation point for a fixed body. The Reynolds number covered the range from 1 000 to 10 000. Afterwards the flow
field past an oscillating cylinder was studied.
In order to explain the results from the experiments a discrete
vortex method has been developed. Finally lift and drag coefficients have been compared with other experiments. The agreement
was acceptable.

Introduction

The wake of a bluff body consists of an alternating vortex
street if the Reynolds number is not too small. Because of this
unsteady effect in the wake one should expect that the position
of the separation point should oscillate too. It is obvious that
there is an important interaction between the body e.g. the circular cylinder and the separated flow. For the determination of
the time-dependent fluid drag and the properties of the flow-induced vibration one needs more details of the flow in the near
wake.

If one looks in the literature one finds out that the information about the near wake is not very minute. Many aspects have
not been discussed at all.

In order to get a better picture of the boundary layer near
the separation point a refined flow visualization method can be
used. It is important to apply an optical system, where there is

nearly no light reflection from the surface of the cylinder. Otherwise one would obtain results with such large errors that one cannot use them.

The second question concerns the theoretical explanation of the whole flow field. Of course it is not possible to integrate the complete Navier-Stokes equation for a reasonable number of cases, because the number of parameters is too large and the Reynolds number is rather big. The computational time would increase too much. Therefore an alternative procedure has to be applied. Here a discrete vortex method starting from the vorticity transfer equation has been developed. Information from the boundary layer theory will be introduced to fulfill the no-slip condition at the surface of the body.

LASER-LIGHT-SHEET TECHNIQUE

For the correct determination of the position of the separation point the laser-light-sheet technique has been used. To avoid the scattering of the laser beam at the surface of the cy-

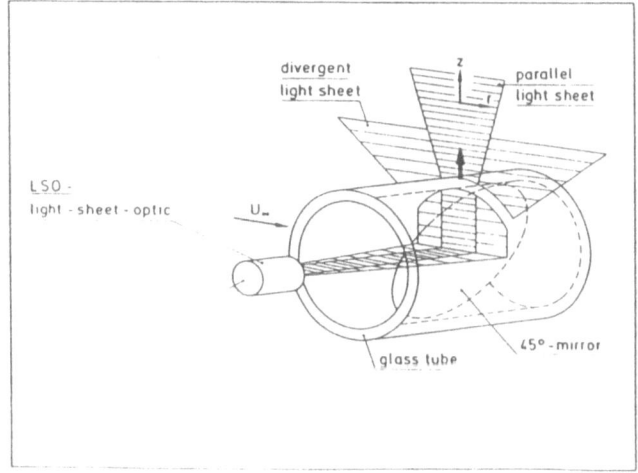

Fig.1 Schematic view of the laser-light set up

linder a particular set up has been built [1]. The laser beam will penetrate the surface of the cylinder from inside. In Fig.1 a schematic view of the optical system has been shown. The laser beam passes through the interior of a glass cylinder. There is a 45°-mirror at the plane where the experimental information will be obtained. The light beam will be reflected by 90° and penetrates the wall of the cylinder and the boundary layer. There will be no scattering of light outside the cylinder. As a consequence the results near the wall are very precise. The turbulence level inside the measuring section was less than 1%. The important result of these experiments is that the separation point for a fixed cylinder does not oscillate although the wake is an unsteady one.

For the oscillating cylinder it is much more difficult to get a realistic picture of the flow near the separation point. Even the definition of the separation point in unsteady flows is not quite clear. Here the definition given by Taneda in [2] is used which means that the separation point is determined as the point at which the streakline will depart from the surface of the wall. In the experiments which have been performed the amplitudes of the vibration were in the range of 3% to 5%, related to the diameter of the cylinder. Otherwise the exposure conditions were too poor. So no spectacular results have been obtained. Further details can be found in [1].

THE DISCRETE VORTEX METHOD

In order to check the experimental features a discrete vortex method has been developed. Instead of integrating the Navier-Stokes equations a different procedure has been used. The starting point was the vorticity transfer equation

$$\frac{D\omega}{Dt} = \nu \, \Delta \, \omega \qquad (1)$$

with ω as the vorticity. For the inviscid case the right hand side can be eliminated.
The complete flow field past a cylinder oscillating perpendicular to the free stream direction will be divided into two layers the far field and the field near the surface of the body. In the far field no viscous effects are present. So the basic equation represents the substantial time derivative for the vorticity ω. The solution can be approximated by a set of discrete point vortices [3]

$$\omega = \sum_{i}^{N} \Gamma_i \gamma(| \vec{r} - \vec{r}_i |) . \qquad (2)$$

$\gamma(r)$ is a distribution function, Γ_i and r_i are the circulation and the position of the i-th vortex. Here a particular partition function given by Leonard [3] has been used

$$\gamma(r_i) = \frac{\sigma^2}{\pi(r_i^2 + \sigma^2)^2} . \qquad (3)$$

σ is proportional to the kernel of a vortex. The Lagrangian description of the flow has the advantage that one does not need to introduce a numerical grid for the integration. The final solution for the flow field can be obtained by Biot-Savart's law

$$\begin{pmatrix} u \\ v \end{pmatrix}_{x,y,t} = \begin{pmatrix} U_\infty \\ 0 \end{pmatrix} +$$

$$+ \frac{1}{2\pi} \int_{-\infty}^{\infty} \int_{-\infty}^{\infty} \begin{pmatrix} y^* - y \\ x - x^* \end{pmatrix} \frac{\omega(x^*, y^*, t)dx^* dy^*}{(x - x^*)^2 + (y - y^*)^2} \cdot \quad (4)$$

In order to fulfill the boundary and initial conditions one may introduce the stream function. One knows that the surface is given by a stream line.

The near field is described by the boundary layer theory. The integration has been carried out by a Warming-Beam procedure. It is quite obvious that the numerical grid which surrounds the whole cylinder will also include the separation point.

The border line between the far field and the near field plays an very important role. If the velocity vector has been directed from the boundary layer to the inviscid field vortices will be eliminated from the calculation. In the other case new vortices will be created.

The important quantities of the whole investigation are the lift and the drag. If all quantities have been obtained it is not difficult to compute also the forces. Quartapelle and Napolitano [4] have derived an analytic expression for lift and drag regarding the circulation of all vortices.

The motion of the vortices will be found by integrating the equation for the velocity

$$\frac{d\ r_i}{d\ t} = \vec{u}_i \ . \tag{5}$$

NUMERICAL RESULTS

There are three important physical parameters, the Reynolds number, the dimensionless frequency of the cylinder and the dimensionless amplitude. First of all the flow past a fixed cylinder has been investigated. As an example the Strouhal number and the drag coefficient versus Reynolds number have been shown in Fig.2 and 3. Comparisons have been made with data from Staubli [5]. In the range of Re = 10^4-10^5 the agreement is very good, if one looks at the Strouhal number S. At larger Reynolds numbers the numerical method showed severe convergence problems. In Fig. 3 the time-average value of the drag coefficient and the RMS-value have been plotted against the Reynolds number. The experimental data for the average value are larger and the RMS-values are smaller. An explanation for this difference could be found in the application of a FFT-analysis of the time signal of

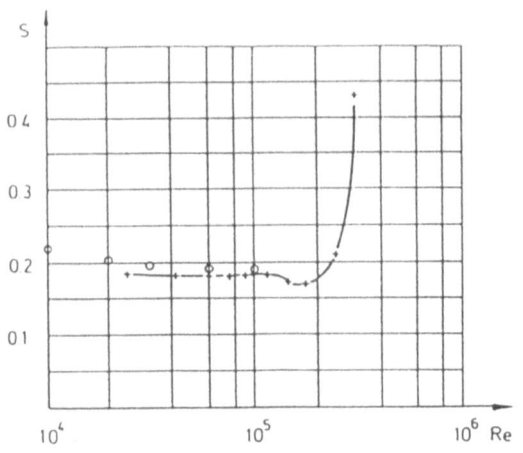

Fig.2 Strouhal
number versus
Reynolds number
+ Staubli [5]
o own results

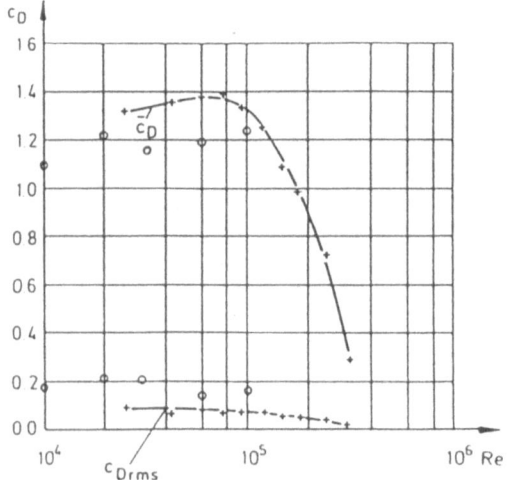

Fig.3 Drag coef-
ficient versus
Reynolds number
+ Staubli [5]
o own results

$\overline{c}_D$ average value

c_{Drms} RMS value

the pressure distribution.

For the oscillating cylinder the time development of the lift coefficient has been approximated by [6]

$$c_L = c_{LK} \cos \omega_K t + c_{LZ} \cos (\omega_Z t + \phi). \qquad (6)$$

ω_K is the frequency of the non-synchronized Karman vortices and ω_Z is the frequency of the vibration of the cylinder. c_{LK} and c_{LZ} are the corresponding amplitudes. ϕ is the phase angle. In Fig. 4 the dimensionless frequency S_K of the non-synchronized Karman vortices has been plotted against A/D which means the relation between the amplitude of the cylinder vibration and the diameter. Various Reynolds numbers have been used. The value of S_Z, the dimensionless frequency ω_Z has been chosen to 0.14. The frequency S_K increases with increasing A/D. The behaviour at small amplitudes for Reynolds numbers near 10^4 might come again from the FFT-analysis. If one compares these results with data of Straubli one finds similar tendencies.

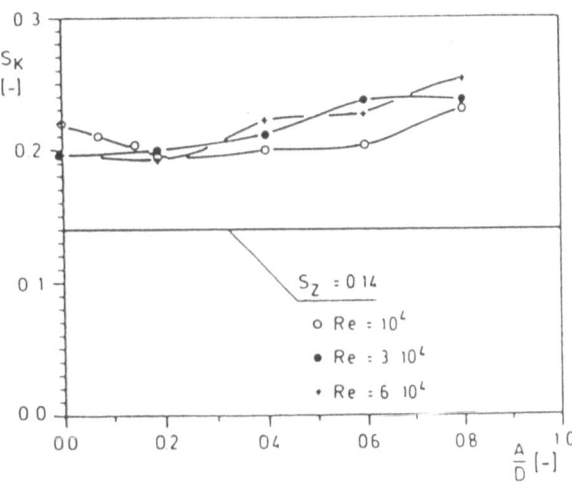

Fig.4 Dimensionless frequency of the non-synchronized Karman vortices versus the relation of the vibration amplitude to the diameter of the cylinder

CONCLUSION

In the first part of this report experimental results for the flow past a circular cylinder have been given. The important remark is that the separation point of the boundary layer at the surface of a fixed cylinder does not oscillate, although an unsteady wake is present. That means that there is no influence from the unsteady effects past the cylinder in front of the separation point.

In the second part a discrete vortex method has been applied in order to obtain information for a modified Strouhal number, the lift and drag coefficient of the oscillating cylinder. With increasing amplitude of the vibration the modified Strouhal number will increase too.

REFERENCES

[1] STÜCKE,P.:"Über die Ablösung der Strömung am querange-strömten Kreiszylinder", Dissertation U. Hannover 1990.

[2] TANEDA,S.: "Visual Study of Unsteady Separated Flows Around Bodies", Prog. Aeron. Sci. 17, (1977) pp.287-348.

[3] LEONARD,A.: "Vortex Methods for Flow Simulation" (Review), J.Comp. Phys. 37 (1980) pp. 289-335.

[4] QUARTAPELLE,L., NAPOLITANO,M.:"Force and Moment in Incompressible Flows", AIAA J.21 (1983) pp.911-913.

[5] STAUBLI,T.: "Untersuchung der oszillierenden Kräfte am querangeströmten schwingenden Kreiszylinder", Dissertation ETH Zürich 1983.

[6] GOTTSCHLICH,M.: "Das Strömungsfeld in der Nähe eines quer zur Anströmung oszillierenden Kreiszylinders" Dissertation U. Hannover 1990.

Vortex Resonant Vibrations of Circular and Square Cylinders

B. Fago, O. Mahrenholtz

Arbeitsbereich Meerestechnik II / Strukturmechanik, TU Hamburg-Harburg
Eissendorfer Str. 42, 2100 Hamburg 90

Summary

The investigation deals with the phenomena of vortex resonant vibrations of circular and square cylinders, normal to the direction of the uniform incident flow. The measurement and observation of the relation between body motion and flow structure and the resulting fluid force are intended to furnish a better understanding of the fluidelastic system and a development of mathematical modelling.

The experiments refer to wind tunnel measurements of the fluid force acting on oscillating circular and square cylinders, latter with the face normal either to the absolute or to the relative flow, illustrating the effect of the motion induced flow separation. In addition, water channel experiments give insight into the fluid-structure interaction of the oscillating square cylinder. The Lissajous figures of fluid force and body displacement are recorded simultaneously with the pattern of the visualized flow.

On the basis of the experimental results the mechanism of some phenomena is explained and numerically simulated. Further, fundamental inconsistencies of topical oscillator models with the natural system and their reduction are demonstrated.

1 Introduction

1.1 Nondimensional parameters

The dimension analysis yields, that the flow induced vibration of the coupled system (fig. 1.1-1), consisting of the flow and a mechanical oscillator with a single degree of freedom

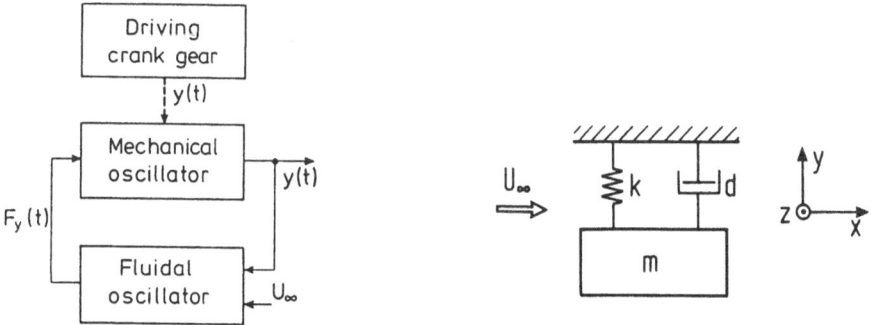

Fig. 1.1-1: Fluidelastic system   Fig. 1.1-2: Mechanical oscillator

and linear spring and damping (fig. 1.1-2), depends on four nondimensional variables. An often used set is:

reduced flow velocity $V_{rn} = \dfrac{U_\infty}{f_n b}$, damping ratio $\delta = \dfrac{d}{2 m \omega_n}$,

(1.1-1)

mass parameter $n = \dfrac{\rho b^2 l}{2 m}$, Reynolds number $Re = \dfrac{U_\infty b}{\nu}$,

with the natural frequency $\omega_n = \sqrt{k/m} = 2\pi f_n$, cylinder height (square section) or diameter (circular section) b, fluid viscosity ν and fluid density ρ.

At 'resonance' the flow velocity is $U_\infty = U_\infty^*$ and the vortex frequency from the fixed cylinder is $f_s = f_n$. To utilize this central configuration the reduced velocity is substituted by the frequency ratio

$$V_{rn} S = \frac{f_s}{f_n} = \Omega_{sn} \quad . \tag{1.1-2}$$

Of course, the Strouhal number $S = f_s b / U_\infty$ depends on Re, but quite weakly. Therefore, with $Re^* = Re(U_\infty^*)$ is in a wide enough range $Re^* - a < Re < Re^* + a$ in good approximation $S(Re) = S(Re^*)$ or $U_\infty = f_s b / S(Re^*)$. Thus, for the coupled system the fundamental parameter set

$$\Omega_{sn}, \delta, n, Re \tag{1.1-3}$$

is used in this paper.

If the cylinder is forced (fig. 1.1-1) by a crank gear to oscillate harmonically with a frequency f_o and amplitude $\hat{y}$, the dimension analysis yields three nondimensional parameters:

reduced flow velocity $V_{ro} = \dfrac{U_\infty}{f_o b}$, reduced amplitude $\hat{Y} = \dfrac{\hat{y}}{b}$, (1.1-4)

and the Reynolds number Re. Again, the reduced velocity is substituted by a frequency ratio, here $\Omega_{so} = f_s / f_o$. So, for the forced oscillation the fundamental parameter set is

$$\Omega_{so}, \hat{Y}, Re . \tag{1.1-5}$$

Dependent variables of the free system are, for instance, the lift force F_y or the lift force coefficient $c_y = 2F_y/(\rho U_\infty^2 bl)$, the motion y or $Y = y/b$, the fundamental frequency of motion f_o or $\Omega_{so} = f_s/f_o$ (in case of the forced system Y and Ω_{so} are fixed) and the phase angle Φ between the fundamental harmonics of displacement and lift force, i. e. between

$$Y = \hat{Y}\cos(\omega_o t) , \quad c_y = \hat{c}_y \cos(\omega_o t + \Phi) \quad . \tag{1.1-6}$$

1.2 Phenomena

An exemplary presentation of the phenomena of a vortex resonant vibration gives a diagram of Parkinson [1], illustrating experimental results of Feng [2]. In the lock-in or synchronization range $1 \lesssim V_{rn} S \lesssim 1{,}38$ the vortex frequency stops to be proportional to $V_{rn} S$, but keeps a virtually constant value just below ω_n. Inside the lock-in range c_y, Y and Φ are strongly increasing, until they jump to smaller (c_y, Y) or higher (Φ) values. On the way back, for decreasing flow velocity, the variables enclose a hysteresis loop.

Particularly interesting is the curve of the phase angle Φ, because it determines the direction of energy flow between fluid and mechanical structure. Fig. 1.1-4 and 3.1-1 show $\Phi(V_{ro})$

166

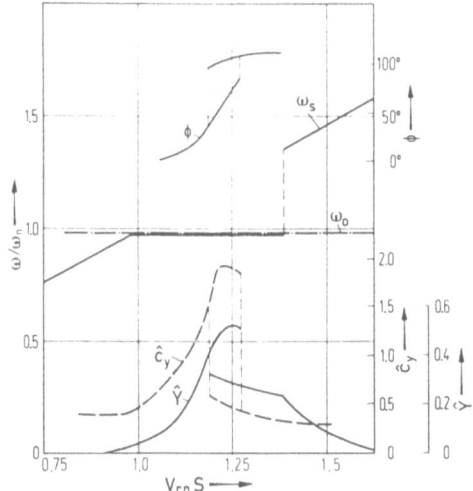

Fig. 1.1-3: Vortex resonance vibration of
a circular cylinder (from [2])

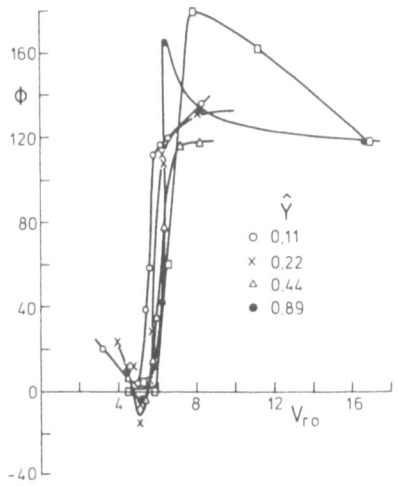

Fig. 1.1-4: $\Phi(V_{ro})$ of a forced oscillating
circular cylinder (from [3])

of a circular and a square cylinder, resp., both with sharply increasing slopes in the
lock-in range, for the square cylinder with remarkable negative values for $V_{ro} < 1/S$.

Beside by vortex shedding the square cylinder can be excited by galloping. Due to the Φ-
curve, if the theoretical velocity for the onset of galloping $\check{V}_{rn}$ is $\check{V}_{rn} < 1/S$, the earliest
possible velocity for oscillations is the resonance velocity $V_{rn} = 1/S$. Coreless and Parkinson
[4] suppose an asynchronous quenching of the galloping excitation by vortex shedding in
the range $V_{rn} < 1/S$.

2 Experimental arrangements

The **wind tunnel** experiments were conducted in a return circuit tunnel with a nozzle diameter of 1,5 m and an open 2,1 m long working section. The turbulence intensity in the
free jet is ca. 1 %. To shield the cylinder from pertubations generated by the surrounding
supports, springs etc., it was mounted inside of a rectangular test section, 790 mm wide
and 1200 mm high. The length l of the model with horizontal axis was 798 mm.

The circular cylinder (b = 100 mm) is made of foam. For this model especially the energy
flow between the fluid and the oscillating body was measured with the 'method of linear
self-excitation', described in [5]. The square cylinder (b = 120) is a shell made of balsa
wood (mass 237 g), supported by two load cells, which are attached to a stiff coaxial
pipe. To get the fluid force, the inertia force was calculated by means of the displacement signal and then subtracted from the total force measured by the load cells.

Fig. 2.-1 shows a sketch of the arrangement for forced oscillations of a square cylinder,
which face plane remains normal to the relative flow with the components U_∞ and $\dot{y}$. The
maximum angle of rotation was 20° with an error less than 2 %. For pure transverse oscillations the swing beam construction was removed. During free oscillations the driving
motor as well as the steel strip were disconnected. The free oscillator had a natural
frequency of 6 Hz. The maximum amplitude was 42 mm, $Re^* \approx 40000$.

167

Model
U∞
Induktive
displacement
pickup
Velocity
(proportional)
articulation
Swing beam
Displacement
articulation
Motor

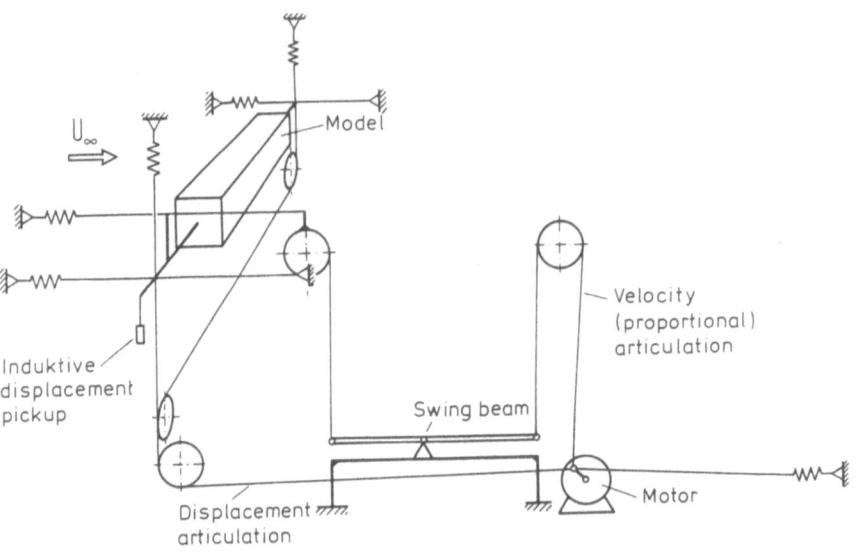

Fig. 2.-1: Arrangement for the combination of a transverse and adapted rotational oscillation

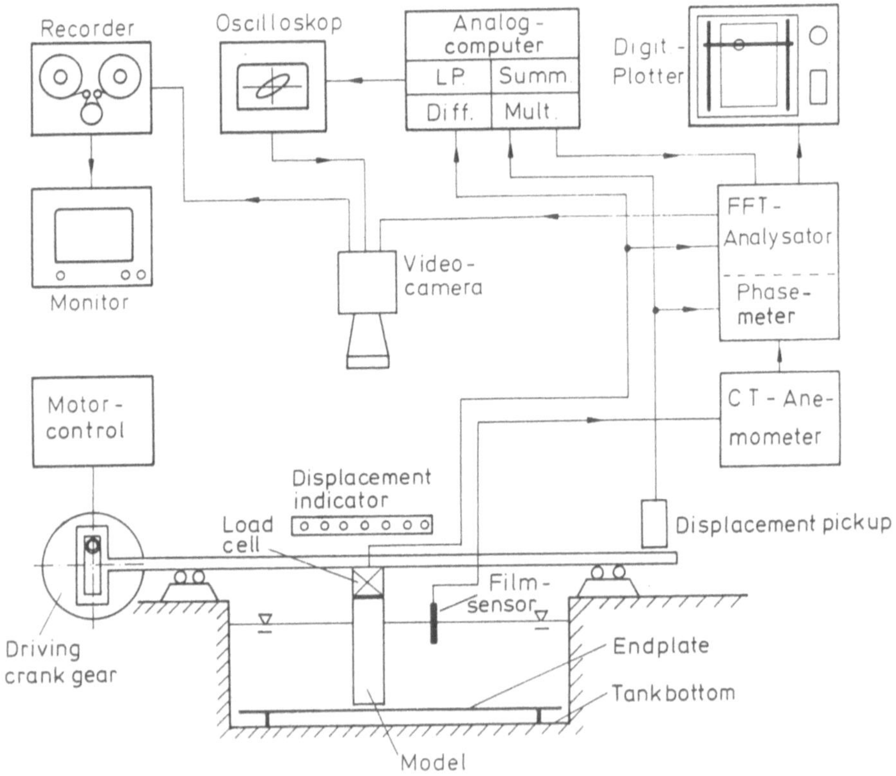

Recorder
Oscilloskop
Analog-
computer
LP.
Summ.
Diff.
Mult.
Digit.-
Plotter
Monitor
Video-
camera
FFT-
Analysator
Phase-
meter
Motor-
control
CT-Ane-
mometer
Displacement
indicator
Load
cell
Displacement pickup
Film-
sensor
Driving
crank gear
Endplate
Tank bottom
Model

Fig. 2.-2: Water channel arrangement

The **water channel** experiments were conducted in a free surface channel with a cross section of 290 x 180 mm (fig. 2.-2). The body is a 30 x 30 mm square cylinder with a vertical axis and 99 mm depth of immersion. By a load cell the cylinder is connected with a supporting beam, driven by a crank gear mechanism. The flow is visualized by aluminium particles swimming on the water surface. The signals of the fluid force (total force minus inertia force of the cylinder mass) and the displacement produce a Lissajous figure, which is recorded together with the flow pattern. To avoid capillary waves, the flow velocity is small (ca. 120 mm/s). The Reynolds number is Re ≈ 3500.

3 Results

3.1 Effects of motion induced circulation

Fig. 3.1-1 gives a comparison of the phase angles $\Phi(\Omega_{so})$ of a forced oscillating square cylinder, firstly in a pure transverse motion, secondly with the face plane remained normal to the relative flow. In case oft the pure transverse motion for all amplitudes $\hat{Y}$ noticeable negative values of Φ can be found in the range $\Omega_{so} < 1$. This energy extraction out of the

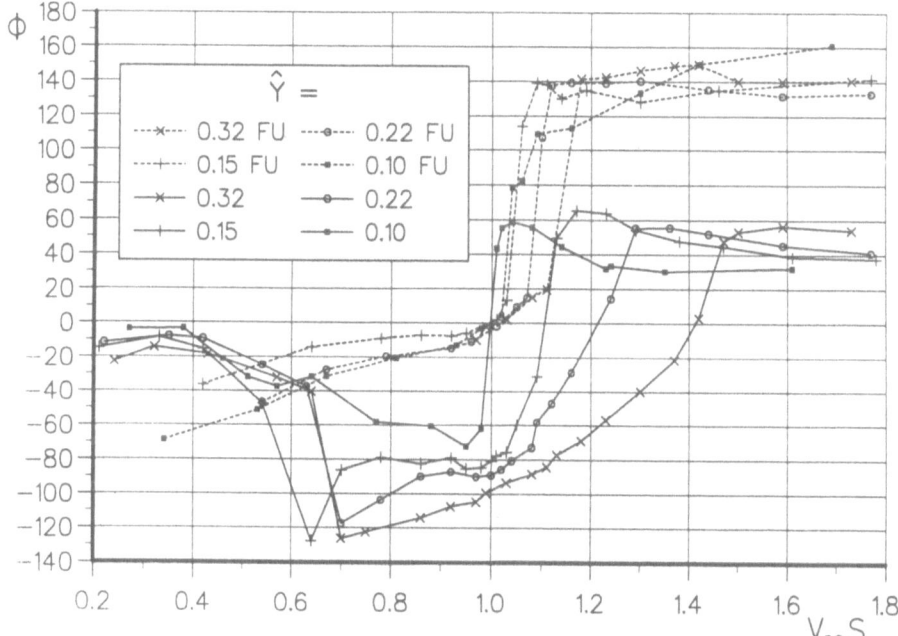

Fig. 3.1-1: $\Phi(\Omega_{so})$, square cylinder in a forced transverse and 'follow up' ('FU') motion

mechanical oscillator is caused by strong motion induced vortices, generated in connection with the variation of the angle of the relative flow on the lee side just behind the lea-ding edge of the body (fig. 3.1-2). Unlike the flow induced vorticies, which follow Strou-hal's law, they have no mutual interaction, but absorb the circulation of the shear layers. So in the range $\Omega_{sn} < 1$ (with $\Omega_{sn} \approx \Omega_{so}$) the flow induced vorticies do not quench the (motion induced) galloping instability, but they are quenched themselves by the effect of body motion.

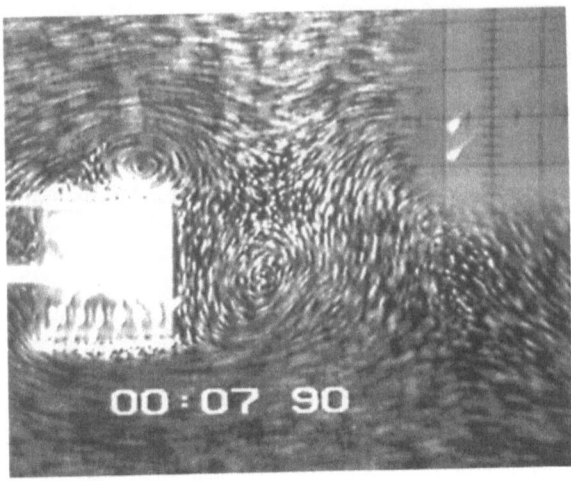

Fig. 3.1-2: Motion induced vorticies ($\hat{Y} = 0{,}125$, $\Omega_{so} = 0{,}55$)

Outside the synchronization-range, close to its boundaries, the frequency f_v, which is measured beside f_o, does not agree with the Strouhal frequency f_s [6,7]. In contrast to the expected behaviour of a phase locked loop, f_v is not pulled nearer to the 'guide frequency' f_o, but moves to $|f_v - f_o| > |f_s - f_o|$.

By the observation of the visualized flow it becomes evident, that the shedding frequency of the flow induced vorticies is really f_s. In addition, motion induced vorticity arrive at the wake with f_o, i. e. with a monotonously varying phase difference to the Strouhal vorticies. At a certain 'unfavourable' phase shift these can no longer incorporate the motion induced circulation and the Strouhal mechanism breaks down [5]. About one period later, at a certain 'favourable' phase angle, it is restored. It was shown by numerical simulation, that the Fourier spectrum of such a regularly interrupted process has dominating components in f_o and $f_v = f_o + k(f_s - f_o)$ with $k \approx 2$, while f_s is faded out. With increasing gap of Ω_{so} to the lock-in boundary, k goes to $k = 1$, so that one gets a curve sketched in Fig. 3.1-3.

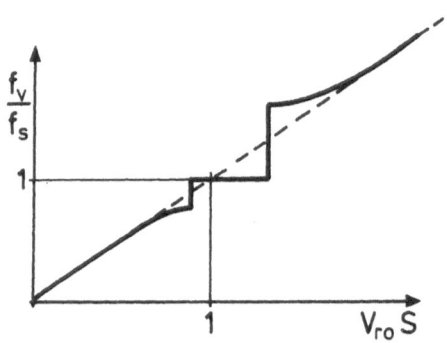

Fig. 3.1-3: Shift of frequency f_v

A further investigated phenomenon is the rapid change of the phase angle Φ and the corresponding energy transfer inside the synchronization range. About this a model, basing on a phase controlled vorticity flow, with encouraging results was developed, which also gives an explanation for the lock-in effect in general [8]. However, the numerical results are not yet satisfactory. A paper to this subject will soon be published by the authors.

3.2 Fundamental remarks to oscillator models

A general form of an oscillator model is

$$Y'' + D_m Y' + K Y = c_y n \frac{1}{(2\pi S)^2} \Omega_{sn}^2 \,, \tag{3.2-1a}$$

$$c_y'' + D_f c_y' + (V_{rn} S)^2 c_y = B Y' \,, \tag{3.2-1b}$$

with $\tau = \omega_n t$, $()' = \partial/\partial\tau$. The various models differ in the formulation of the terms D_m, K,

D_f and B. If a crank gear takes the place 'of the viscoelastic suspension of the mechanical oscillator, the fluidal component of the fluidelastic system is the same as before. This means, it is described by the same equation (3.2-1b). Its transformation into the proper time $\lambda = \omega_o t$ leads to

$$\overset{\circ\circ}{c}_y + D_f \overset{\circ}{c}_y + \Omega_{so}^2 c_y = R(\Omega_{so}, \hat{Y}) \tag{3.2-2}$$

with $(\,)^\circ = \partial/\partial\lambda$. The function R represents a general form of the right side, which can only depend on Ω_{so} and $\hat{Y}$, according to the dimension analysis.

The quality of eq. (3.2-2), the mathematical model of the fluidal oscillator, immediately becomes apparent in its ability to describe the experimental results, obtained with the *forced* oscillating cylinder. These experiments show (fig. 1.1-4, 3.1-1), that the phase angle Φ at resonance ($\Omega_{so} = 1$) depends on $\hat{Y}$. However, the phase difference γ (considering the fundamental harmonics) between c_y and the right side of (3.2-2) at $\Omega_{so} = 1$ is always 90°, independently of the formulation of D_f. Thus, the phase shift $\Phi(\hat{Y})$ at $\Omega_{so} = 1$ has to be generated within the right side R (fig. 3.2-1). Due to $\Phi = \varepsilon - \gamma$ and $\gamma(\Omega_{so} = 1) \equiv \pi/2$, the relation $\Phi(\Omega_{so} = 1) = f(\hat{Y})$ results in $\varepsilon(\Omega_{so} = 1) = f(\hat{Y})$. In consequence R has to be a nonlinear function of orthogonal derivatives of Y.

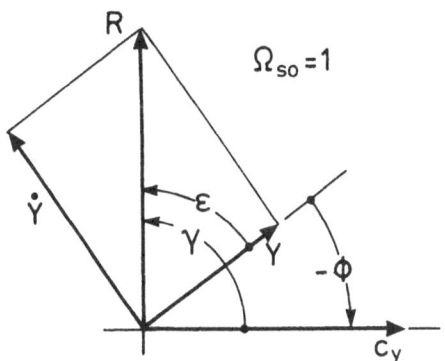

Fig. 3.2-1: Phase differences

The recent development in the research of oscillator models is given by Berger [9]. Berger found, that the reproduction of fundamental features of the fluidelastic system requires

- a coupling term B, which has to be a nonlinear function of Y',
- a *fluiddynamic* contribution to the mechanical damping D_m, depending on $\hat{Y}$ and acting in addition to c_y.

As pointed out before, the first condition seems to be not adequate. The second one obviously implies, that (3.2-1b) is an insufficient description of the fluid force, because there has to be another fluid force besides it. From the fluiddynamic point of view, it is sensible (and at the same time it presents a solution of the latter problem), that the lift force is not the fluidal oscillator itself, but it is a variable, depending on the actual fluidal oscillator, i.e. on the flow field. In a first examination the fluctuating wake may be represented by a variable Γ, characterizing the time dependent circulation of the vortex trail, described by the Rayleigh equation or with a term $\tilde{D}_f$ of higher order. So it is possible to create a set of equations

$$Y'' + D_m Y' + KY = c_y n \frac{1}{(2\pi S)^2} \Omega_{sn}^2 \,, \tag{3.2-3a}$$

$$c_y = f\left(\Gamma, \frac{\Gamma'}{\Omega_{sn}}, Y, \frac{Y'}{\Omega_{sn}}, \frac{Y''}{\Omega_{sn}^2}\right), \tag{3.2-3b}$$

$$\Gamma'' + \tilde{D}_f \Omega_{sn} \Gamma' + \Omega_{sn}^2 \Gamma = \Omega_{sn}^2 R\left(Y, \frac{Y'}{\Omega_{sn}}, \frac{Y''}{\Omega_{sn}^2}\right) \tag{3.2-3c}$$

with $\qquad K = 1, \quad D_m = 2\delta \,, \quad \tilde{D}_f = f\left(\frac{\Gamma'}{\Omega_{sn}}\right), \quad \Gamma = \tilde{\Gamma}/(U_\infty b) \,,$ (3.2-3d)

171

$$\tilde{\Gamma} = \oint \vec{v}\,d\vec{s} \quad \text{(vortex circulation)} \quad , \tag{3.2-3e}$$

which allows to assemble the lift force taking into account effects of the wake and the body motion, now at least without formal inconsistencies.

To give an example, the vortex resonant vibrations of the square cylinder in the windtunnel were calculated. On the basis of the experimental results, which were measured with the forced oscillating body and considering the parameters of the viscoelastic suspension of the cylinder, the following approach was used:

$$Y'' + 2 \cdot 0{,}0047\, Y' + Y = \frac{0{,}00106}{(2\pi\, 0{,}135)^2}\, \Omega_{sn}^2\, c_y \quad , \tag{3.2-4a}$$

$$c_y = 0{,}644\, \Gamma - 1{,}73\, \frac{Y''}{\Omega_{sn}^2} \quad , \tag{3.2-4b}$$

$$\Gamma'' - 0{,}04\, \Omega_{sn} \Gamma + 0{,}053\, \frac{1}{\Omega_{sn}}\, \Gamma'^3 + \Omega_{sn}^2\, \Gamma = R \tag{3.2-4c}$$

$$\tag{3.2-4d}$$

$$R = \frac{10}{\left(\left(\frac{Y'}{\Omega_{sn}}\right)^2 + Y^2\right)^{1{,}3}} \left(-\frac{24}{5}\, Y^4 Y'' + \frac{80}{\Omega_{sn}}\, Y^2 Y'^3 + \frac{136}{\Omega_{sn}^3}\, Y Y'^3 Y'' + \frac{0{,}395}{\Omega_{sn}^3}\, Y' Y''^2 + \frac{57{,}8}{\Omega_{sn}^5}\, Y'^3 Y''^2 \right).$$

Fig. 3.2-2 shows the measured and the calculated curves of the amplitude $\hat{Y}$ for increasing and decreasing velocities U_∞. At the turning point, before U_∞ was decreased, in the experiment as well as in the calculation the cylinder's oscillation was shortly stopped.

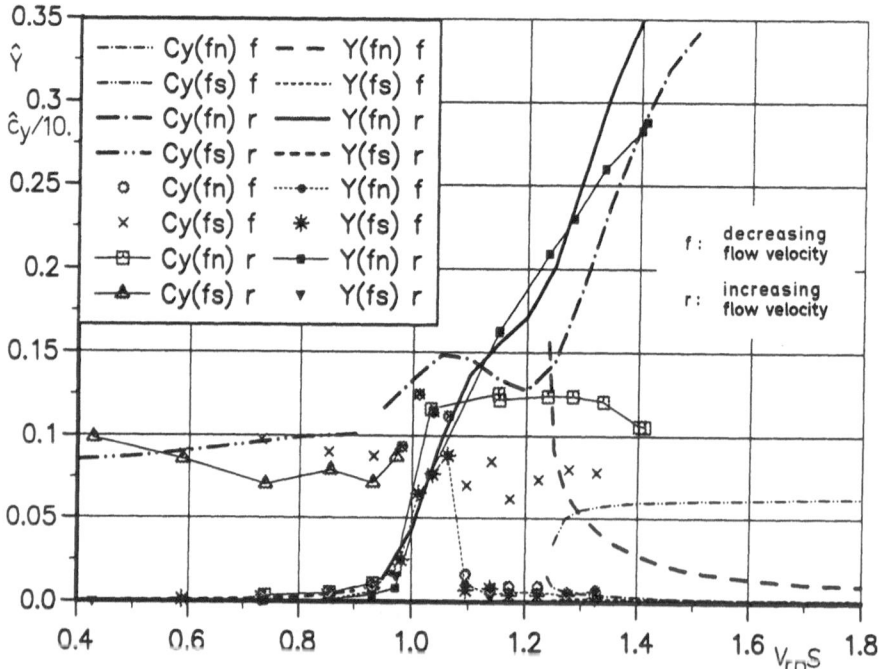

Fig. 3.3-2: Measured (with marker) and calculated (no marker) oscillation of a square cylinder

If the equations (3.2-4b), (3.2-4c), (3.2-4d) are transformed into the proper time $\lambda = \omega_0 t$, they represent the forced oscillating system. Fig. 3.3-3 shows the calculated curves of Φ, which are (to some extent) fitted to the experimental results in fig. 3.1-1. Additionally curves of Φ for a coupling term $B = 2{,}5\,\Omega_{so}$ or right side $R = 2{,}5\,\Omega_{so}Y'$ are shown. Of course, all curves go through $\Phi = 0$, $\Omega_{so} = 1$, if R is a (even higher order) function of Y' only.

By the respective power of Ω_{sn} in the equations (3.2-4b), (3.2-4c), (3.2-4d) is achieved, that the free oscillating system depends on Ω_{sn} and not on Ω_{so} and the forced oscillating system depends on Ω_{so} and not on Ω_{sn}, according to the dimension analysis.

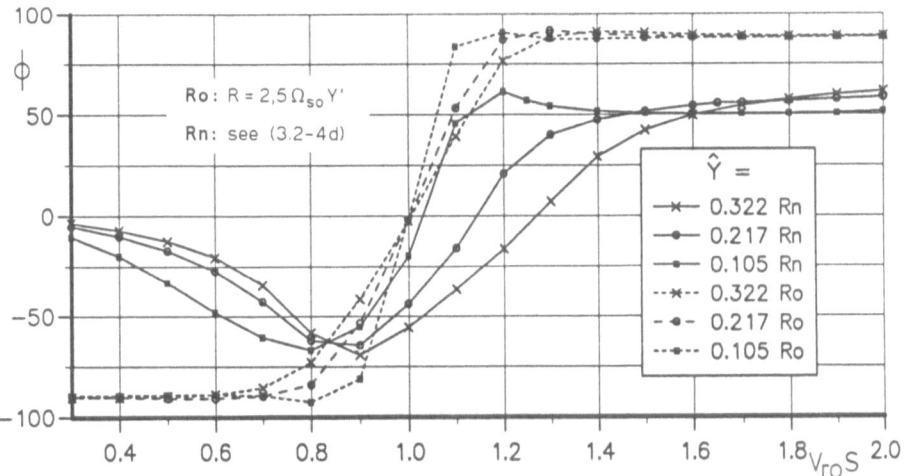

Fig. 3.3-3: Calculated curves of Φ for the forced oscillating square cylinder

References

[1]   Parkinson, G.V.: Mathematical Models of Flow-Induced Vibrations of Bluff Bodies.
      In: Naudascher, E. (Ed.), Flow-Induced Structural Vibrations. Springer-Verlag, Berlin, Heidelberg, New York 1974.

[2]   Feng, C. C.: The Measurement of Vortex Induced Effects in Flow Past Stationary and Oscillating Circular and D-Section Cylinders. M. A. Sc. Thesis, U. British Columbia, 1968.

[3]   Bearman, P. W., Currie, I. G.: Pressure-Fluctuation Measurements On an Oscillating Circular Cylinder. J. Fluid Mech. **91**, 4 (1979), 661-677.

[4]   Coreless, R.M., Parkinson, G.V.: A Model of the Combined Effects of Vortex-Induced Oscillation and Galloping. J. Fluids and Structures **2** (1988), 203-220.

[5]   Mahrenholtz, O.: Fluidelastische Schwingungen. ZAMM **66** (1986)1, 1-22.

[6]   Staubli, Th.: Untersuchung der oszillierenden Kräfte am querangeströmten, schwingenden Kreiszylinder. Dissertation, T.H. Zürich, 1983.

[7]   Fago, B., Mahrenholtz, O.: Entwicklung eines strömungsmechanisch fundierten Modells zur Beschreibung eines fluidelastischen Schwingers im Bereich der Wirbelresonanz. DFG-Bericht Ma 358/37-2, 1986.

[8]   Fago, B., Mahrenholtz, O.: Wirbelresonanz. DFG-Bericht Ma 358/37-4, 1988.

[9]   Berger, E.: Zwei fundamentale Aspekte wirbelerregter Schwingungen.
      DFG-Bericht Be 343/12, Berlin, 1984.

Experimental Analysis of the Unsteady Turbulent Boundary Layer Flow with Separation on Turbine Blades

G.H. Dibelius
Institute for Steam and Gas Turbines, RWTH Aachen
Templergraben 55, D - 5100 Aachen, Germany

E. Ahlers
BMW Rolls Royce Aero Engines GmbH
Carl-von-Linde-Str. 25, D - 8044 Lohhof, Germany

ABSTRACT

The influence of periodically unsteady perturbations on the turbulent flow along the suction side of turbine blades is investigated in a test rig. The blade suction side is represented by a flat plate. The pressure profile typically encountered in a turbine blade channel is generated by a curved wall opposite to the flat plate. The angle of the divergent part of the test section and hence the pressure can be increased to induce flow separation on the flat plate. For the simulation of wakes being shed from upstream blade rows the incoming flow is periodically disturbed by a wake generator. A 2D-LDV with high spatial resolution is used to measure averaged and fluctuating components of the velocity inside the boundary layer flow. By Fourier analysis of the measured time related velocity distributions the stochastic and periodic parts of the overall turbulence are identified.

The structure of the boundary layer flow is investigated for both steady and unsteady flow conditions. With the periodic wake flow the separation is shifted downstream as compared to the steady flow situation. Conclusions derived from the experimental results for the theoretical understanding are discussed in particular with respect to turbulence modelling.

INTRODUCTION

Bladings of turbomachinery are designed for high aerodynamic loads. In compressors as well as in turbines areas with rising pressure are formed along the profile contour. The pressure increase is related directly to the flow deflection achieved by a blade row and thus to the aerodynamic load. If a flow separation

174

occurs under __high load conditions__ due to a high adverse pressure gradient, the pressure rise downstream of that location is only small. The irregular velocity distribution influences the flow to the following blade row and aerodynamic losses increase significantly. The lay-out of bladings for __low aerodynamic loads__ requires additional stages inside the turbomachine causing higher costs. Therefore the bladings of modern turbomachines have to be designed to work very close to flow separation. For an off-design point also the flow behaviour with beginning flow separation is of interest.

Calculation procedures used so far for the aerodynamic lay-out of turbomachine bladings do not consider this flow structure adequately, mainly for two reasons:

1. Usual turbulence models are based on an isotropic turbulence structure, whereas in turbomachines the character of turbulence is strongly anisotropic.

2. In a system of stationary and moving blade rows the flow to one particular blade is unsteady due to the wakes generated by the blade row located upstream. The wakes are distributed periodically onto several blade channels, thus generating areas with reduced velocity and increased turbulence level that are transported downstream with the main flow.

The unsteady and anisotropic flow structure in turbomachines has to be implemented in new calculation procedures. Experimental investigations give the base for a better theoretical understanding of flow phenomena in turbomachine bladings. Therefore, at the Institute of Steam and Gas Turbines experimental work on this subject has been initiated by simulating the cascade flow in a special test rig.

TEST APPARATUS

The measurements were performed in a low-speed open circuit wind tunnel described in more detail in [1,3]. The test section is shown in __Fig.1__. The blade suction side is represented by a flat plate of L = 550 mm length. The pressure profile typically encountered in a turbine blade channel is generated by a curved wall opposite to the flat plate. Flow separation on the curved top wall is prevented by a highly twodimensional suction system. The Reynolds number based on the length L of the flat plate (equivalent to the chord of the blade) is $Re_L = 3,7 \cdot 10^5$ for all measurements presented here.

The angle of the divergent part of the test channel can be increased up to 10^0 simulating various aerodynamic loads of the blade. The pressure recovery in the diffusor increases with rising diffusor angle α_D. For the diffusor angle $\alpha_D = 10^0$ the pressure increases strongly up to the flow separation point and only slightly downstream of that location, __Fig.2__.

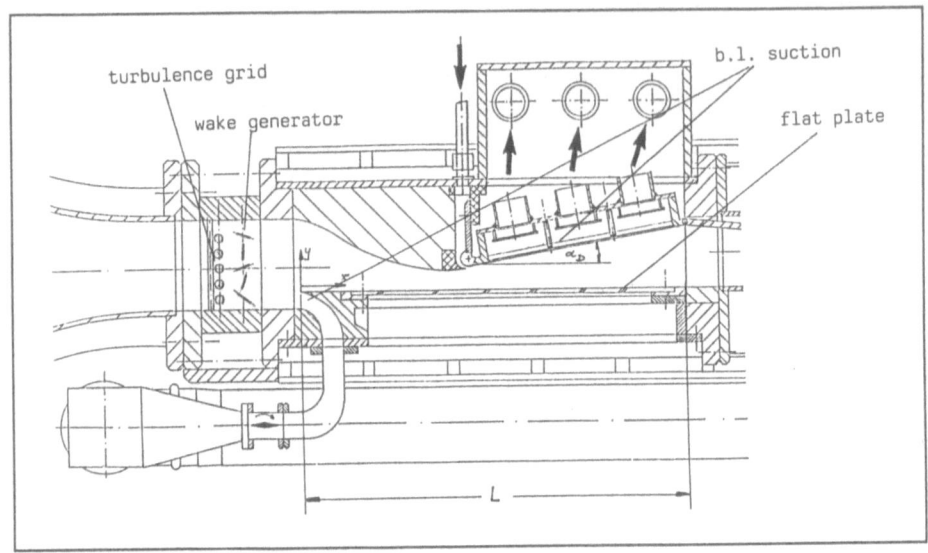

Fig.1: Test section

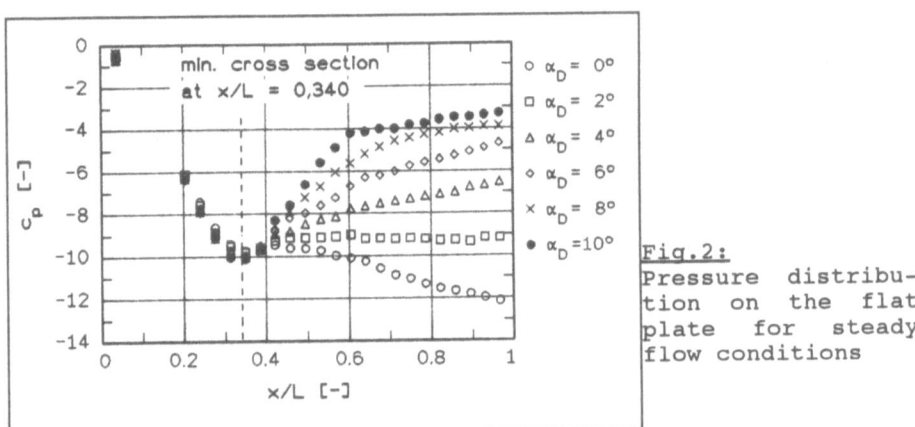

Fig.2:
Pressure distribution on the flat plate for steady flow conditions

For ensuring defined starting conditions for the boundary layer flow on the flat plate the boundary layer developped in the contraction nozzle is removed before the leading edge of the test plate. Laminar-turbulent transition is enforced by a trip edge at a streamwise location $x/L = 0,1$. Tripping the boundary layer in a highly accelerated flow the possibility of boundary layer relaminarisation has to be taken into account. Therefore, the velocity profiles at the location of minimum channel cross section were checked for both steady and unsteady flow conditions. In both cases the boundary layer at that location and hence over the whole length of the flat plate is turbulent [1,3].

In turbomachines the wakes of blade rows located far upstream are equalized by mixing processes and cause an increase of stochastic turbulence in the flow. The wakes of the blade row located directly upstream superimpose periodic fluctuations of the local velocity and the flow angle. In order to simulate these conditions, a turbulence grid and a wake generator are installed in the entry cross section of the test rig upstream the flat plate.

The turbulence grid consists of cylindrical bars perpendicular to each other. Varying the geometrical parameters of the grid the turbulence level of the flow was raised to values between 2% and 7%. The variation of the stochastic turbulence level shows no significant effect on the flow separation in the diffusor [3].

A wake generator consisting of five flat profiles rotating with adjustable speed was used for simulating the periodical flow perturbations of an upstream blade row [1,3]. Each profile causes a large velocity wake with increased turbulence level in the upright position of highest drag, **Fig.3a**. Adjacent profiles being coupled by a gear box are positioned with an angular deviation $\Delta\varphi$ to each other in order to keep the channel blockage constant while rotating. During rotation of the generator the wakes change their position normal to the wall surface, **Fig 3b**, thus behaving very similarly to wakes generated by moving blades. However, the influence of the flow angle changing inside a wake is not simulated correctly.

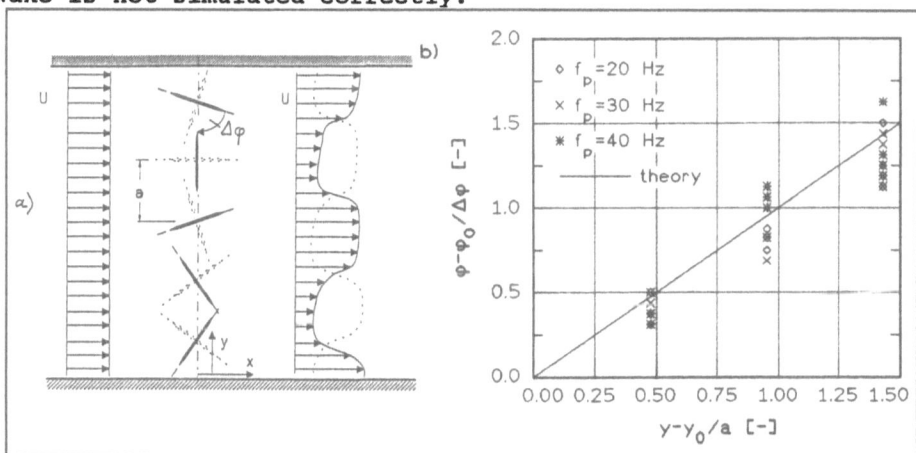

Fig.3: Wake generator   a) principle   b) movement of the wake normal to the plate surface

A 2D-LDV system with high spatial resolution described in detail in [2,3] has been used to measure averaged and fluctuating components of the velocity inside the boundary layer as close to the wall as y = 0,05 mm. Thus boundary layer parameters as well as the wall shear stress were determined from the measu-

red velocity profiles. The flow was seeded with either Di-2-
Ethylhexyl-Sebacate (DES) or silicon oil droplets with a particle size of about 1 μm. The overall measurement accuracy is
estimated to be better than 2% for the velocity and boundary
layer parameters, better than 5% for turbulent fluctuations and
the wall shear stress and better than 8% for turbulent velocity
correlations such as the Reynolds shear stresses.

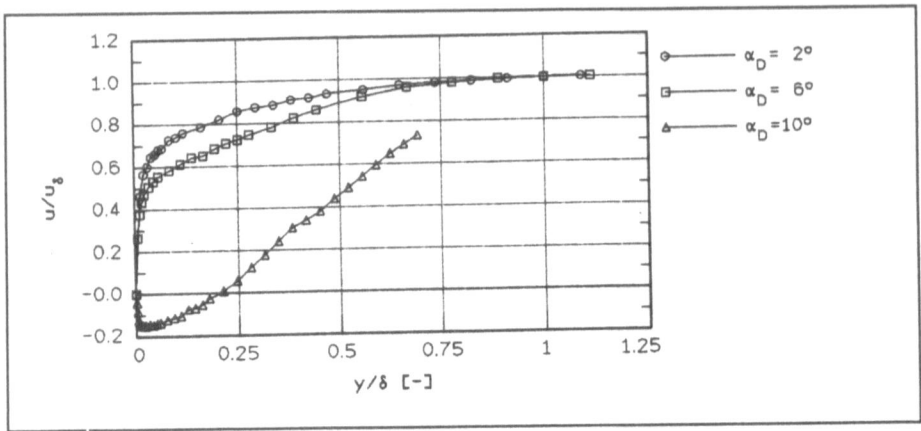

Fig.4: Velocity profiles in the boundary layer at a streamwise
location x/L = 0,931 for steady state conditions

TEST RESULTS

Influence of pressure gradient The influence of the adverse
pressure gradient on the flow in the diffusor has been investigated by varying the diffusor angle α_D between 0^0 and 10^0. The
pressure distribution on the flat plate is shown in **Fig.2** above.
The velocity profiles at a streamwise location x/L = 0,931 are
given for the three diffusor angles 2^0, 6^0 and 10^0 in **Fig.4**. The
increase of the diffusor angle causes a loss of kinetic energy
in the boundary layer. For α_D = 10^0 the flow detaches and a
backflow region is formed in the vicinity of the wall. The wall
shear stress becomes zero at a streamwise location x/L = 0,7 and
is negative downstream of that location, **Fig.5**. As a result of
reduced kinetic energy the shape factor H_{12} of the boundary
layer increases downstream of x/L = 0,6 reaching a value of 3 at
the separation point and about 7 at the end of the flat plate,
Fig.6. The results compare well with experimental data given in
literature.

Influence of the unsteady flow The influence of unsteady flow
conditions on the flow separation has been investigated at a
diffusor angle of 10^0. The pressure recovery in the diffusor is
higher and more uniform under unsteady flow conditions compared
to the steady state case, **Fig.7**. There is a downshift of flow
separation and stabilization of the flow. The velocity profiles

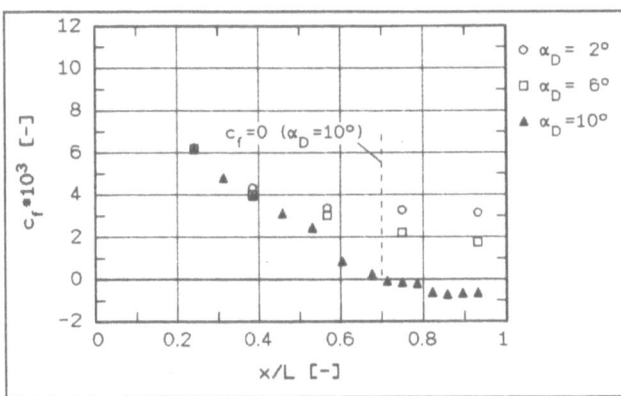

Fig.5:
Wall shear stress
coefficient c_f on
the flat plate for
steady state condi-
tions

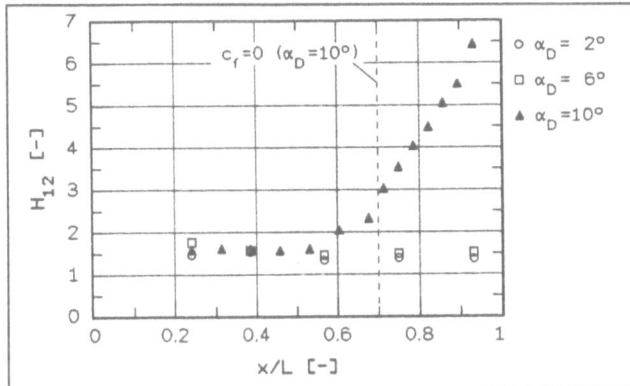

Fig.6:
Shape factor H_{12} on
the flat plate for
steady state condi-
tions

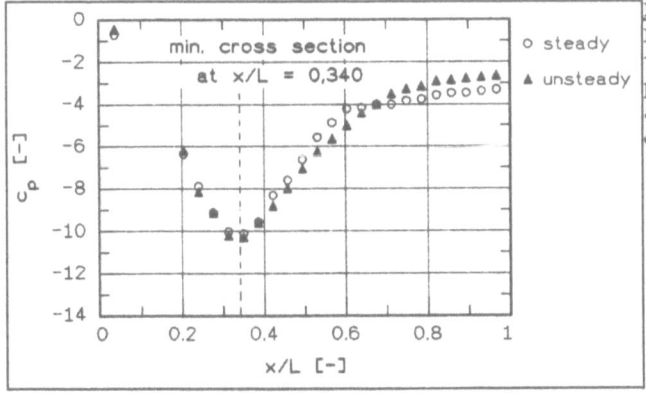

Fig.7:
Pressure distribu-
tion on the flat
plate for steady
and unsteady flow
conditions
$\alpha_D = 10^0$

are compared for both flow situations in **Fig.8** at two streamwise
locations. Upstream as well as downstream of flow separation the
kinetic energy inside the boundary layer is higher in case of
periodically disturbed flow. Due to the downshift of flow separation the backflow region is smaller compared to the steady
state case. For unsteady flow conditions the wall shear stress

is higher at every streamwise location and becomes zero at x/L = 0,84, **Fig.9**. Hence the flow separation has been shifted downstream 14% related to the length L of the flat plate. The increase of the shape factor H_{12} downstream of x/L = 0,6 is lower in case of the unsteady flow, **Fig.10**. Like in the case of undisturbed flow, it reaches a value of about 3 at the separation point.

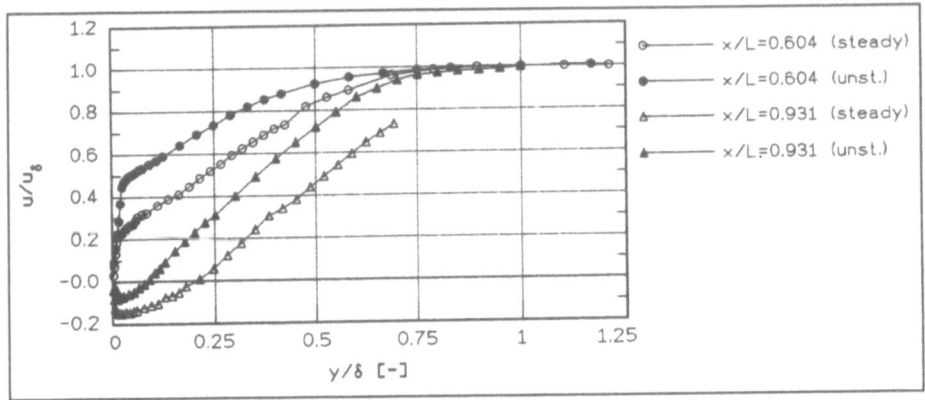

Fig.8: Velocity profiles in the boundary layer for steady and unsteady flow conditions: $\alpha = 10^0$

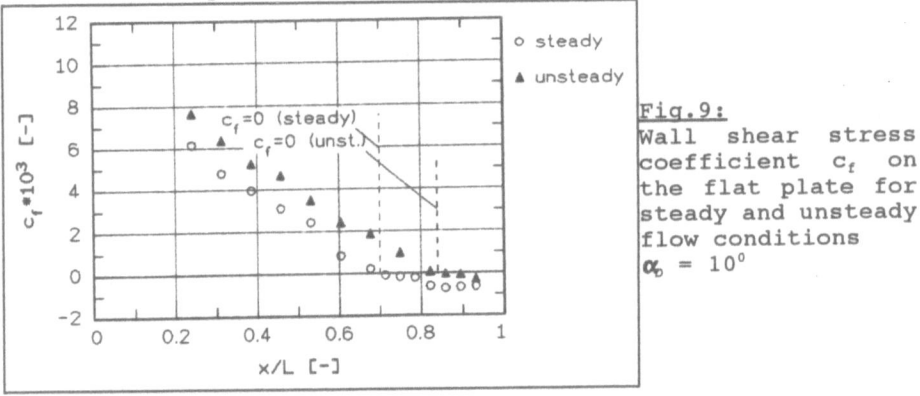

Fig.9: Wall shear stress coefficient c_f on the flat plate for steady and unsteady flow conditions $\alpha_0 = 10^0$

The reason for the significant downshift of flow separation for unsteady flow conditions is the energization of the boundary layer flow by the periodical perturbations being converted into stochastic turbulence. This is derived from the detailed analysis of the periodic and stochastic parts of velocity fluctuations [3]. A compilation of the results is shown in **Fig.11**. The stochastic fluctuations especially in the main flow (u-) direction rise with increasing axial position on the flat plate whilst the periodic fluctuations decrease simultaneously. Due to this effect the energy transport between core flow and boundary layer is enhanced and the flow separation is moved downstream compared to the steady state case.

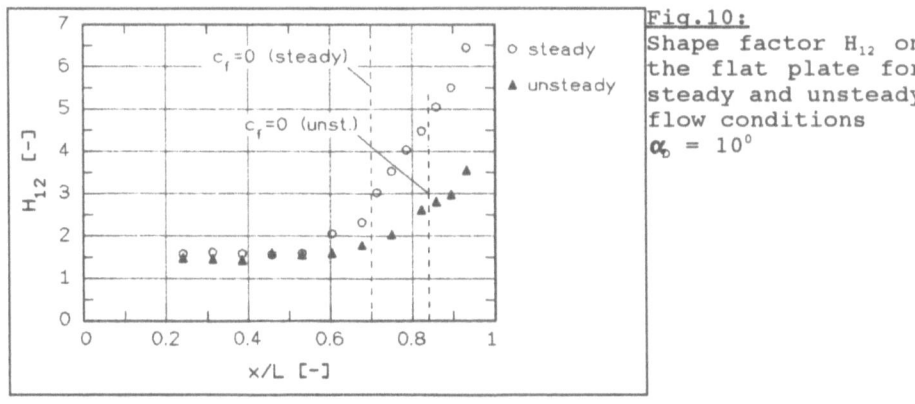

Fig.10:
Shape factor H_{12} on the flat plate for steady and unsteady flow conditions $\alpha_0 = 10^0$

The flow structure inside the separated flow region is analyzed in detail in [3,4] for both steady and unsteady flow conditions. The velocity distribution in this area can be described by a universal dimensionless function, if the actual velocity u is related to the maximum backflow velocity u_N and the wall distance y is related to the position y_N of this maximum [3,4]. The dimensionless wall distance y/δ obviously is not dependent on the Reynolds number Re_N calculated from the velocity u_N and the distance x_N to the separation point $(c_f = 0)$, Fig.12.

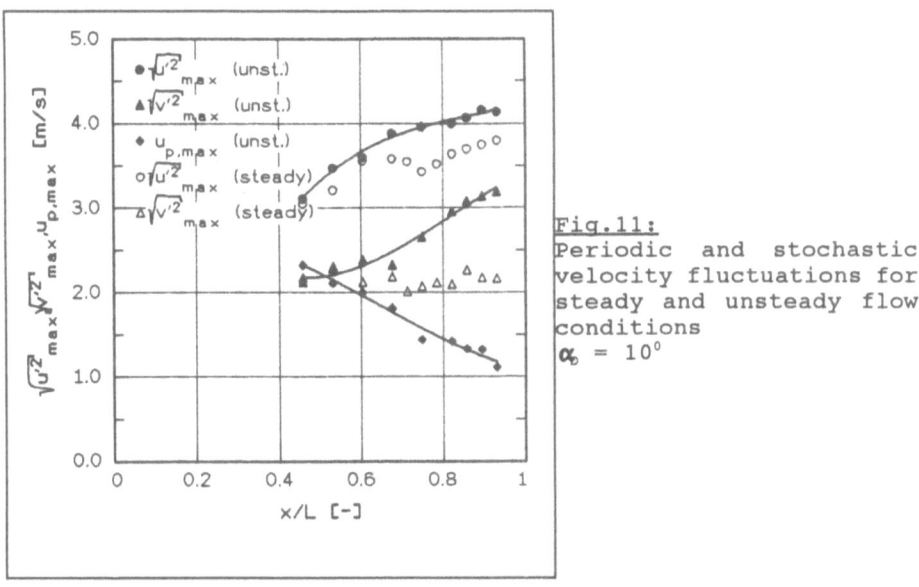

Fig.11:
Periodic and stochastic velocity fluctuations for steady and unsteady flow conditions $\alpha_0 = 10^0$

Four different turbulence models were examined by comparing calculated and measured values [3,4]. The best results in the area of separated flow were obtained using zero or one equation turbulence models (Cebeci/Smith, Johnson/King) compared to more complicated procedures like the k-ϵ-model which do not consider non-equilibrium effects inside the boundary layer. As an example the calculated and measured wall shear stress distribution on the

flat plate is shown in **Fig.13** for steady and unsteady flow
conditions.

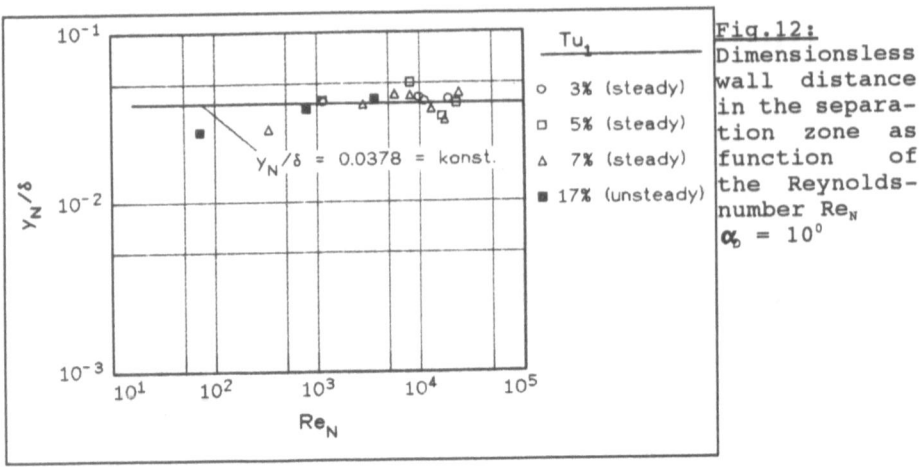

Fig.12:
Dimensionless
wall distance
in the separa-
tion zone as
function of
the Reynolds-
number Re_N
$\alpha_b = 10^0$

$y_N/\delta = 0.0378 = \text{konst.}$

Tu_1
○ 3% (steady)
□ 5% (steady)
△ 7% (steady)
■ 17% (unsteady)

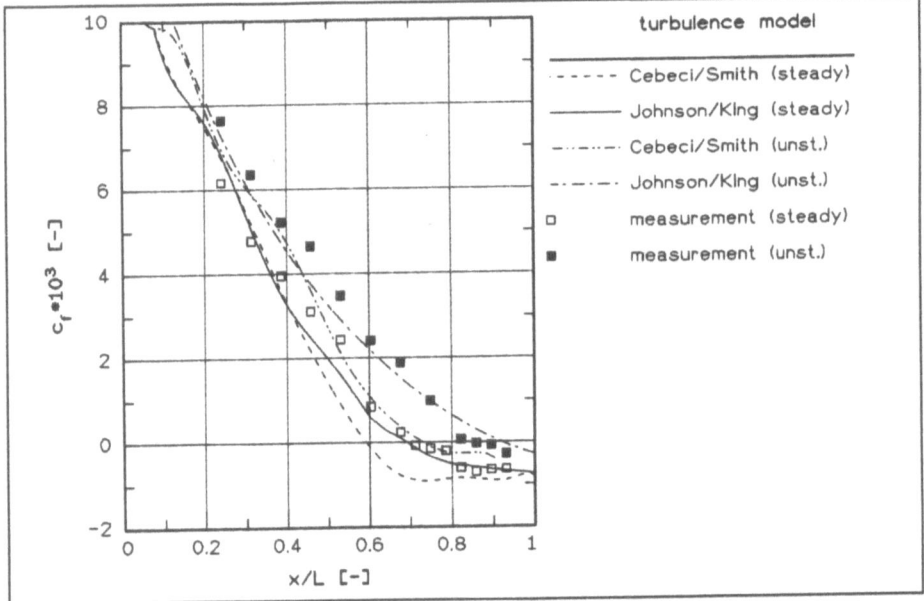

turbulence model

----- Cebeci/Smith (steady)
——— Johnson/King (steady)
—·—·— Cebeci/Smith (unst.)
—··—··— Johnson/King (unst.)
□ measurement (steady)
■ measurement (unst.)

Fig.13: Comparison of measured and calculated wall shear stress
distributions on the flat plate for steady and unsteady
flow conditions; $\alpha_b = 10^0$

CONCLUSIONS

The periodically unsteady wake flow in a turbomachine causes
a significant downshift of the separation zone on the blade

182

suction side and thus has a stabilizing effect on the blade channel flow. The downshift is associated with an increased energy transport between core flow and boundary layer due to the conversion of periodic into stochastic parts of the velocity fluctuations. The stochastic turbulence in turbomachines is anisotrop for steady as well as for unsteady flow conditions. This is true especially for the boundary layer flow [3].

For the theoretical prediction of the flow in blade channels in particular the turbulence modelling is of interest. Higher order turbulence models like the k-ϵ-model fail especially in the vicinity of flow separation. Using existing turbulence models the best results are obtained applying the method by Johnson/King in which the Reynolds shear stress and non-equilibrium effects inside the boundary layer are modelled.

The perfect simulation of the blade channel flow in a simple test rig is not possible. In our case the inability of the wake generator to change the flow angle in the wakes and the artificially enforced laminar-turbulent transition of the boundary layer make a verification of the experimental results necessary. Therefore additional measurements have to be performed in a real turbomachine. Nevertheless the experimental results obtained in the test rig are a valuable test case for the improvement of calculation procedures.

ACKNOWLEDGEMENT

The support of this work by the DFG in the frame of this Schwerpunktprogramm "Physik abgelöster Strömungen" has been greatly appreciated.

REFERENCES

[1] Dibelius, G.H., Ahlers, E.: Influence of Periodically Unsteady Wake Flow on the Flow Separation in Blade Channels, ASME paper 90-GT-253, Brussels, 1990

[2] Dibelius, G.H., Ahlers, E.: LDV Measurement Technique for Boundary Layer Flow with Separation, 5th Int. Symposium on the Application of Laser Techniques to Fluid Mechanics, Lisbon, 1990

[3] Ahlers, E.: Einfluß von Druckgradienten und instationärer Gitteranströmung auf die turbulente Strömungsablösung im Schaufelkanal hochbelasteter Turbinen, PhD-thesis, RWTH Aachen, 1992

[4] Bohn, D., Dibelius, G.H., Sucharski, Z.: Comparison between Experiment and Prediction for a Subsonic Turbulent Separated Boundary Layer, 37th ASME Int. Gas Turbine & Aeroengine Congress and Exposition, Cologne, 1992

FLOW SEPARATION IN ARTIFICIAL HEART VALVES

Klaus Affeld, Klaus Schichl*, Andreas Ziemann

Biofluidmechanik Labor, UKRV, Freie Universität Berlin
Hermann-Föttinger-Institut, Technische Universität Berlin*
1 Berlin, Germany

INTRODUCTION

Flow separations are most unwelcome in the flow through artificial heart valves as they are in most technical flows. The bulk flow properties as the resistance of the valve or its closing behaviour are not so much influenced by flow separations as another phenomenon which is unique to blood flow - the generation of a thrombus. This phenomenon which is normally useful by closing leaks in the vessel system becomes harmful, when a mechanical device such as an artificial heart valve is implanted. In areas of stagnant flow the blood solidifies and can impede or even completely close off the flow. At a rate of 3 to 5 % the patients die each year of complications related to this undesired interaction of the blood with the implant. From detailed in vitro experiments one knows the flow conditions, which are relevant for the generation of a thrombus. The first condition is the flow of blood through a volume of high shear stress, either in a jet or at a wall, which exceeds $\tau = 100$ N/m^2 and lasts for more than 30 ms [1]. This high shear stress activates the cells, which are responsible for a thrombus formation, the platelets. The second condition for the generation of a thrombus is the presence of an area of recirculation close to a wall of a foreign material. The recirculation is found in a flow separation and allows a concentration of platelets. If they are activated by the mechanical shear stress and then accumulated densely enough and in addition placed in the vicinity of foreign material to which they can attach, then the formation of a thrombus is likely. Investigations of the flow through artificial heart valves in the past have been concentrated on flow properties as resistance or turbulence, however, investigations of the flow in regard of the thrombus generation have been scarce, and it was the aim of this project to illucidate the role of the flow in thrombus generation.

METHODS AND RESULTS

ENLARGED MODEL - SIMILARITY LAWS

The flow through artificial heart valves has a great degree of complexity: the flow is usually three dimensional, instationary and periodic; in addition, it has flow separations, and often a laminar turbulent transition. The Laser Doppler Velocimetry (LDV) has enabled us to investigate extensively these fluid mechanics [2,3,4]. However, as successful as the LDV method is, it has the great disadvantage that the velocity can only be measured at one particular location at a time. The picture of the flow must be composed from many velocity vectors, which in addition are averaged vectors. Information is lost in this process. An overview of the flow can be obtained by the use of flow visualization, which has been used for heart valve flow studies since they were first began. Hitherto, however, they have lacked great detail, mainly due to the small size of the valves and the high velocity of the fluid. Thus, the methods to visualize the flow render an integrated picture but lack the precision of the LDV method. Most methods of flow visualization require a low velocity of around 5 cm/s in water. At higher velocities dyes disperse too quickly and particles are difficult to light properly. Velocities of 150 cm/s which are typical for flow in an artificial heart valve are much too high for good flow visualization. A sufficiently low velocity can be achieved, however, with upscaling while observing the similarity law. The number of greatest importance is the Reynolds number. A scale of 10 to 1 and the use of water as model fluid allows the reduction of the velocity in the model flow to 1/25 th of the real one, producing a maximum velocity of about 7 cm/s. For this purpose a water tunnel has been built for a model valve of a diameter of 220 mm which is mounted in a transparent model of the aortic root. Unlike a conventional watertunnel this one produces a pulsatile flow as is required for artificial heart valves. An even and laminar flow to the model valve is established by a free floating wall which separates still water from a settling chamber from the disturbed returning fluid. This wall also serves to monitor the flow. The displacement of the piston is the same as the displacement of the water and so through differentiation the water flow can be calculated. The fluid is moved by a propeller driven by a computer controlled electric DC motor, which is controlled so that the displacement of the piston follows the physiological curve. With u being the velocity, d the diameter of the valve and ν the kinematic viscosity we obtain:

$$Re = \frac{u \cdot d}{\nu} \ . \tag{1}$$

With a model scale of 10:1 for the geometry and 1:2.54 for the kinematic viscosity - (water, 20°C: blood, 37°C) - we obtain a reduction of the velocity of 1:25.4 and a time expansion of 254:1. This means that a systole of 300 milliseconds lasts in the model more than 76 seconds thereby giving the flow a slow motion appearance. The second number to be observed is the Strouhal number (S):

$$S = \frac{d}{u\,t} \ . \tag{2}$$

This similarity is automatically kept, if the Reynolds-similarity is observed for each time step, i.e. if the flow curves of original valve and its model are properly scaled. Another number to be considered is the Archimedes number (Ar). With ρ_{fluid} being the specific density of the fluid, $\rho_{occluder}$ that of the valve occluder and g the gravity, the Archimedes number is defined as:

$$Ar = \frac{\rho_{occluder} - \rho_{fluid}}{\rho_{fluid}\,u^2}\,d\ g \ . \tag{3}$$

The Archimedes number takes into account the buoyancy of the occluder. In the real valve the specific weight of the occluder is about 1.6 and differs considerably from that of the blood. This has little influence on the valve's movements because the velocities are high

185

and the dynamic pressure easily overcomes the influence of its buoyancy. In the enlarged model this cannot be said because the dynamic forces are so small that the buoyancy plays an important role. With the model scale inserted in the formula one obtains the specific weight of the model occluder as $\rho_{occluder}$ = 1.000079 times the specific weight of water. This is practically equal to that of the water. To achieve this density the model occluder is made of polyethelene, which is lighter than water and weighted with lead to obtain a neutral buoyancy.

Some artificial heart valves are not made of a rigid but of flexible material. Thin and pliable membranes form the occluder like the leaflets in the natural heart valve. To model such a valve, the fluid forces have to be in the proper relation to the elastic forces which resist the bending of the leaflet. For this case a similarity law can be obtained:

$$K = \frac{\rho}{E} \frac{u^2 d^2}{s^3} ,$$

4)

E is the modulus of elasticity of the material, d a characteristic length, u the fluid velocity, and s the thickness of the leaflet. The leaflets of the natural valve are extremly flexible [5]. Artificial heart valves with leaflets of polyurethane with a thickness of 0.15 to 0.22 mm have been developed [6]. If the enlarged model valve is made of this material and the model law - equation 4 - is applied one obtains the same thickness for the model leaflet, a condition which obviously is easy to meet. This valve has leaflets that are bent in only one axis. However, if the leaflets have a double curvature a change of its shape can lead to folding and buckling, and in this case the relation of thickness and length has be kept constant. One must then fabricate the leaflet from a material with a different modulus of elasticity. The enlarged model has to be much softer, for the model relation of ten to one obtains that the modulus of the model leaflet is only a thousendth of the original. Since the modulus of the natural valve is already low, this condition certainly is difficult to achieve. This clearly indicates the limitations of this model technique.

Unfortunately, however, this is not the only limitation of the enlarged model. Another one is: one cannot model the fluid. Blood is a non-Newtonian fluid and there are succesful attempts to model it with water and polymeric additives, such as Separan® and Xanthem® [7]. Over a wide range of shear this model fluid imitates well the rheology of blood, showing a major deviation only in the low-shear range. Since the enlarged flow model allows for the detection of fine details of the flow in small areas, where the shear is low, it is desirable to have a non-Newtonian fluid. However, it has not been achieved to model the blood for the enlarged flow model. Modeling of non-Newtonian fluids in general is difficult and even may be impossible [8]. A simple calculation may illustrate this difficulty. If one assumes a simple law for a shear dependant viscosity such as:

$$\eta = a \dot{\gamma}^b$$

(5)

one obtains laws, which model well blood and also its model fluid water-Separan. This is true only for a model scale of 1:1. However, if one scales the fluid to a ten times enlarged model flow one needs a fluid, which has a lower viscosity than water at larger shear rates. This obviously cannot be achieved. For the precise modeling of flow separations in the very low shear range one has to go back to the real size flow model and use a model fluid or even blood itself.

Once all the limitations are considered, the enlarged valve being fabricated and the proper movement of the occluder being established, the flow can be made visible. To this end a great variety of flow visualization techniques have been developed. From these methods, the most appropriate ones have been selected and applied.

186

VISUALIZATION AND QUANTIFICATION OF FLOW SEPARATIONS

Two methods have mainly be used to visualize the flow, the particle method and the fluorescent dye method. In the particle method small plastic (polystyrene) spheres with a diameter of 0.2 to 0.4 mm are added to the water. The flow is then lit with a light plane so that only a section of the flow becomes visible. The spheres move with the flow and show the path lines. In photographs taken with an extended exposure time, each particle produces a streak the length of which is proportional to the velocity. Figure 1 shows the center line section of the flow through the model of the Björk-Shiley Monostrut valve. The jet through the major orifice is visible as well as the flow separations behind the monostrut, occluder-disk and at the valve ring. The flow is visualized in great details, which one could not detect in a real size flow model. The concentration of the flow through only one orifice is ob-

Figure 1. Flow through a model of the Björk-Shiley Monostrut valve. The drawing on the left depicts the arrangement of the occluder and the light plane. The flow comes from the left. Behind the occluder a large flow separation becomes visible. The flow through the upper orifice is reduced. This phenomenon is found also in case of clinical application in real blood flow.

served also in clinical cases [9]. Radioopaque dye is injected in the aortic root by the cardiologist and gives a shadow in the x-rays, thus making the bloodflow visible. The same phenomenon as shown in figure 1 appears. This assures that the flow in the enlarged model is comparable with the real blood flow.

The use of a videorecorder adds a further feature - the slow motion effect permits the observation of the flow in its development and the repetition of the observation as often as needed. The timescale of 254 to 1 results in a camera speed which is 25 times 254 equal to 6350 frames per second, a speed, which corresponds to that of a high speed camera. Using these video recordings it is possible to quantify the flow with methods of image processing. From the displacement of the particles between different video frames the velocity field can be calculated, which is a PIV (Particle Image Velocimetry) method. The result of this process is described in detail in [10,11,12,13].

The particle method renders an overview of the flow and permits to quantify the flow-field as well. It also gives an indication of the numerous flow separations of the heart valve flow. These flow separations play an important role in the long term performance of the

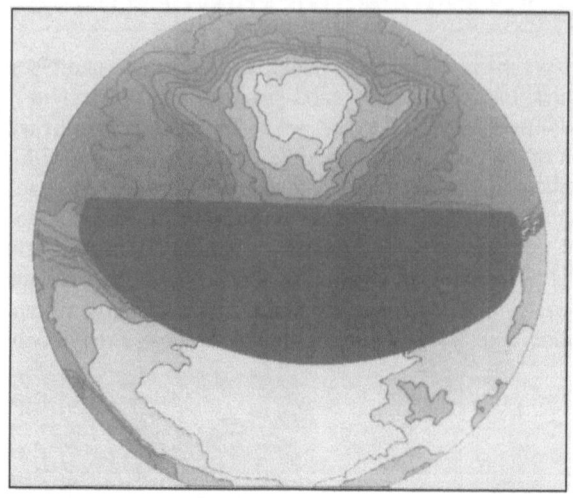

Figure 2. The application of the fluorescent dye method shows the residence time - depicted dark - immediatly downstream of the ring of a Björk-Shiley Standard valve

heart valve in the patient. A flow separation in the bloodstream is inevitably connected with a stagnant flow area and it is known since many years that such a stagnant flow is associated with the danger of thromboembolic complications. How can a stagnant flow be made visible and discriminated from the rest of the flow? The appropriate method is coloring a part of the fluid with a dye. The dye is infused into the model of the aortic root and thoroughly mixed. The space immediately downstream of the valve is stirred to assure a homogeneous concentration of the dye. This initial separation of the clean portion from the dyed fluid provides an optical distinction between the fluid from the previous cycle and the fluid from the new cycle. If there is no flow separation, the new fluid will completely replace the old fluid forming a distinct front. However, in case of a flow separation the old fluid will remain in some areas and stay there for a while, the dye discriminating it from the new fluid. Uranin® is used as a dye. This substance is a watersoluble dye and flares up in yellow-green tinge when lighted up. Areas which are not lit remain transparent. This important property makes the unlit dye invisible, so that the dye between the observer and the light plane does not obscure the view. This permits us to look into the center of the flow which is undisturbed by the dye surrounding the center. As the valve opens the clear fluid displaces the dyed fluid except in those regions where separations occur. We observe flow separations right at the ring, at the aortic sinuses and behind the struts. The grayscale of the video frames can be corrected to be proportional to the concentration of the dye. One further can compute the average concentration at every part of a cross-section and from this compute an average residence time of a particle in the flow separation. Figure 2 shows such a graph of the residence time. The area analized is just behind the ring, the dark element is a part of the occluder, which protrudes the light plane. The valve is a Björk-Shiley Standard valve, which has two struts to hold the occluder. Both generate flow separations, which induce a longer residence time and thus show in the picture. The clear distinction between dyed und clear fluid permits to obtain a clear picture of the mixing layer. From this one can make the turbulence of the mixing process visible, an observation, which has not yet been made in a real scale model. The main frequency of the turbulence in this valve is found to be 272 Hz, calculated to the real size flow.

The flow through the valve is instationary and so are the flow separations. The development of the flow separations can be observed and recorded with the video technique. Figure

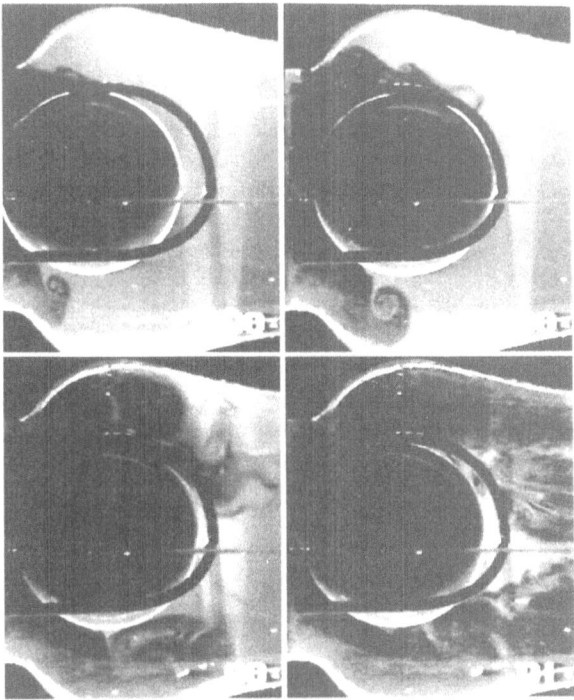

Figure 3. Four frames from a video taken of the flow through a Starr-Edwards heart valve. The flow comes from the left. Flow separations are found behind the ball and partially behind the valve ring. The geometry of the aortic root favours the generation of flow separations, since it is expanding in it cross section in the vicinity of the valve. With a designed duct flow separations can be avoided, however, this has a limited application only.

3 shows 4 frames from the opening of a Starr-Edwards heart valve, this type of valve has a ball functioning as the occluder. The frames are 40, 49, 64 and 82 milliseconds apart from the beginning of systole (300 ms duration). The dyed fluid is in the ball valve faster replaced by the dark fluid than in a valve with a disk occluder. One reason for this is that the ball has a much larger volume than a disk occluder valve and consequently it displaces more fluid. As one would expect, there is an extended flow separation at the downstream side of the ball. This flow separation is greatly influenced by proper fitting of the voluminous ball into the aortic root. This proper fit often may have not been achieved, having the consequence of flow separations and thromboembolic complications. Thus the ball valve has slowly disappeared from the clinical practice. However, some cases are reported from patients with ball valves having an extraordinary longivety. On the basis of these flow studies we assume that these patients accidentally had a very good fit of vessel and ball. It is theorized, that the small or favourably shaped flow separations were influencial on the lack of thrombus generation and thus helped the patient to sustain the implant well.

This leads to an important question: has the influence of the vessel geometry on flow separations sufficiently been considered? We think it has not. The flow around the occluder is usually looked at as a flow around an airfoil, which is surrounded by an infinitely extended fluid. The influence of the pressure along the x-axis is only little considered. In the valve flow, the structure which extends into the flow is subjected to a pressure which in general increases downstream, because the aortic root has a larger cross section in the vicinity of the valve. Thus we have a ducted flow with increasing static pressure, a flow separation is nearly inevitable. In engineering application one would avoid such flow, instead, one would decrease the cross section and accelerate the flow. We cannot change the human anatomy and it certainly is difficult surgically to change the shape of the aorta and thus create a better duct. However, there are applications where one has the freedom to design the duct - in artificial bloodpumps the occluder and the duct are subject to design. Up to now little effort has been made to match the duct and the occluder of an artificial valve. Instead, artificial heart valves made originally for the replacement of the diseased natural valve are put into an artificial duct without considering the basic laws of ducted flow. A ball valve in a duct specifically designed for it in order to minimize the energy loss has been investigated [14, 15]. Their ducted valve is intended to be used clinically as a conduit. Such a device is guiding the bloodflow from the apex of the heart to the aorta descendens, bypassing the diseased natural aortic valve. However, the important issue of flow separation has not been adressed. In an attempt to create a valve with little or no flow separation a duct has been calculated and fabricated for a ball as occluder [16, 17]. The main idea is to generate an accelerated flow, which can be achieved by giving the duct a downstream decreasing cross-section. Flow visualization studies made with these ducted valves showed, that flow separation can be avoided nearly totally and thus gives hope for the design of a valve with superior performance. The application of this idea for an artificial cardiac valve however is difficult, since the aortic root has a given anatomy. In the implantable ducted ball valve, the above mentioned conduit, flow separations can be avoided, because one here again has the freedom of designing the duct, which surrounds the occluder.

DISCUSSION

The flow through an artificial heart valve is very complex. It is instationary-periodic, three-dimensional, with moving boundaries, with a laminar turbulent transition and in addition the fluid is non-Newtonian. Thus making heart valve flow much more complex than typical engineering flows. No wonder it has so far resisted a complete clarification, despite the sophisticated methods, which have been applied. However, with new approaches one can clarify some of the still unsolved questions. A new flow model makes use of an enlarged geometry and of similarity laws. This permits the application of many flow visualization methods using particles and dye. The resulting still-pictures and video-pictures show a high resolution in space and time, which previously was unobtainable. With this model it is possible to locate and quantify the areas of flow separations, which are likely sites for a thrombus to grow. Pathologists have indeed found thrombi at these loci through the examination of artificial heart valves which have been implanted in patients or which have been subjected to bloodflow in artificial bloodpumps. This gives an indirect proof that areas of thrombus generation and areas of flow separation coincide. This is an encouraging result, because there are many ways to modify the flow. With the enlarged model we have a way to objectively check the modifications and to predict the thrombogenicity of a valve.

As convincing the correspondence of residence time and thrombus generation may appear, these calculations of the platelet's residence time are not yet a quantification of the thrombogenicity of an artificial valve. Still, it would be very valuable if one could predict its thrombogenicity before a newly developed valve leaves the laboratory and prior to any animal experiments. In additon to the residence time we need to know the average stress history of a platelet. This means that for our flow studies we will have leave the Eulerian way of looking at the flow in favour to the Lagrangeian way, i.e., the observer does not look at the flow from the outside, but looks at an individual particle and so to speak travels with the fluid. Further, one has to quantify the flow separations better, paying attention to their properties in regard of a chemical reaction. This will be most likely a combination of a specific shear rate, which bridges the distance of the platelets and provides a time for a reaction. Both are qualities, which can be assessed with methods developed in this project.

REFERENCES

1) Wurzinger L., Opitz R., Wolf M., and Schmid-Schönbein H., 1985, Shear Induced Platelet Activation, Biorheology.

2) Affeld K., Pszolla H., Lehmann B. and Mohnhaupt R., 1979, Measurement of the Flowfield behind Artificial Heart Valves with the Help of the Laser-Doppler-Effect, Proc. ISAO II: 439-441.

3) Pszolla H., Affeld K., Lehmann B., and Mohnhaupt A., 1979, Messung des Geschwindigkeitsfeldes hinter einer künstlichen Herzklappe (Björk-Shiley-Ventil) mit dem Laser-Doppler Anemometer, Biomed. Technik, 24, Erg.-Band.

4) Reul H., 1989, Qualitätssicherung von Herzklappenprothesen, "Abschlußbericht BMFT Projekt 01 ZQ 0141".

5) Affeld K., Schmidt S., and Bücherl E.S., 1973, Form- und Festigkeitsuntersuchungen von Aorten- und Pulmonalklappen des Rindes, Jahrestagung der Deutschen Gesellschaft für Biomedizinische Technik: 85-88.

6) Jansen J., WillekeS., Reiners B., Harbott P., Reul H., and Rau G., 1991, New J-3 Flexible-Leaflet Polyurethane Heart Valve Prosthesis with Improved Hydrodynamic Performance, The International Journal of Artificial Organs, 14, 10: 655-660.

7) Liepsch D.W., Levesque M., Nerem R.M., and Moravec S.T., 1988, Correlation of Laser Doppler Velocity Measurements and Endothelial Cell Shape in a Stenosed Dog Aorta, Vascular Endothelium in Health and Disease, Plenum Publ. Corp.: 43-50.

8) Astarita G., 1978, Scale-up Problems Arising with Non-Newtonian Fluids, Journal of Non-Newtonian Fluid Mechanics, 4: 285-298.

9) Taenzer L., 1991, "Konvektiver Fluidaustausch bei künstlichen Herzklappen", Diplomarbeit, Technische Universität Berlin 1991. Unpublished M.A. - thesis.

10) Affeld K., Walker P., and Schichl K., 1988, Novel Flow Visualization to Detect Sites of Thrombus Formation at Artificial Heart Valves, Proc. ESAO, Brno: 91-100.

11) Affeld K., Walker P., and Schichl K., 1989, The Use of Image Processing for the Investigation of Artificial Heart Valve Flow, Proc. ASAIO: 294-298.

12) Affeld K., Walker P., and Schichl K., 1990, Upscaling as a Tool in Biofluidmechnics - Demonstrated at the Artificial Heart Valve Flow, in: "Biomechanical Transport Processes", F. Mosora et al., ed., Series A: Life Sciences, 193, Plenum Press, New York.

13) Affeld K., Schichl K., Ziemann A, Flow Model Studies of Heart Valves, 1992, Plenum Press, New York, in press

14) Gentle R., 1983, Minimizing of Pressure Drop Across Heart Valve Conduits: a Preliminary Study, Life Support Systems, 1, 4: 263-270.

15) Tansley G.D., 1988, "Numerical Analysis of Turbulent, Non-Newtonian Fluid Flow through Heart Valve Conduits", PH.D. - thesis, Trent Polytechnic, Nottingham, U.K.

16) Vondran T., 1991, "Untersuchung der Umströmung einer Starr-Edwards-Kugelherzklappe im Modell", Diplomarbeit, Technische Universität Berlin. Unpublished M.A. - thesis.

17) Svejda D., 1992, "Untersuchung der Strömungsablösung an zwei künstlichen Herzklappen", Diplomarbeit, Technische Universität Berlin. Unpublished M.A. - thesis.

Large-eddy Simulation of Turbulent Flow
Over Sharp-edged Obstacles in a Plate Channel

H. Wengle and H. Werner

Institut für Strömungsmechanik und Aerodynamik, LRT, WE7,
Universität der Bundeswehr München, D-8014 Neubiberg, Germany

1 Introduction

The major goal of our work within the framework a priority research program of the
German Research Society (DFG) on 'Physics of Separated Flows' was to apply the
solution concept of large-eddy simulation (LES) to turbulent flow over and around
flow obstacles with sharp edges and corners. For engineering applications of LES, it
is essential to consider (a) high Reynolds number flow, (b) three-dimensionality of the
mean flow and (c) non-periodic boundary conditions in the main flow direction.

Two examples of flow obstacles have been considered: the first one is a single cube,
creating a fully 3D mean flow field and making it necessary to provide the statistics
by time-averaging only, and the second example is a single square-rib, the corresponding (nominally) 2D case, with a (nominally) homogeneous direction in the lateral flow
direction which usually can be used to improve the statistics. The flow obstacles are
mounted on the bottom of a plate channel. This choice of flow situation makes it easier
to provide proper inflow boundary conditions from a LES of a fully developed channel
flow with corresponding Reynolds number. We selected a Reynolds number of about
50000 (based on mean bulk velocity and obstacle height, which is equal to the channel
half width) for the case of a cube and $Re = 42500$ for the case of a square-rib. This
choice of Reynolds numbers was determined by the corresponding experiments which
have been carried out by Larousse, Martinuzzi and Tropea [1] and by Dimaczek, Tropea
and Wang [2].

Earlier related work on LES of turbulent flow over a *periodic* arrangement of cubes
in a simulated atmospheric boundary layer has been published by Murakami, Mochida
and Hibi [3], results for the flow over a *periodic* arrangement of square ribs in a channel
have been presented by Kobayashi, Kano and Ishihara, [4]. More detailed results from
our own work on flow over a *single* square-rib and over a *single* cube can be found in
[5],[6],[7],[8] and [9].

2 Mathematical models, solution technique and evaluation of the statistics

The governing equations describing the resolvable flow quantities are derived following the 'volume balance method' of Schumann [10].

The subgrid scale (SGS) stresses, arising from the nonlinear convection terms, are
evaluated by the Smagorinsky-Lilly model (with $c_1 = 0.1$) which relates the SGS stresses
to the grid scale (GS) velocity field via an eddy viscosity model.

The governing equations are solved numerically on a staggered and non-uniform
grid using second order finite-differencing in time and space (explicit leap-frog for time

discretization, central differencing for convection terms and for time-lagged diffusion terms). The problem of pressure-velocity coupling is solved iteratively (point-by-point relaxation).

The *direct* results from LES are the time-dependent and three-dimensional fields for the GS quantities of the three velocity components and the pressure. By time-averaging of the instantaneous flow field, the *mean* flow field is obtained, and as soon as the mean velocity field has reached stable (i.e. time-independent) values, the fluctuating velocity field can be evaluated. Finally, from the fluctuating fields, the root-mean-square values, e.g. for velocity, vorticity and pressure fluctuations can be calculated, as well as the Reynolds-stresses and other statistics desired.

3 Inflow and wall boundary conditions

At the inflow section, we used at each time step the instantaneous flow field of a LES result of the corresponding (fully developed) channel flow (see figure 1). Boundary conditions at horizontal and vertical walls were specified by assuming that at the grid points (P) closest to the wall, (a) the instantaneous velocity components tangential to the wall (u_P, v_P) are in phase with the instantaneous wall shear stress components (τ_{ub}, τ_{vb}) and (b) the instantaneous velocity distribution is assumed to follow the linear law-of-the-wall $u^+ = z^+$ for $z^+ \leq 11.81$, and for $z^+ = z_m > 11.81$ it is continued by a power-law description of the form $u^+ = A(z^+)^B$ (with A=8.3 and B=1/7). The velocity components tangential to a wall at the grid point next to the wall (u_P, v_P) can be related to the corresponding wall shear stress components by integrating the velocity distribution over the height of the first grid element, and the resulting expression can be resolved *analytically* for the wall shear stress component. From an application point of view this procedure offers the advantage that the averages $< \tau_{ub} >$ and $< u_P >$ are *not* required to evaluate the instantaneous wall shear stress component, e.g. from $\tau_{ub} = u_P < \tau_{ub} > / < u_P >$. This is important in flow situations in which these variables may be slowly varying in time and, in addition, numerical problems are avoided using this relation in reattachment regions. In consequence of the experimental results from Ruderich and Fernholz [11] we have abandoned the use of the logarithmic-law-of-the-wall.

4 Discussion of results

The results shown in this paper are made dimensionless using a reference height $L_{ref} = H$ (cube height) and a reference velocity $U_{ref} = U_b$ (mean bulk velocity, fig.1). Note, that in our nomenclature, Z is the coordinate *normal* to the walls of the plate channel.

LES offers the great benefit that it provides insight into the time-dependent and three-dimensional large-scale structure of a turbulent flow field. A view at the *instantaneous velocity field* in a vertical plane through the center of the cube (fig. 2) exhibits an extremely complicated flow field. The interaction of different processes like the development of a three-dimensional shear layer, the reattachment of flow on the bottom plate behind the flow obstacle, the recirculation of highly turbulent flow and its reentrainment into the free shear layer takes place within a spatial regime which is significantly smaller compared to the case of turbulent flow over a square-rib: the mean recirculation length is about 2.0 for the flow over a cube, and about 7.0 for the square-rib (measured from the back side of the flow obstacle). Strong horizontal fluctuations in the lateral direction close to the walls can be observed in the small recirculation regimes in front and on top of the sharp-edged obstacles and in the large recirculation regime behind the flow obstacles. The creation of strong horizontal fluctuations by splashing down of tongues of fluid material can be seen in figure 5(above). An ideal way of presenting time-dependent results would be a video movie, but, in a printed medium, the only way is to present a series of snapshots from flow regions of particular interest, see e.g. in [8].

The *mean structure* of the flow field (e.g. mean velocity, mean vorticity, second order statistics, mean enstrophy, mean helicity) can be provided by time-averaging only if no homogeneous direction is available to improve the statistics. For the results shown here for the case of turbulent flow over a cube (fig. 1 and 3), we started averaging for the first-order statistics after 60 reference times $T_{ref} = U_{ref}/L_{ref}$; the first order statistics was sufficiently stable after about 100 reference times (taking one sample at every 40th time step), and samples for the second order statistics have been collected over the latest 300 reference times. *One* reference time T_{ref} is equivalent to the time a tracer particle needs to travel with bulk velocity U_b over a distance of *one* obstacle height H (fig.1). Fig. 3 shows a comparison of LES results for first and second order statistics with the experimental data of Martinuzzi in [1] for the case of turbulent flow over a cube. For the case of turbulent flow over a square rib, a corresponding comparison between LES results and the experimental data of Dimaczek in [2] is given in figure 4. For the latter case, the experiments indicate a three-dimensional *mean* structure immediately in front and above the top face of the square rib. If the nominally homogeneous direction is *not* used to collect samples for the statistics, the LES results confirm this observation (see figure 5 below).

From a research programm carried out over several years, the major results only can be discussed on the limited space given here. Further results and details, e.g. on the theoretical background of LES, on the numerical method used (stability, errors, effect of different compuational grids) as well as an extensive collection of LES results for the case of turbulent flow over a square rib can be found in [9]. Additional results for the case of a square rib (distribution of inclination angles of vorticity vectors, integral length scales, effect of different computational grids) have been presented in [7], and additional results for the case of turbulent flow over a cube (snapshots of instantaneous velocity and vorticity fields, mean recirculation lengths, vorticity, enstrophy, helicity) are given in [8].

5 Conclusions

The most valuable result from a large-eddy simulation is the time-dependent and three-dimensional large-scale structure of a turbulent flow field, and a lot of additional effort seems to be necessary to present such data to the research community in a proper way. It is always very helpful and, at least for some time in the future, it will be necessary to compare LES data with corresponding experimental data for the first order and, most important, for the second order statistics. The two methods will be able to supplement each other, and it is to be expected that LES will develop to a valuable tool for a sufficiently accurate prediction of complex turbulent flows of engineering interest.

For flow cases without any homogeneous directions, large amounts of computing time have to be spent to provide sufficiently stable second order statistics. For example, for the results of turbulent flow over a single cube (fig. 3), about 500 CPU hours on a

CRAY/Y-MP were necessary and the number of grid points used for this case should be considered to be a minimum to reach satisfying agreement with the experiment.

The case of turbulent flow over a cube in a plate channel at high Reynolds number represents a flow problem which is very well suited for testing and validating numerical simulation techniques and turbulence models.

The existence of a 3D *mean* flow structure close to a square rib, as observed in the experiments [2], can be confirmed by LES if time-averaging only is used to provide the *mean* flow field (i.e. if the nominally homogeneous (lateral) direction is *not* used to improve the statistics). Quite surprisingly, the case of turbulent flow over the 2D obstacle turned out to be the more difficult case of the two flow problems (a single 3D cube, a single 2D square rib) considered in the research program.

References

[1]   LAROUSSE, A., MARTINUZZI, R., TROPEA, C.: "Flow around surface-mounted three-dimensional obstacles", Proc. 8th Symposium on Turbulent Shear Flows, Sept. 9-11, Technical University of Munich, Munich, Germany (1991).

[2]   DIMACZEK, G., TROPEA, C., WANG, A.B.: "Turbulent flow over two-dimensional, surface-mounted obstacles: plane and axisymmetric geometries, Proc. 2nd European Turbulence Conference, August 30 - September 2, 1988, Berlin, Germany, in: Advances in Turbulence 2, pp. 114-121, Springer, Berlin (1989).

[3]   MURAKAMI, S., MOCHIDA, A., HIBI, K.: "Three-dimensional numerical simulation of air flow around a cubic model by means of large eddy simulation", J.Wind Engng. and Ind. Aerodynamics, 25, 291-305 (1987).

[4]   KOBAYASHI, T., KANO, M., ISHIHARA, T.: "Prediction of turbulent flow in two-dimensional channel with turbulence promoters", Bull. JSME,Vol. 28, No. 246, 2948-2953 (1985).

[5]   BAETKE, F., WERNER, H. WENGLE, H.: "Computation of turbulent flow around a cube on a vector computer", Proc. 6th Symposium on Turbulent Shear Flows, Sept. 7-9, Toulouse, France (1987).

[6]   WERNER, H., WENGLE, H.: "Large-eddy simulation of turbulent flow over a square rib in a channel", Proc. 2nd European Turbulence Conference, August 30 - September 2, 1988, Berlin, Germany, in: Advances in Turbulence 2, pp. 418-423, Springer, Berlin (1989).

[7]   WERNER, H., WENGLE, H.: "Large-eddy simulation of turbulent flow over a square rib in a channel", Proc. 7th Symposium on Turbulent Shear Flows, August 21-23, 1989, Stanford University, USA (1989).

[8]   WERNER, H., WENGLE, H.: "Large-eddy simulation of turbulent flow over and around a cube in a plate channel", Proc. 8th Symposium on Turbulent Shear Flows, September 9-11, 1991, Technical University of Munich, Munich, Germany (1991) in: Turbulent Shear Flows 8, Springer, Berlin (1992).

[9]   WERNER, H.: "Grobstruktursimulation der turbulenten Strömung über eine querliegende Rippe in einem Plattenkanal bei hoher Reynoldszahl", Ph.D. Thesis (in german), Technical University of Munich, Germany (1991).

[10]   SCHUMANN, U.: "Subgrid scale model for finite difference simulations of turbulent flows in plane channels and annuli", J.Comp.Phys.18, 376-404 (1975)

[11]   RUDERICH, R., FERNHOLZ, H.H.: "An experimental investigation of a turbulent shear flow with separation, reverse flow, and reattachment", J.Fluid Mech. 163, 283-322 (1986).

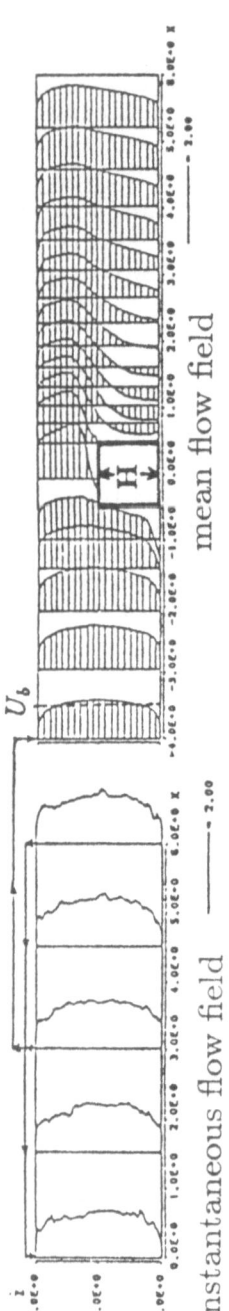

U_b

H

mean flow field

instantaneous flow field

Fig. 1: Geometry of computational domain and inflow boundary conditions

left: channel flow $NX * NY * NZ = 54 * 92 * 58$ grid points
right: flow over cube $NX * NY * NZ = 144 * 92 * 58$ grid points

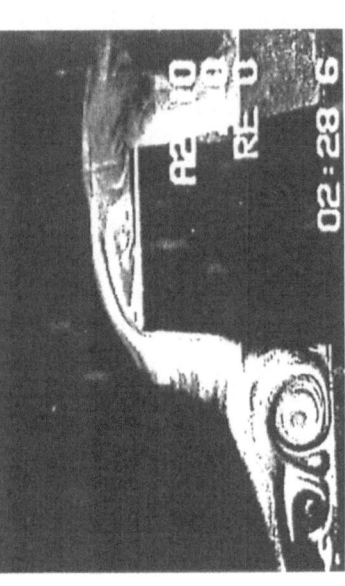

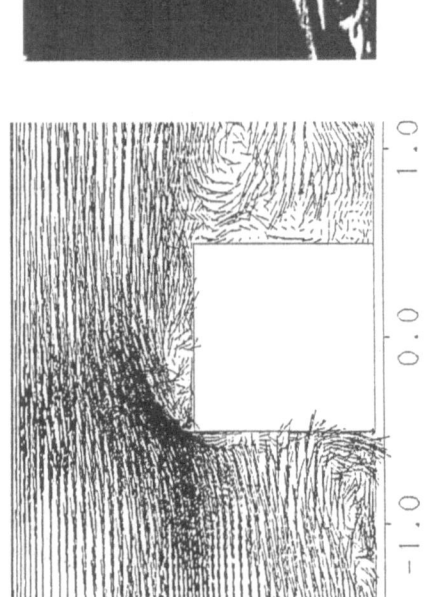

large-eddy simulation

experiment by Martinuzzi [1]

Fig. 2: Instantaneous flow field in a vertical plane
through the center of the cube

196

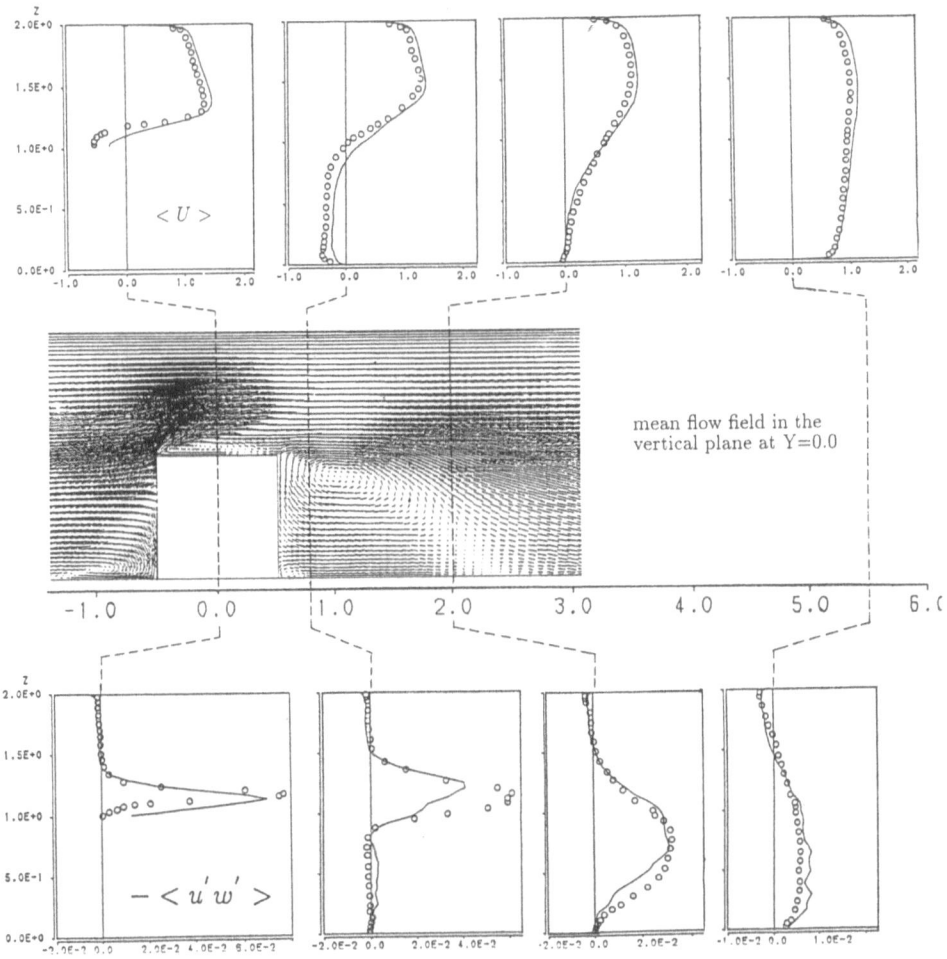

**Fig. 3: Comparison of LES with experiment
for turbulent flow over a cube**

above: vertical profiles of mean U-velocity component $< U >$
below: vertical profiles of Reynolds shear stress $- < uw >$
o o o: experimental data from Martinuzzi [1]

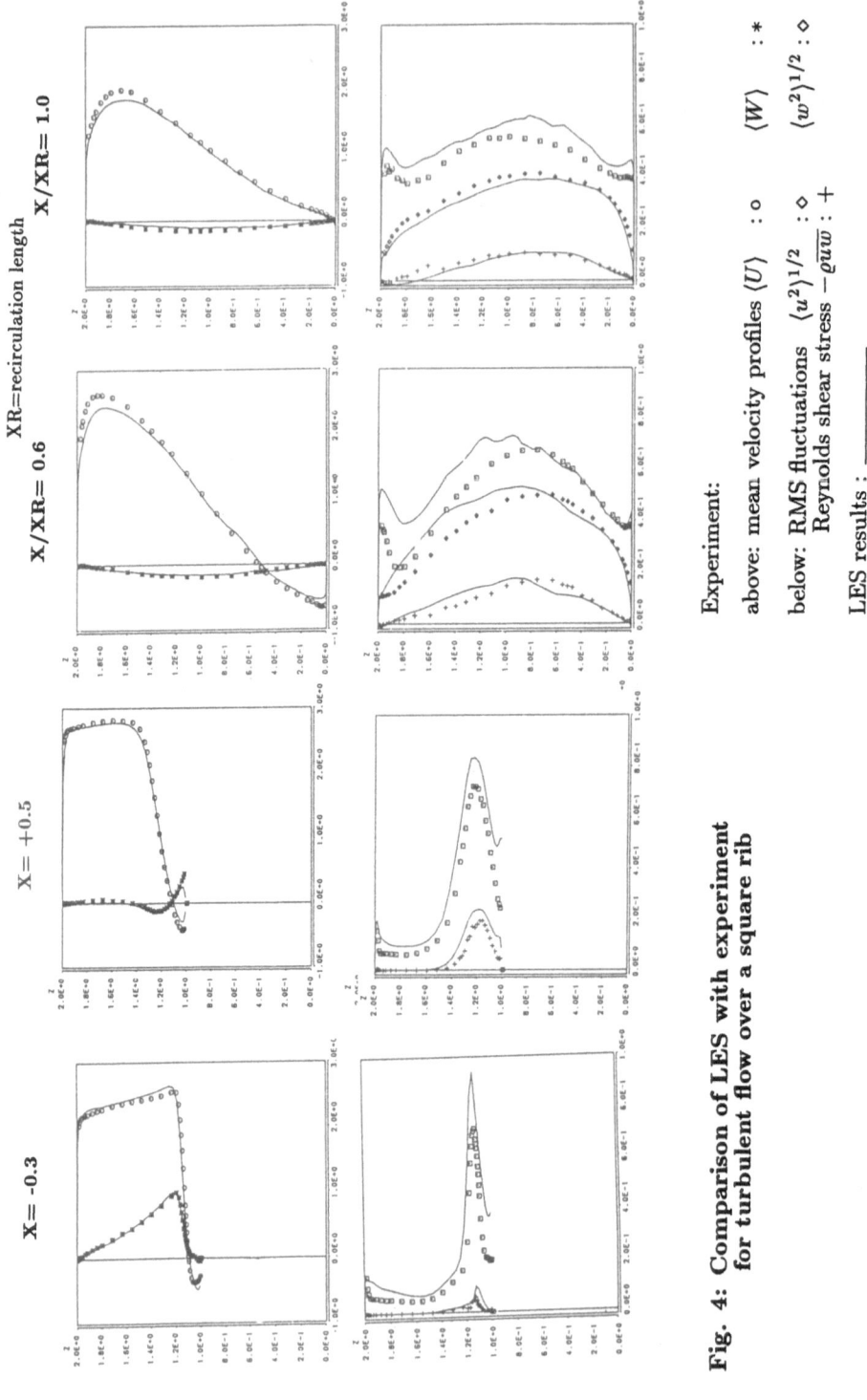

Fig. 4: Comparison of LES with experiment for turbulent flow over a square rib

Experiment:

above: mean velocity profiles $\langle U \rangle$: o $\langle W \rangle$: *

below: RMS fluctuations $\langle u^2 \rangle^{1/2}$: ◇ $\langle w^2 \rangle^{1/2}$: ◇

       Reynolds shear stress $-\varrho \overline{uw}$: +

LES results : ——

X= -0.3

X= +0.5

X/XR= 0.6

X/XR= 1.0

XR=recirculation length

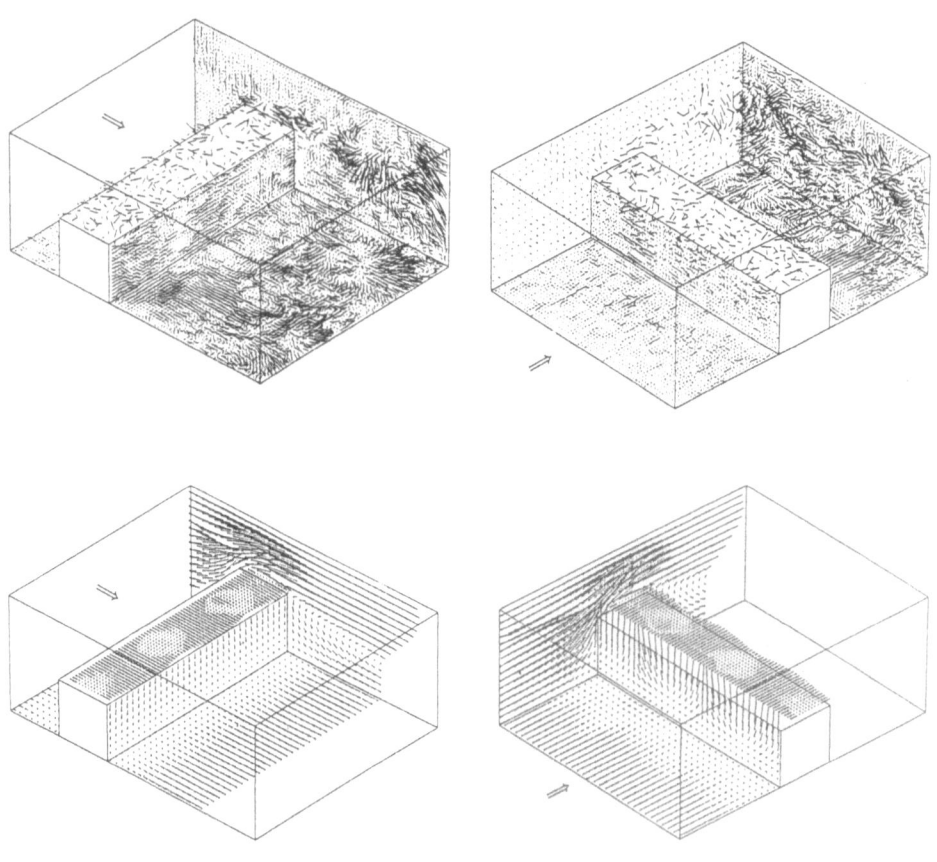

Fig. 5: Perspective views of *fluctuating* **velocity field (above)**
and of time-averaged *mean* **velocity field (below)**

Acknowledgements

This work has been supported by the German Research Society (DFG), Grant No.
497/5 (Römer/Wengle). We gratefully acknowledge the support by the computing center
of the Universität der Bundeswehr München, by the Leibniz Computing Center of the
Bavarian Academy of Sciences, and by Convex Computer GmbH (GS Bayern).

On the separated flow behind a swept backward-facing step

H.H. Fernholz, G. Janke , M. Kalter, M. Schober

Hermann-Föttinger-Institut, Technische Universität Berlin, Strasse des 17. Juni 135
D-1000 Berlin 12

Abstract

Nominally two-dimensional flows can be strongly affected by three-dimensional flow effects. The influence of these three-dimensionalities is investigated in the reverse-flow region downstream of a swept backward-facing step by varying the Reynolds number and the sweep angle. Prandtl's independance principle was found to hold for the ratio of reattachment length and step height up to sweep angles of about 40° within a certain Reynolds-number range. The principle also holds for the skin-friction component normal to the step.

A brief summary is given in the appendix of an improved method to determine skin-friction distributions by means of oil-film interferometry. Comparisons of different measuring techniques for the skin friction and turbulence measurements in the flow field at different sweep angles will be published separately.

Introduction

The interaction of a controlled three-dimensional wall flow with a strong reverse-flow region was investigated using as a model the flow over a straight, swept rearward-facing step. The sweep angle could be varied between 0° and 50° resulting in an aspect ratio defined as the ratio of the spanwise step length b and the chordwise lenght x_r (i.e. normal to the step) of the reverse-flow region, between 14 and 33. In a co-ordinate system attached to the model (fig. 1), the three-dimensionality can be viewed as a superimposed spanwise flow component.

Only very few studies of this model flow exist. Mostly related is the phenomenological study of the swept backward facing step geometry by Selby [8],

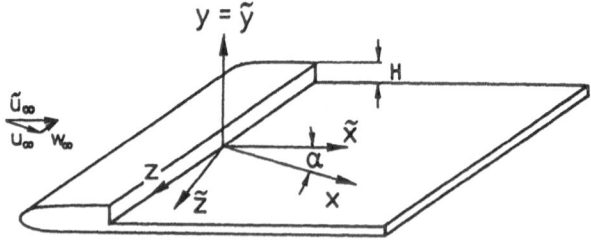

Fig. 1 Experimental configuration

consisting mainly of oil-flow visualizations and surface pressure measurements. Barkey Wolf [3] studied the effect of sweepback on a separation bubble formed at a blunt plate with a maximum sweep angle of $45°$, measuring the mean and fluctuating surface pressure, the skin friction, and the mean velocity, the latter with a three tube yaw probe. Sutton [9] showed a few flow visualization pictures of the near-wall flow of the swept normal plate with a splitter plate. McCluskey et. al. [5] did oil-flow visualizations and took pulsed-wire measurements in the swept lying-T-configuration.

The continuity and momentum equations for the quasi two-dimensional flow in the coordinate system given in fig.(1) are

Chordwise flow:

$$\frac{\partial \bar{u}}{\partial x} + \frac{\partial \bar{v}}{\partial y} = 0$$

$$\bar{u}\frac{\partial \bar{u}}{\partial x} + \bar{v}\frac{\partial \bar{u}}{\partial y} = -\frac{1}{\rho}\frac{\partial \bar{p}}{\partial x} - \frac{\overline{\partial u'^2}}{\partial x} - \frac{\overline{\partial u'v'}}{\partial y} + \nu\left(\frac{\partial^2 \bar{u}}{\partial x} + \frac{\partial^2 \bar{u}}{\partial y}\right)$$

$$\bar{u}\frac{\partial \bar{v}}{\partial x} + \bar{v}\frac{\partial \bar{v}}{\partial y} = -\frac{1}{\rho}\frac{\partial \bar{p}}{\partial y} - \frac{\overline{\partial u'v'}}{\partial x} - \frac{\overline{\partial v'^2}}{\partial y} + \nu\left(\frac{\partial^2 \bar{v}}{\partial x^2} + \frac{\partial^2 \bar{v}}{\partial y^2}\right)$$

Spanwise flow:

$$\bar{u}\frac{\partial \bar{w}}{\partial x} + \bar{v}\frac{\partial \bar{w}}{\partial y} = -\frac{\overline{\partial u'w'}}{\partial x} - \frac{\overline{\partial v'w'}}{\partial y} + \nu\left(\frac{\partial^2 \bar{w}}{\partial x^2} + \frac{\partial^2 \bar{w}}{\partial y^2}\right).$$

They show that the coupling between the chordwise and the spanwise flow is solely due to Reynolds stresses. The infinetely yawed backward-facing step is thus a shear-driven flow, as noticed by Pierce & McAllister [6]. In laminar flow the equations are decoupled and Prandtl's independence principle holds. This principle seems to carry over, at least partially, even to the turbulent separated flow. Up to sweep angles of $30°$ to $40°$ the independence principle applies to the chordwise reattachment length, the chordwise skin-friction component, the chordwise velocity profiles in the separation bubble, and to the static pressure ([3],[8]).

Experimental apparatus

Figure 2 shows the experimental setup. The model plate was installed in sidewall slots in the centre of the testsection (50×49cm) of an open-circuit suction wind tunnel with a turbulence level of 0.1 %. The step configuration consisted of a nose piece whose upstream half had an elliptical cross section. The nose piece was fastened to the test plate, so that a 5 mm step was formed. The test plate was 12 mm thick and contained a section with a computer controlled eccentric double-turntable with exchangeable plugs, allowing to reach every point on the wall within a circle of 260 mm diameter. One of the plugs was equiped with a traverse gear for probe movements normal to the wall.

The investigation started with the visualization of the near-wall flow using TiO_2 particles in an oil/kerosene mixture. Velocity measurements were taken with X-wire and normal wire probes. The mean and fluctuating part of the skin-friction vector was measured with a wall pulsed-wire and a wall hot-wire

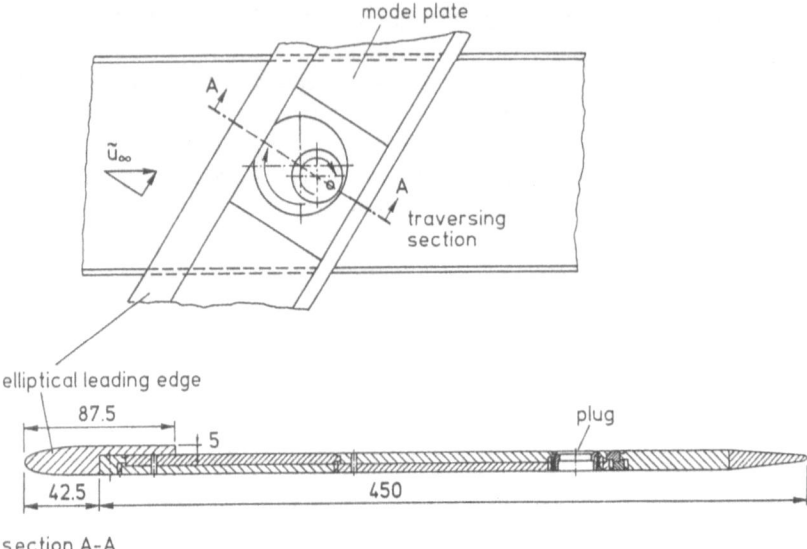

Fig. 2 Model plate with eccentric double turntable

probe. The flow visualization in the main test section was complemented by visualizations in a smoke tunnel and in a water channel.

The boundary layer upstream of the step was laminar. The Reynolds number Re_α defined with the step hight H and the component U_∞ of the flow normal to the step was varied in the preliminary oil-flow visualizations. For more detailed surveys at sweep angles of $0°$, $30°$ and $50°$, it was later kept constant at the highest possible Reynold number $Re_\alpha = 5040$.

Results and discussion

Figure 3 shows the distance of the reattachment line normal to the step for different sweep angles, as taken from oil-flow visualizations. Without sweep a two-dimensional middle region with a spanwise invariant reattachment distance can be clearly recognized and distinguished from the two side regions where the lenght of the reattachment region is reduced. This is an indication that in these regions an additional pressure-driven spanwise flow exists, which is due to the interaction of the separated flow with the boundary layer on the wind tunnel side walls (see [5]). As the sweep angle increases, the spanwise extent of the upstream side region increases and for the case of $30°$ sweep the reattachment line is parallel to the step only over a very short region. Since small parallel reattachment regions could also be produced with rather small aspect ratios, the spanwise invariance of the static pressure on the wall appeared to be a better check for the two-dimensionality of the flow. For $50°$ sweep the reattachment length is again uniform over about 75% of the span, though smaller. So far we have no explanation for the different behaviour at the $30°$ sweep angle.

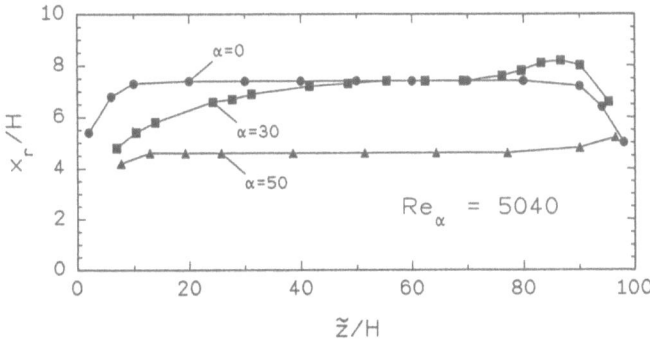

Fig. 3 Chordwise reattachment length along the span

Figure 4 shows the reattachment length plotted against the sweep angle for a slightly different configuration. (4 mm step height). In the Reynolds-number regime covered by this experiment the reattachment length is reduced with increasing Reynolds number (see e.g. [2]). Due to the limited width of the test section, at lower Reynolds numbers two-dimensional regions along the span did not always exist. It can be seen, however, that if the ratio of tunnel width over reattachment length x_r is larger than about 15, there are two-dimensional regions for all sweep angles, and, in agreement with the results of Selby [8], up to angles of about 40° the independence principle applies to x_r/H.

For higher sweep angles the reattachment length is considerably reduced, and at $\alpha = 55°$ it reaches a limiting value of about $x_r/H = 4.5$ that is independent of the Reynolds number in the range covered here.

The distributions of the wall static pressure coefficient against x/x_r (not shown) for 0° and 30° sweep differ hardly from each other. But the reduced length of the reverse-flow region for 50° sweep is accompanied by a higher peak-pressure rise shortly downstream of reattachment. The level of c_p remains higher downstream, which is due to the fact that the sidewalls of the wind-tunnel are not shaped like the streamlines for infinite sweep.

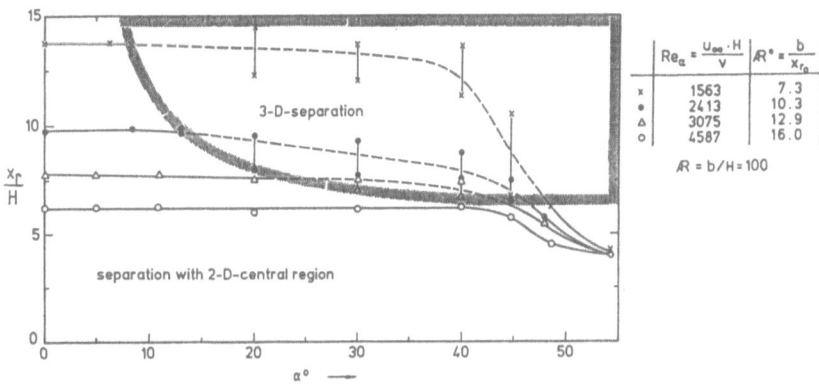

Fig. 4 Reattachment length as a function of Reynolds number and sweep angle

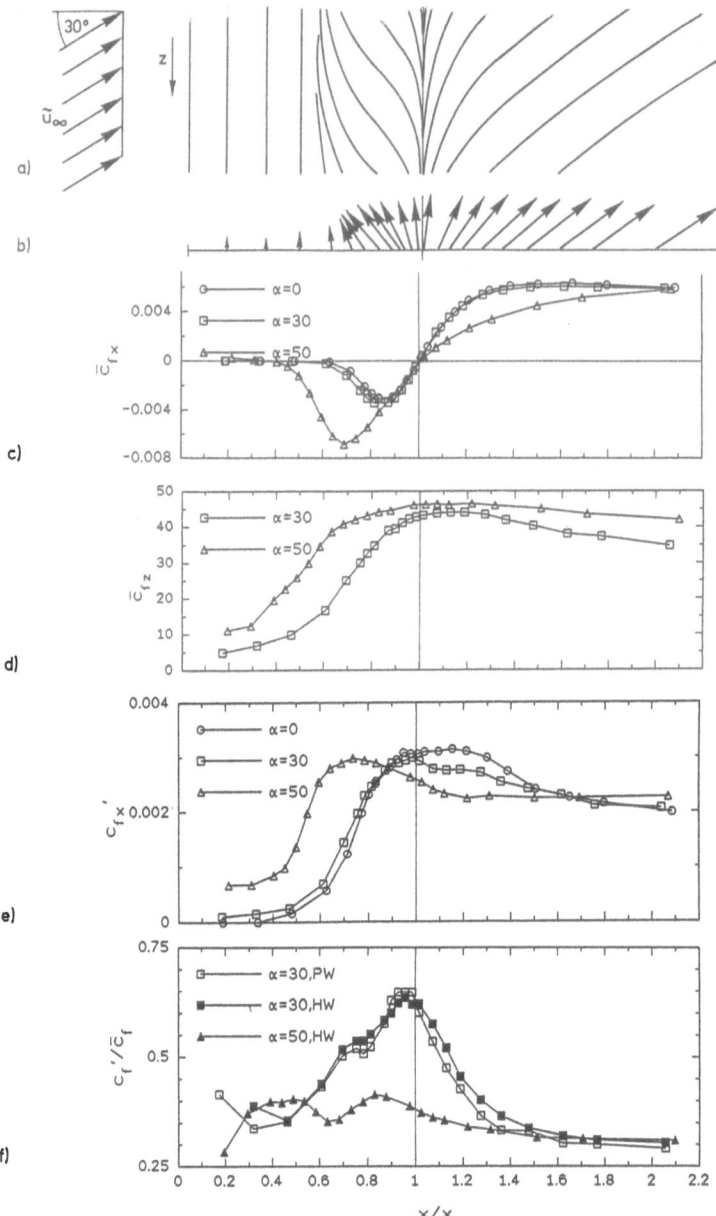

Fig. 5 Skin-friction distribution in the two-dimensional region
a) skin-friction lines, b) vektor plot of mean skin-friction vector,
c) chordwise component of skin friction coeficient $\bar{c}_{fx}$ =2 $\tau_{wx}/\rho U^2_\infty$,
d) spanwise component of skin friction coefficient $\bar{c}_{fz}$ =2H $\tau_{wz}/\rho\nu W_\infty$,
e) fluctuating part of chordwise skin-friction componen c'_{fx} =2 $\tau'_{wx}/\rho U^2_\infty$,
f) turbulence level of skin friction in streamwise direction

Figure 5 displays the chordwise (5c) and spanwise (5d) components of the mean skin-friction coefficient for three sweep angles and the resulting vector plot (5b) and wall streamlines (5a) for the 30° case. The unusually high values of the chordwise $\bar{c}_{fx}$ - values are due to the small Reynolds numbers, where $\bar{c}_f$ is also high for a flat plate turbulent boundary layer. At 0° and 30° sweep $\bar{c}_{fx}$ is zero over about half the reverse-flow region. This is indicative of a region with little kinetic energy near the step, which is confirmed by the fluctuating parts of the normal (5e) and streamwise (5f) components.

In figure 5c the spanwise skin-friction component $\bar{\tau}_{wz}$ is normalized by $\rho \nu W_\infty/(2H)$. Then the curves collapse in the case of laminar flow, because in this case the spanwise velocity distributions of different sweep angles should be distinguished only by this factor But this is no longer true for turbulent flow of these high sweep angles.

Although, at 30° sweep the chordwise flow component looks exactly like that in a separated flow without sweep, the absolut value of the skin friction has substantial values in the reattachment region, so that hot wires could be used in this region. The value $c_f'/\bar{c}_f$ reaches 65 % (fig. 5f). The wall pulsed wire and the wall hot wire agree well despite this high turbulence level (fig.7), probably because the small v-component in the vicinity of the wall only causes small errors in the hot-wire values.

Mean flow profiles (not shown here) for the flow cases 0° and 30° sweep agree well with each other on the high speed side of the reverse-flow region, where also at 0° sweep the hot wire measurements do not suffer from rectification errors. Taking into account the close agreement between the mean skin friction measurements of the 0° sweep and the 30° sweep case, it can be concluded, that the chordwise component of the mean flow field does not change either. Thus, in agreement with the results of Barkey Wolf [3], the independence principle applies to the chordwise mean-flow component.

Conclusions

Up to at least 30° the chordwise mean flow component ist independent of the spanwise flow.

Up to 40° the reattachment length x_r is independent of the spanwise flow.

For higher sweep angles the chordwise reattachment length is reduced by at least two step heights, which corresponds to a reduction of about 35 % compared to the unswept case. Following the modell of Chapmann (1959) (see [1]), i.e. the reattachment length is determined by the balance of pressure induced backflow and entrainment of the separated shear layer, this reduction could be caused by an increased entrainment due to an additional instability of the chordwise shear in the separated shear layer.

Acknowledgements

We would like to thank the DFG for financial support of projects (Fe 43/33, 1-5)

References

1. Adams, E.W. & Johnston, J.P. 1988 Effects of the separating shear layer on the reattachment flow structure Part 1: Pressure and turbulence quantities. Exp. Fluids 6, 400-408

2. Armaly, B.F. Durst, F., Peireira, J.F.C., Schönung, B. 1983 Experimental and theoretical investigation of backward-facing step. J. Fluid. Mech., 127 , 473-496

3. Barkey Wolf, F. D. 1987 Swept and unswept separation bubbles.
   Ph.D. Dissertation University of Cambridge

4. Kalter, M. 1991 Untersuchung der Turbulenzstruktur in der Strömung hinter einer schrägen rückwärtsgewandten Stufe. Diplomarbeit, Hermann-Föttinger-Institut, TU Berlin

5. McCluskey, F. Hancock, P.E., Castro, I.P. 1991 Three-dimensional separated flows. 8th Symp. Turb. Shear Flows, TU Munich, Sept. 9-11, 1991

6. Pierce, F.J. & McAllister J.E. 1983 Measurements in a pressure-driven and a shear-driven three-dimensional turbulent boundary layer. In " Three -dimensional turbulent boundary layers, Eds. H.H. Fernholz & E. Krause, Springer Verlag Berlin, 44-54

7. Schober, M. 1991 Experimentelle Untersuchung der Umbildung einer abgelösten, turbulenten Strömung hinter einer rückwärtsgewandten, schrägen Stufe. Diplomarbeit, Hermann-Föttinger-Institut, TU Berlin

8. Selby, G.V. 1982 Phenomenological study of subsonic turbulent flow over a swept rearward-facing step. Ph.D. Dissertation, University of Delaware

9. Sutton, E.P. 1983 Experiments on a flow with swept separation and reattach-ment of a boundary layer. In " Three-dimensional turbulent boundary layers, Eds. H.H. Fernholz & E. Krause, Springer Verlag Berlin, 44-54

10. Tanner,L.H. & Blows, L.G. 1976 A study of the motion of oil films on surfaces in air flow. with application to the measurement of skin friction. J. of Phys. E.: Sci. Instrum. 9, 194-202

Appendix: Oil-film interferometry

The possibility of measuring the skin friction from the movement of inter-ference fringes of a thin oil film has first been realized by Tanner & Blows (1976). As shown in fig. 6 for the case of the normal backward-facing step, an oil film is applied as an extremely thin layer (2 μm) to a smooth solid wall (e.g. glass), so that it does not change the flow on top of it, but is only driven by the time invariant mean skin-friction distribution $\underline{\tau}(x,z) = \{\overline{\tau}_x, \overline{\tau}_z\}$ of this flow. In this case the temporal development of the oil-film surface h(x,z,t) can be described by

$$\frac{\partial h}{\partial t} = - \frac{1}{2\eta} \text{ div } (h^2 \, \underline{\tau}) \qquad (1)$$

where div() is a two-dimensional operator and η is the dynamical viscosity of the oil. From this equation it can be concluded that the surface of the oil film moves in a slow motion exactly like a near-wall material surface of the mean fluid flow above the oil film (fig 6.b). Along skin-friction lines, i.e. the lines that are tangential to the skin-friction vector, equ. (1) can be solved for τ by integration and a general analytical solution can be found for h.

As shown in fig. 6a, only a sodium lamp and a blackened glass plate are needed to visualize the contour lines of the oil-film surface h(x,z,t) via interferometry.

Along the skin-friction co-ordinate s an s-t-diagram (fig. 6c) of the film movement is recorded by repetitively reading only one line of a properly adjusted video image into an image processor. The s-t diagram enables one to

- find characteristics of equ. (1), which are identical with the paths of the particles that make up the oil-film surface (fig. 6c)
- identify the contour lines t(h,s) at later stages as similarity solutions of (1), which are spread out or drawn together by characteristics (fig. 6c)
- infer the wall-shear stress (fig. 6d) and the streamlines of the outer flow in the vicinity of the wall (fig. 6b) even in cases of highly varying skin friction.
- monitor the development and movement of jumps, as contained in the general solution of equ. (1)

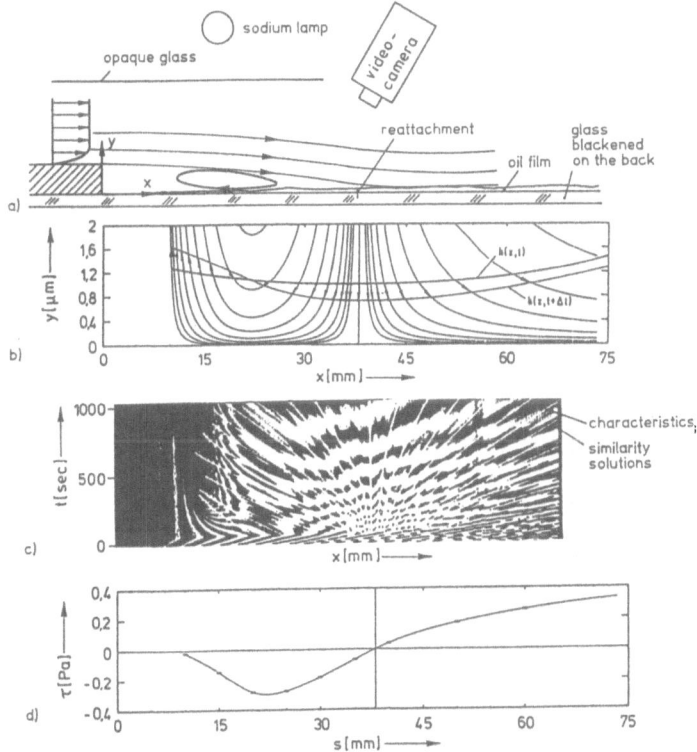

Fig. 6 Example for a setup and results of oil-film interferometry
a) experimental setup, b) near-wall streamlines, c) s-t diagram, d) skin-friction distribution

Flow around finite lengthed cylinders at low Reynolds number: End effects and their origins

Holger Eisenlohr and Helmut Eckelmann

Institut für Angewandte Mechanik und Strömungsphysik
der Universität Göttingen
Bunsenstraße 10, D-3400 Göttingen

Summary

There is experimental evidence that the appearence of discontinuities in the Strouhal number Reynolds number range 50–160 depends on the cylinder boundaries and the aspect ratio. The discontinuity at Re ≈ 90 observed by Tritton [22] is therefore a three-dimensional effect. The discontinuities in the Strouhal number are accomplished by sudden changes in the shedding angle and by vortex splitting. The discontinuities and slanted shedding disappear if the cylinder boundaries are decoupled from the rest of the cylinder by using "end-cylinders". Furthermore, the base pressure and velocity profiles of the wake near an end plate are presented.

1 Introduction

The flow around cylinders is one of the classical problems of fluid mechanics and after over 100 years of research since Strouhal [20] it should be expected that there are few or no open questions left. Indeed, the drag of a cylinder in a flow was already calculated by von Kármán in 1911 [13] which caused his name to be associated with the vortex street since then. However, it is characteristic of a non linear fluid mechanical system that despite its geometrical simplicity, almost each analysis adds new information which does not necessarily complement the old. For example, in recent years numerical solutions of the flow around a cylinder have been published [2] which have added to a debate about the reliability of experimental data. Yet, it is also difficult to believe that all important parameters are taken into account by the numerics.

Experimental data has proved to be contradictive, one of the best examples is the debate over the so-called Tritton discontinuity [22]. Whereas Roshko [17] found the Strouhal number to depend continuously on the Reynolds number in the "regular" (50 – 160) as well as the "irregular" (300 – 2000) Reynolds number ranges, Tritton had observed discontinuities at Reynolds numbers of about 90.

The results of Gerich [9],[10] show that the Strouhal number, i.e. the shedding frequency, is not necessarily constant across the span of the cylinder, but might be different near the cylinder ends in the laminar, as well as the irregular [19] Reynolds number range. This "end effect" could also be observed by Ayoub and Karamcheti [1] near the free end of a cylinder at Reynolds numbers of the order of 10^5.

Some of the conflicts in experimental data can no doubt be attributed to the finite length of real cylinders, as the parameter aspect ratio is often not considered. Even worse, in theoretical examinations it is completely ignored when the flow considered as two dimensional, implying an infinitely long cylinder. Any doubts that a finite length will influence the flow were already dispelled by Wieselberger [25]. He showed that the drag coefficient of a cylinder depends on its aspect ratio!

The examinations presented in this paper were carried out with the aim of understanding the implications which arise from the fact that real cylinders have a finite length, i.e. how important is the parameter aspect ratio? Is it linked to the Tritton discontinuity? What causes the shedding frequency to be lower near the cylinder end than when far away from it? Why is the drag coefficient belonging to a long cylinder different from the one measured for a short one? How long must a cylinder be for the flow to be considered two dimensional? Is there a method of manipulating a cylinder so that even a short cylinder can lead to a 2d-flow? The results will be limited to the regular range $50 \leq Re \leq 160$.

2 Experimental Methods

The experimental techniques used were flow visualization, hot-wire anemometry and surface pressure measurements. The coordinate system was chosen so that x is the mean flow direction, y is perpendicular to the cylinder axis and the flow direction and z is parallel to the cylinder axis. The origin was located at one of the cylinder ends. The flow visualization was based on the continuously operating smoke-wire, an improvement over the regular smoke-wire which only operates intermittently introduced by Gerich [11]. An oil or, even better because less persistant and healthier, "disco fog fluid" flows onto the smoke-wire through a hollow needle. A strong current (about 3 A) heats the resistance wire so that the fluid evaporates only to condense again into a smoke sheet consisting of droplets of about 1 μm in diameter. The diameter of the smoke-wire is 0.1 mm so that in the velocity range where the system works, about $0.6-3$ m/s, no vortex street is produced by the smoke-wire itself. As the oil droplets which continuously feed the wire are driven by gravity, the wire must be placed vertically inside the wind tunnel. To have either a section or a top view of the flow around the cylinder, it must be mounted horizontally or vertically respectively. Very little light is scattered by the smoke sheet, especially at higher velocities, so the most important factor in producing satisfactory pictures is the illumination. For the pictures presented here, two electronic flash units were used in parallel. It is also important to make sure the wind tunnel windows and the background are kept out of the direct light, otherwise the smoke sheet will become weak in contrast.

Standard hot-wire anemometers (DANTEC 55M10) were used with 5 μm sensors. The output was linearized with 55D10 units when necessary. The linearizers were adjusted to allow for the deviation from King's law at low velocities. The probe was mounted on a traversing mechanism which could be computer driven. At the same time integrating digital voltmeters (PREMA 6031S) collected the data which was passed on to the computer through an IEEE interface. With this setup it was possible to measure automatically velocity profiles as well as profiles of the mean fluctuation. Frequency was determined with the help of a Nicolet Scientific FFT Analyser (Model 660A). Because the FFT analyser always averages over a certain period of time, frequency was also measured using a phase-locked-loop circuit which tracks the input with a voltage controlled oscillator. The voltage from the secondary oscillator is a measure of the frequency so that the variation of the shedding frequency with time could be seen. In addition, as phase jumps in the time signal from the hot-wire are due to vortex splitting, these could also easily be detected [7].

The surface pressure measurements necessitated a compromise between the size of the pressure tap and the range of dynamic pressure. To measure at a given low Reynolds number it is possible to use a rather large cylinder at very low velocities, or a smaller cylinder at higher velocities. It was found that hollow cylinders of 1 and 1.5 mm diameter with a hole in the surface of 1/10 this size yielded satisfactory results. The cylinder could be turned to measure the base pressure or the stagnation pressure, or even the complete distribution around the cylinder. It could also be moved back and forth, thus moving the pressure tap closer to or further away from the wall or an end plate. The manometer used was an MKS Instruments type 398HD Baratron with a full scale reading of 1 $mmHg$. At the lowest pressures measured the instrument drift was frequently checked and for a reproducible reading integration times of several minutes were necessary.

3 Observed Phenomena

As briefly mentioned in the introduction, the end effects manifest themselves in a variety of different phenomena. The best known is the susceptibility of the shedding frequency to different end conditions. Various other observations will also be presented in this section.

Figure 1 is taken from [14] and shows the Strouhal number as a function of the Reynolds number. The measurements were taken in the plane of symmetry, midway between the cylinder ends. Cylinders of different diameter and aspect ratio were used. The left hand plot shows the data following a continuous curve, similar to the one originally proposed by Roshko [17]. It is also very similar to data from Williamson [26] and

Hammache and Gharib [12]. It will be seen later, that the data follows a continuous curve when the shed vortex axes are parallel to the cylinder. End cylinders [4] were used to promote parallel shedding for the data presented here. The wake of a cylinder in which vortices are shed parallel to the cylinder is shown in a flow visualization on Figure 2a.

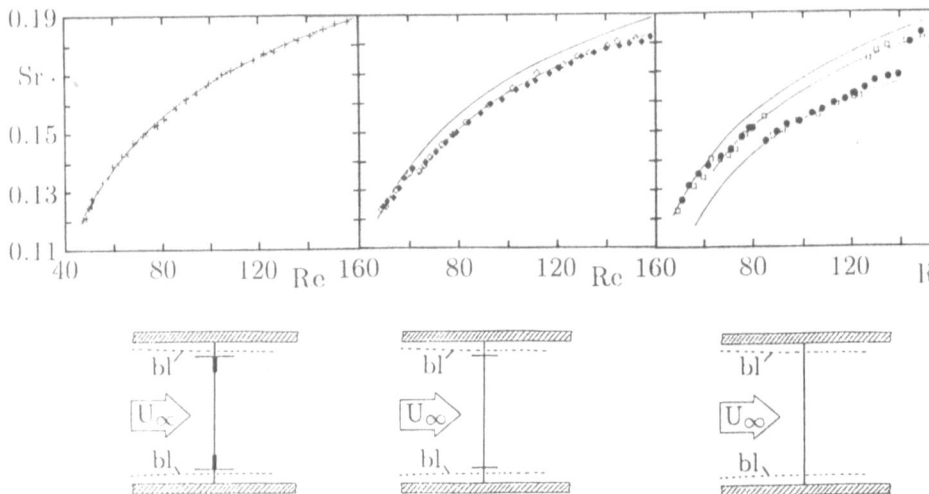

Fig. 1   The Strouhal number at midspan position in the wake of a long (aspect ratio > 100) cylinder as a function of the Reynolds number for different end conditions. The end conditions are sketched on the bottom. In the left hand diagram, the cylinder is bounded by end cylinders and the curve is continuous. The middle diagram shows data when the cylinder is bounded by end plates. There is a discontinuity at $Re \approx 65$. In the right hand diagram the cylinder is placed directly between the wind tunnel walls where a large boundary layer is present. There are several discontinuities.

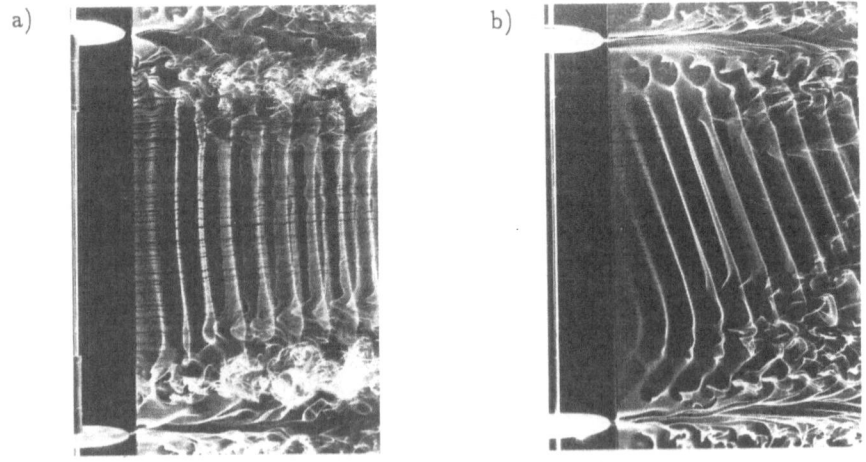

Fig. 2   Smoke-wire flow visualization of the wake of a cylinder, $Re \approx 130$. a) With end cylinders, vortex shedding is parallel; b) With end plates only, vortex shedding is slanted.

210

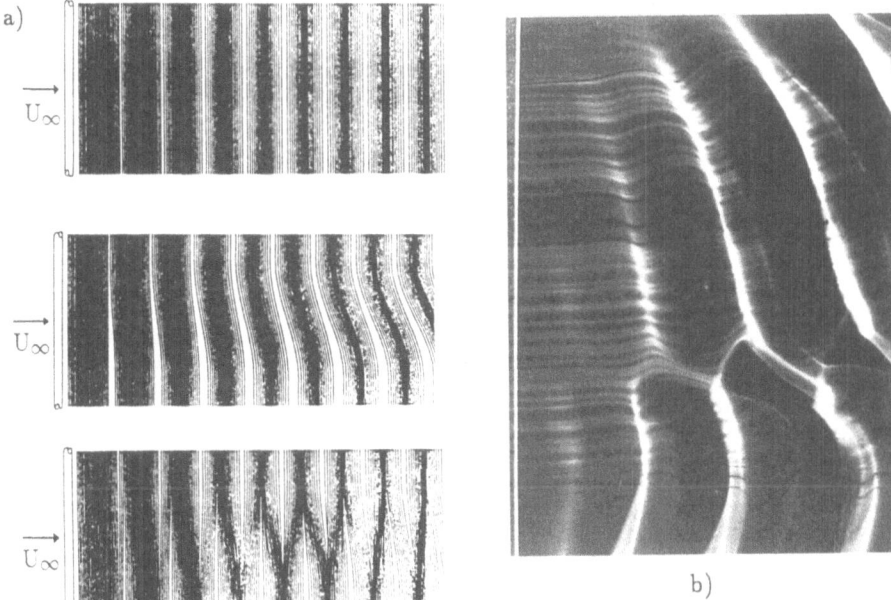

Fig. 3 a) Sketch of the vortex splitting mechanism. Vortex lines separate from the cylinder to agglomerate in vortices. In the top sketch the frequency is constant across the span. If this is not the case, the shedding angle grows more and more as time goes by (middle), or vortex splitting occurs (bottom);
b) Vortex splitting in the wake of a thin flat plate. The trailing edge can be seen on the far left. The smoke wire is upstream and on the side of the leading edge. The Reynolds number based on the plate thickness is 200.

End plates were used as cylinder boundaries in the middle graph of Figure 1. For this boundary condition, vortex shedding is slanted, Figure 2b. The data falls below the line which approximates the data of the left hand graph. This means the shedding frequency is lower, despite the fact that the same cylinders as in the top case were used. It is therefore an effect of the end plates, which extends over a distance much larger than can be expected. Another point to note is that at Reynolds numbers below about 65, the data from the end cylinders and that from the end plates is identical. At such low Reynolds numbers, the end effect does not come into play. Such a discontinuity at $Re \approx 65$ was also found by Williamson [26].

For the data in the right hand graph of Figure 1, the cylinder extended all the way to the wind tunnel walls. Again vortex shedding is slanted. Instead of a discontinuity only at $Re \approx 65$, there are now two further jumps in the curve. One is a transition from a higher frequency mode to a lower one, while the other jump is exactly the reverse. The Reynolds number at which these transitions take place is dependent on the aspect ratio. Tritton [22] was the first to observe this phenomen and it provoked some controversy in the literature. Some of the mechanisms which can lead to such discontinuities are different modes of vortex formation [23], non-uniformities in the oncoming flow [8], turbulence level in the oncoming flow [3], windows of chaos [18] and vibration of the cylinder [24]. Williamson [27] shows the importance of three-dimensional mechanisms and attributes the discontinuities to changes in the shedding angle which in turn is linked to the shedding frequency. König, Eisenlohr and Eckelmann [15] were able to show that various shedding angles are always present across the span of the cylinder for certain types of boundary conditions. For example, when the cylinder is bounded by end plates, there is a cell of lower shedding frequency near the end plate [10]. In this cell the shedding angle is larger than outside it. This can be seen on the picture of the visualized flow, Figure 2b. When the cylinder ends in the wind tunnel walls, there are sometimes three or perhaps even more cells of different angle and shedding frequency [15]. When the Reynolds number changes, the cell boundaries move across the span. If they meet in the plane of symmetry in mid-span, they disappear and a probe placed at that location detects a discontinuity.

At the cell boundaries, by definition, two shedding frequencies meet. Due to the conservation of vorticity, it is not possible for a vortex to "end" in the flow. Instead, a mechanism called vortex splitting takes place [4],[5]. This process is illustrated in Figure 3a, where vortex lines shedding from the cylinder are drawn. The lines agglomerate to form vortices on their way downstream. In the top sketch, the shedding frequency is constant across the span of the wake: shedding is parallel. In the middle sketch the frequency is slightly different across the span. The vortex axes must become more and more slanted as time goes by. Finally, vortex splitting as shown on the bottom takes place. Figure 3b shows the visualized flow in the wake of a bluff body where vortex splitting is taking place.

It was shown in both the frequency measurements and the flow visualization that the geometry of the cylinder ends could be felt many hundreds of diameters from the end itself. It was therefore also of interest to see if and how the wake velocity profile was also influenced [6]. Figure 4 shows the wake profile as measured with a hot-wire anemometer. Near the end plate the velocity defect is much larger than further away from it. The node, i.e. the cell boundary between two shedding frequencies, can be made out at $z/d \approx 9$. Figure 5 shows the centerline velocity profile extracted from Figure 4 in conjunction with the Strouhal number and the base pressure. Stable values for the velocity and pressure are reached only for $z/d > 20$. The fact that the base pressure shows a z-dependancy which is not monotonously increasing near the cylinder end indicates that it is *influenced by the vortices in the wake*. There are no geometrical obstacles in the flow which could lead to a pressure dependance as shown. It is clear that pressure, velocity and frequency are correlated.

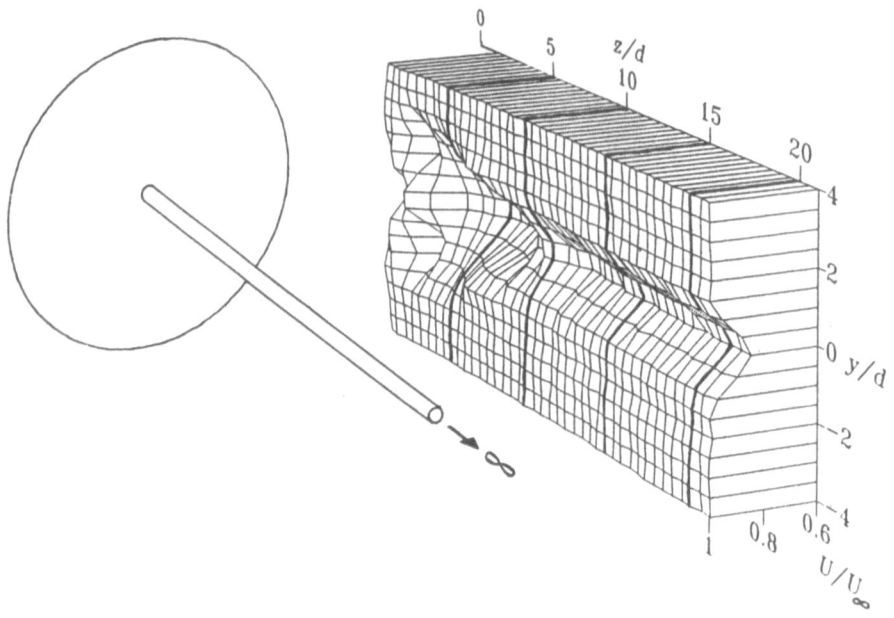

Fig. 4   Perspective view of the wake velocity profile near an end plate, $Re = 130$, probe at $x/d = 10$. Note that the profile is drastically distorted even at quite some distance from the end plate.

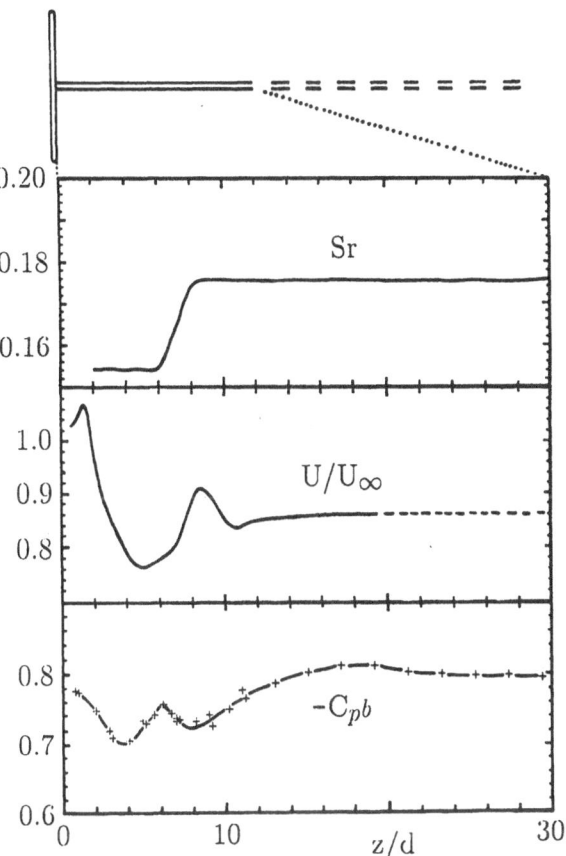

Fig. 5 Strouhal number, mean flow velocity at $y/d = 0$ and base pressure as a function of distance from the end plate. Note how the three values are correlated: they are intimately linked together.

4 Origin of the end effect

A possible explanation of how the velocity profile is distorted and how the vortices in the wake can feed back upstream was derived from the visualized flow shown in Figure 6a. A sketch of the vortex lines associated with the formation of vortex loops is shown in Figure 6b. Near the end of the cylinder, the vortices from either side of the cylinder short circuit. In this way, the necessary condition that vortex lines must always be closed is satisfied. This model was already proposed by Taneda [21], but its implications for the existance of an end effect are only now becoming clear.

All vortices induce a velocity field around them which can be calculated with the help of Biot-Savart's law. When vortices become curved, the induced velocity field becomes focussed. This is the reason why vortex rings are able to propagate for great distances: the ring is constantly feeding its forward motion from the velocity field induced by the rotation. A similar motion can therefore be expected from the loops sketched in Figure 6b. The sign of rotation is such that the loops near the cylinder ends will induce a velocity field which acts upstream against the oncoming flow. This will cause the vortex loops to convect downstream at a lower velocity than the convection velocity of the vortices which form at a distance from the cylinder ends. This in turn explains why the velocity profile is distorted the way it is: the velocity defect is much more pronounced near the ends than far away from them.

The distorted velocity profile also will cause the shedding frequency to vary across the cylinder span. Cells of different shedding frequency form. Noack, Öhle and Eckelmann [16] have shown how the formation of cells is a process which can be derived from relatively simple models. The slanted shedding is also a direct result of the wake profile. A vortex axis must become tilted as it extends through regions of differing convection velocities. There is, however, a mechanism which stops the tilt angle from growing forever larger: vortex splitting. Unfortunately, it is not yet known what defines the angles which can be observed.

a)
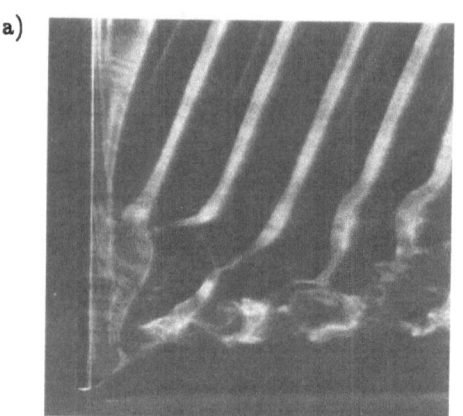
b)
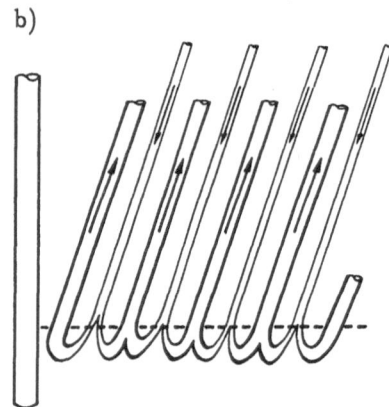

Fig. 6 a) Flow visualization of the flow around a cylinder with a free end at $Re = 140$. Vortex splitting can be observed, as well as the separation of vortex loops; b) Sketch of the formation of vortex loops near the free end of a cylinder. The flow is coming from the left hand side.

5 Conclusions

Different geometries at the cylinder ends give rise to different values of the Strouhal number as a function of the Reynolds number. The Tritton discontinuity is due to a spanwise movement of cells of different frequency. At the cell boundaries, vortex splitting takes place. The appearance of cells is due to the implications of Helmholtz's Law: the vortex lines near the cylinder end must be closed. Vortex loops which separate from the cylinder form. The velocity field induced by the loops gives rise to a distortion in the wake profile which leads to a region of lower shedding frequency as well as slanted shedding.

References

[1] Ayoub, A.; Karamcheti, K.: An experiment on the flow past a finite circular cylinder at high subcritical and supercritical Reynolds numbers, J. Fluid Mech. **118**, 1-26, 1982.

[2] Braza, M.; Chassiang, P.; Ha Minh, M.: Numerical study and physical analysis of the pressure and velocity fields in the near wake of a circular cylinder, J. Fluid Mech. **165**, 79-130, 1986.

[3] Berger, E.; Wille, R.: Periodic flow phenomena, Annu. Rev. Fluid Mech. **4**, 313-340, 1972.

[4] Eisenlohr, H.; Eckelmann, H.: Vortex splitting and its consequences in the vortex street wake of cylinders at low Reynolds numbers, Phys. Fluids A **1**, 189-192, 1989.

[5] Eisenlohr, H.; Eckelmann, H.: Visualization of three-dimensional vortex splitting in the wake of a thin flat plate, *Flow Visualization V*, R. Režniček, editor, 338-343, Hemisphere, New York 1990.

[6] Eisenlohr, H.: Ein kurzer oder ein langer Zylinder, worin liegt der Unterschied für die Kármánsche Wirbelstraße?, Mitt. Max-Planck-Inst. f. Strömungsforschung, Nr. 98, Göttingen 1990

[7] Eisenlohr, H.; König, M.; Eckelmann, H.: Eine Methode zur Erkennung von Frequenzzellen im von Randeffekten beeinflußten Zylindernachlauf, Inst. f. Luft- und Raumfahrt, Mitt. Nr. 260, 77–84, TU Berlin 1991

[8] Gaster, M.: Vortex shedding from slender cones at low Reynolds numbers, J. Fluid Mech. **38**, 565-577, 1969.

[9] Gerich, D.: Einfluß von Endscheiben und freien Enden auf den Nachlauf von quer angeströmten Zylindern, Diplomarbeit, Inst. für Angewandte Mechanik und Strömungsphysik, Universität Göttingen, 1979.

[10] Gerich, D.; Eckelmann, H.: Influence of end plates and free ends on the shedding frequency of circular cylinders, J. Fluid Mech. **122**, 109-121, 1982.

[11] Gerich, D.: Über den kontinuierlich arbeitenden Rauchdraht und die Sichtbarmachung eines Übergangs vom laminaren zum turbulenten Nachlauf, Max-Planck-Institut für Strömungsforschung, Bericht 104, Göttingen 1987.

[12] Hammache, M.; Gharib, M.: A novel method to promote parallel vortex shedding in the wake of circular cylinders, Phys. Fluids A **1**, 1611-1614, 1989.

[13] Kármán, T. v.: Über den Mechanismus des Widerstandes, den ein bewegter Körper in einer Flüssigkeit erfährt, Nachr. der K. Ges. d. Wissensch. zu Göttingen, Math.-phys. Klasse, 1911.

[14] König, M.; Eisenlohr, H.; Eckelmann, H.: The fine structure in the Strouhal-Reynolds number relationship of the laminar wake of a circular cylinder, Phys. Fluids A **2**, 1607-1614, 1990.

[15] König, M.; Eisenlohr, H.; Eckelmann, H.: Visualization of spanwise cellular structure of the laminar wake of wall-bounded circular cylinders, Phys. Fluids A **4**, to appear May 1992

[16] Noack, B.R.; Ohle, F.; Eckelmann, H.: On cell formation in vortex streets, J. Fluid Mech. **227**, 293-308, 1991.

[17] Roshko, A.: On the development of turbulent wakes from vortex streets, NACA Report 1911, 1954.

[18] Sreenivasan, K.R.: Transition and turbulence in fluid flows and low-dimensional chaos, "Frontiers in Fluid Mechanics", S.H. Davis and J.L. Lumley, eds., Springer Berlin, 1985, 41-66.

[19] Stäger, R.; Eckelmann, H.: The effect of end plates on the shedding frequency of circular cylinder in the irregular range, Phys. Fluids A **3**, 2116-2121, 1991.

[20] Strouhal, V.: Über eine besondere Art der Tonerregung, Ann. der Physik und Chemie, N.F. Bd. V, 1878.

[21] Taneda, S.: Studies on wake vortices (I). An experimental study on the structure of the vortex street behind a circular cylinder of finite length, Rep. Res. Inst. Appl. Mech. **1**, 131-143, 1952.

[22] Tritton, D. J.: Experiments on the flow past a circular cylinder at low Reynolds numbers, J. Fluid Mech. **6**, 547-567, 1959.

[23] Tritton, D. J.: A note on vortex streets behind circular cylinders at low Reynolds numbers, J. Fluid Mech. bf 45, 203-208, 1971.

[24] Van Atta, C.W.; Gharib, M.: Ordered and chaotic vortex streets behind circular cylinders at low Reynolds numbers, J. Fluid Mech. **174**, 113-133, 1987.

[25] Wieselberger, C.: Neuere Feststellungen über die Gesetze des Flüssigkeits- und Luftwiderstandes, Phys. Zeitschr. **22**, 321-328, 1921.

[26] Williamson, C.H.K.: Defining a universal and continuous Strouhal-Reynolds number relationship for the laminar vortex shedding of a circular cylinder, Phys. Fluids **31**, 2742-2744, 1988.

[27] Williamson, C.H.K.: Oblique and parallel modes of vortex shedding in the wake of a circular cylinder at low Reynolds numbers, J. Fluid Mech. **206**, 579-627, 1989.

ON THE FLOW AROUND FINITE CIRCULAR CYLINDERS IN TURBULENT SHEAR FLOWS

N. Hölscher, H.-J. Niemann

Ruhr-Universität Bochum, Institut für Konstruktiven
Ingenieurbau
Aerodynamik im Bauwesen
Universitätsstr. 150, D-4630 Bochum, Germany

SUMMARY

The mean flow pattern around a finite span circular cylinder
reveals several stationary flow separations. The unsteady flow
is dominated by the effect of the approaching turbulent shear
flow and by an alternating vortex shedding, which is significantly altered under these three-dimensional conditions.

INTRODUCTION

The flow around finite span bluff bodies in turbulent shear
flows is always three-dimensional, accompanied by flow separations, running in various stationary vortex systems which
interact with each other. The associated separation cavities
are not bounded by a closed surface of mean streamlines. Instead, the mean recirculating flow entrains fluid into the near
wake, which is convected downstream by means of wake vortices.
The unsteadiness of such kind of flows is in principle due to
velocity fluctuations in the near wake and to a periodic vortex
street which forms further downstream. Additional fluctuations
are introduced in the case of a turbulent approach flow. In
most cases this contribution is dominant and obscures the
separation induced flow features. The present experimental
study deals with the flow around finite span circular
cylinders, which are totally immersed in a simulated atmospheric boundary layer flow. Besides the mean flow pattern,
mainly the flow dynamics and its effect on surface pressure
fluctuations are investigated.

BOUNDARY LAYER CHARACTERISTICS

All experiments are conducted in a boundary layer wind tunnel.
Following Counihan's method, different disturbing bodies provide a sufficient simulation of atmospheric flow conditions,
[1]. The mean velocity profile of a flow over rural to urban

terrain is approximated by an empirical power law with a profile exponent of $\alpha \approx 0.18$, a wall friction velocity $u_\tau /u_\delta \approx 0.055$, a roughness length $z_0 \approx 1.34$ mm and a boundary layer height $\delta \approx 1.20$m. Maximum turbulence intensities are in the order of $I_u \approx 20$ % with respect to the longitudinal component. Reynolds stresses correspond to correlation coefficients less than $\varrho_{uw} \lesssim -30\%$. Kármán spectra are fitted to spectral distributions of the velocity fluctuations to obtain principal length scales as $^x L_{ii}$, $i = u$, v, w. Additionally, an appropriate statistical model about the spatial and temporal structure of the simulated shear flow has been developed, adapting the theory of homogeneous isotropic turbulence to measured two point statistics, [2].

MEAN FLOW PATTERN

By means of oil film technique, surface flow patterns on finite circular cylinders with smooth surface have been visualized for a variety of different aspect ratios and upstream boundary layer conditions. Further smoke visualizations reveal some ideas about the instantaneous flow behaviour, [3]. Surface pressure measurements are confined to those configurations, which are typical for the mean flow pattern, [4]. In order to prevent effects of the Re-number, all but one pressure measurements are conducted on models with a rib roughness.

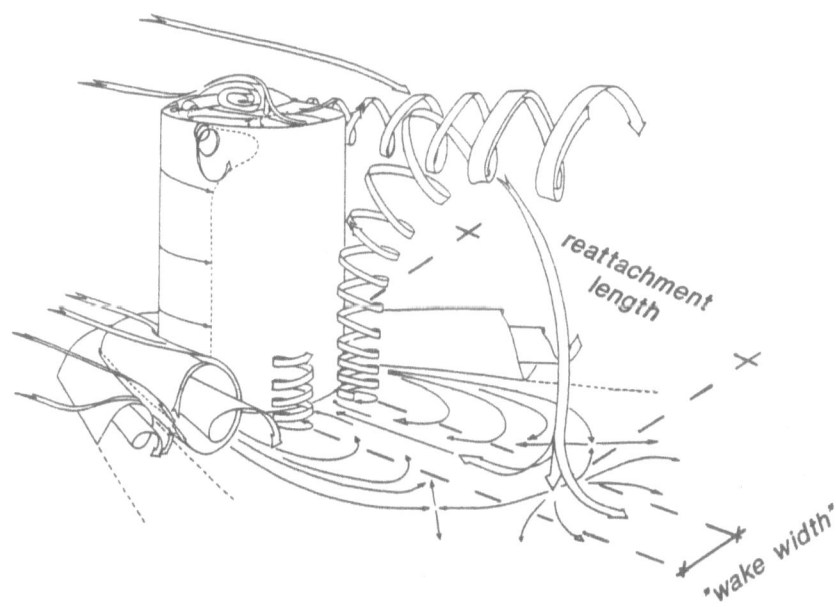

Fig. 1: The three-dimensional flow pattern

Similar to Hunt's sketch [5] about the flow around cubes, fig. 1 illustrates the time averaged structure of the separated flow. The boundary layer approaching the cylinder forms in the plane of symmetry a stagnation point at a relative model height of ≈85%. As surface pressure distributions confirm, Fig. 2, this position is approximately constant for all cylinders investigated, [6]. According to the spanwise pressure gradient in the stagnation line, the flow subdivides into an upward part, separating at the curved leading edge, and a downward flow, which is oblique to the cylinder's axis. Upstream the base of the cylinder, an increasing pressure causes the flow to separate, [7]. A horseshoe vortex system with two main vortices builds up and trails downstream. Positions of both, primary and secondary separation line and thereby size and strength of the involved vortices are significantly sensitive to the approaching flow conditions. Tendentiously, the higher the shear the greater the total separated flow region. Particularly, the primary vortex grows until the secondary vortex is completely displaced. Increasing turbulence and Reynolds number seems to prevent early separation. For the parameter investigated, the horseshoe vortices measured in the plane of symmetry are $L_1/R \approx 0.7 \div 0.9$ and $L_2/R \approx 1.4 \div 1.6$ (distances between lines of separation and the cylinder front). In the spanwise direction the horseshoe vortex reaches up to $z/H \approx 0.2$. Local surface pressures exceed the velocity pressure at the same level, $C_p(z/H \leq 0.2) > 1$, fig. 2.

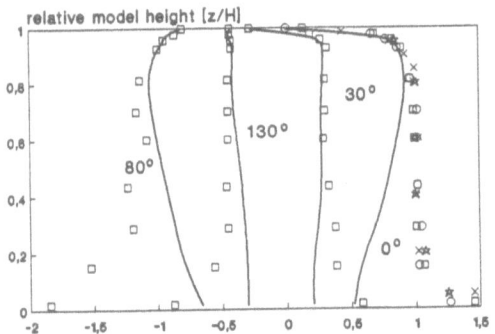

	*	X	O	□	⊖
α	0.18	0.18	0.14	0.18	0.18
δ/H	2.	2.	3.	3.	3.
smooth	x				
rough		x	x	x	x
	$c_p(q_z)$				$c_p^*(q_H)$

Fig. 2: Mean surface pressure distributions along the span of finite circular cylinders

On top of the cylinder, the leading-edge separation forms a sickle shaped flow region with two counter-rotating vortices. Due to the turbulent momentum flux, the approaching flow reattaches. This is accompanied by a strong pressure recovery. A reversal flow covers the top surface. The associated reattachment line is curved upstream and meets the spanwise flow out of the leading edge vortices presumably in a node and saddle-point of separation. Obviously, fluid out of the leading edge vortices, but mainly out of the reattached downstream flow lift off at the circumference. Thereby, the top surface pressures close to the trailing edge are greater than the pressures on the superficies. The separated shear layers are dragged along, roll up and form a conical tip vortex on each side with horizontal axis, eventually accompanied by two additional smaller vortices.

Outside the end regions, for moderate aspect ratios H/D≳ 2, flow separates nearly independent of height. In between the free shear layers, close to the rear face of the cylinder two vertical stationary vortices develop. Additional fluid is entrained into these vortices from the recirculating wake flow. There is a critical aspect ratio where the lateral distance of the vertical wake vortices is maximum. Figure 3 summarizes the characteristic length scales of the flow separations from a finite span circular cylinder.

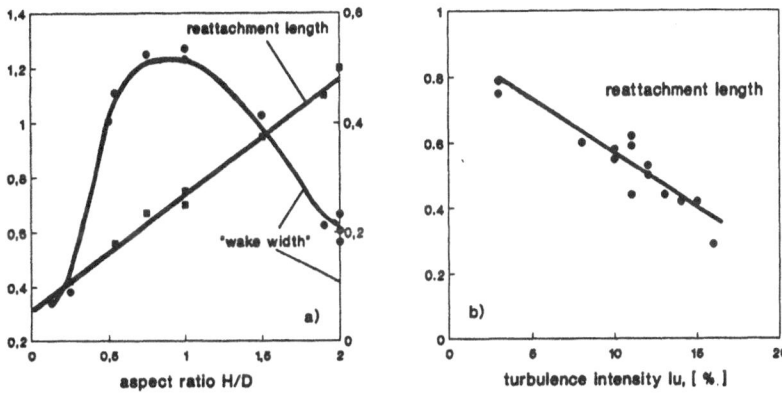

Fig. 3: Characteristic length scales of the flow separations
(flow pattern a) on the ground plate, b) on the top,
for scale definitions see fig. 2)

FLOW DYNAMICS

For a finite circular cylinder, but still in uniform flow, the regular vortex street is significantly disturbed, [8]. It is widely accepted that for tall cylinders with H/D≳ 7, quasi two-dimensional vortex shedding occurs only below ≈4·D from the tip due to the effect of the free end. However, the Strouhal number is considerably reduced compared to an infinite cylinder, [9].

For the flow in the tip region there are two interpretations. One assumes a pair of stationary vortices, arising as a coupling of the blow-down flow from the free-end and the separated up-wash flow around the cylinder's surface, [10]. The other interpretation follows the idea of alternating vortex shedding with a frequency decreasing stepwise as the free end is approached, [11]. This notion is supported by the observation that the lift fluctuations have a maximum close to the tip.

For small aspect ratios no spanwise variation of the separation frequency is discernible, [12]. However, the axial coherence of the vortex shedding is effectively destroyed by the recirculating wake flow. Therefore, the lift fluctuations are significantly smaller than the drag fluctuations, which are primarily due to an oscillation of the wake. Its characteristic frequency is by a factor of ≈ 1.3 larger than that of the vortex shedding process.

Further decorrelation is introduced in the case of a turbulent boundary layer flow. The vortex shedding is modulated into a random process. Particularly, the spectral peak of lift fluctuations is drastically reduced and replaced by random narrow-band fluctuations. With decreasing aspect ratio, a horizontal vortex, separating from the trailing top edge, joins the flow, which separates from the flanges. Instead of a Kármán-type vortex, an arch-type vortex establishes. In this arrangement vortices trail downstream symmetrically. For a circular finite span cylinder in subcritical flow a critical aspect ratio of $H/D \approx 2.5$ is quoted, [13]. This value is reduced for increasing Re-number, [14] for square cylinders. In the present investigation with an aspect ratio of 2.0, no arch-type vortex was observed, since the Reynolds number was $Re(u_B) \approx 3 \cdot 10^5$, (u_B = mean velocity at the top of the cylinder).

Typical features of pressure fluctuations on short circular cylinders in simulated atmospheric boundary layer flow are:

• Due to the approaching turbulent flow, maximum pressure fluctuations occur at the stagnation line. Except for the end regions, the vertical distribution of the rms-pressures is approximately proportional to the profile of turbulence intensity $I_u(z)$, fig. 3. Around the circumference, pressure fluctuations decreases slowly, with a slight maximum close to the position of minimum pressure. There is no significant rms-peak at the mean position of flow separation. In the wake rms-values drop drastically to about 50% of the fluctuations in the stagnation region. This indicates, that the influence of the impinging turbulence on the wake is quite weak. At the free end of the cylinder, local fluctuation maxima occur due to the tip vortex. On the top, maxima are localized in that area, where the mean recirculating flow and fluid out of the leading edge separation whirl together. Close to the base rms-values are more or less constant even in the separated flow region.
• The normalized pressure spectra $S_{cp}(f)/\sigma^2$ vary with height and circumferential position. The effect of height disappears

approximately using a non-dimensional frequency with $^xL_u(z)$ as reference length. This does not apply to the spectral peaks induced by vortex shedding, which depend on the diameter D. On the windward side the pressure spectra are an attenuated form of the velocity spectrum regarding the high frequency range. On the other hand, low frequency pressure fluctuations exceed the amplitudes of the velocity spectrum. These effects may be caused by the distortion of turbulent eddies. Further down-stream, the influence of the oncoming turbulence is superim-posed by vortex shedding and finally decreases further in the wake.

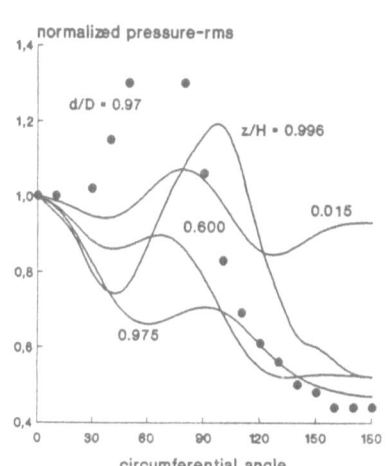 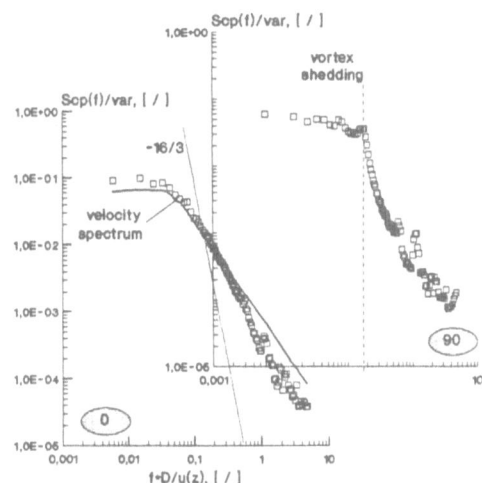

Fig. 3: RMS-values and spectra of fluctuating pressures
(H/D= 2.0, δ/H= 3.0, ^xL_u /D= 2.0, α= 0.18,
Re(u$_H$)= 3·10^5, rib roughness, spectra at z/H= 0.6)

• Pressure fluctuations, induced by an asymmetrical vortex shedding, result in negative correlations between pressures on the cylinder sides. On the contrary and in particular, fluc-tuations, induced by large scale turbulence, are positively correlated. Furthermore cross correlation functions are symmetrical with respect to the time lag 0. Accordingly, in the frequency domain the quadratur spectrum is negligible. Remar-kably, close to the regular vortex shedding frequency the phase is shifted up form 0 to its maximum π and keeps constant until a sharp drop to the original value, Fig. 4. Whereas the zero phase is caused by the approaching turbulence, a phase of π is related to quasi-periodic vortex shedding. This interpretation is confirmed by the coherence function, which at first decreases continuously from its low frequency limit of almost unity. In the narrow frequency band dominated by alternating flow separation, the coherence exhibits a significant increase. By means of such cross-spectral analysis it has been found, that the bandwidth of separation induced pressure fluctuations is nearly constant between φ≈ ±45° ÷ ±175°. In the stagnation or base pressure region, such a distinct shape of the phase spectra is not discernible, so that no significant influence of the vortex shedding is expected. Fig. 4 reveals a slight spanwise variation of the bandwidth of quasi-periodic vortex shedding.

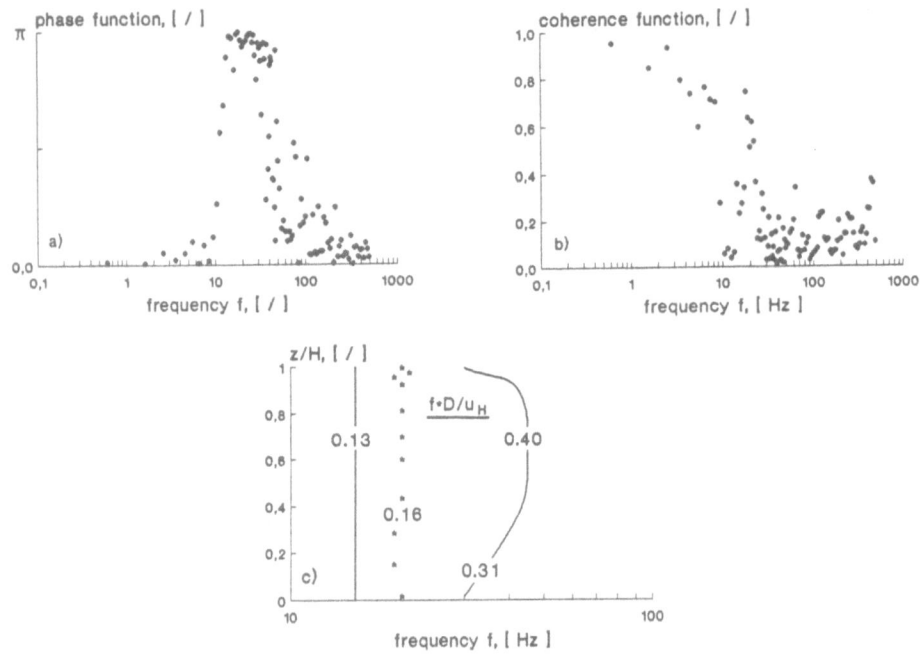

Fig. 4: Influence of vortex shedding on surface pressure fluctuations at symmetrical positions, a) phase, b) coherence function, c) bandwidth, (u_H= 20.75 m/s, D= 0.2 m, z/H= 0.6, φ= ±120°, α= 0.18, rib roughness)

• The spectral contribution of vortex shedding is shown in fig. 5. The partial spectra are unsymmetrical with respect to the peak value. This peak occurs always at the lower limit of the frequency band. The associated Strouhal number is constant across the cylinders surface Sr= $f \cdot u_H$/D≈ 0.16. The rms-pressures induced by vortex shedding relative to the total rms-pressures yield a maximum of ≈45% nearby φ≈90°. In the wake, this contribution levels to ≈35%. Except for the end regions, where a constant value of ≈20% has been found, these relations and furthermore the Strouhal-peak of the partial pressure spectra are more or less independent of height. For taller structures it is known, that with increasing turbulence intensity the peak value decreases and the bandwidth increases with respect to height, [15]. For the circular cylinder in-vestigated (H/D= 2.0), such a relation could not be confirmed.

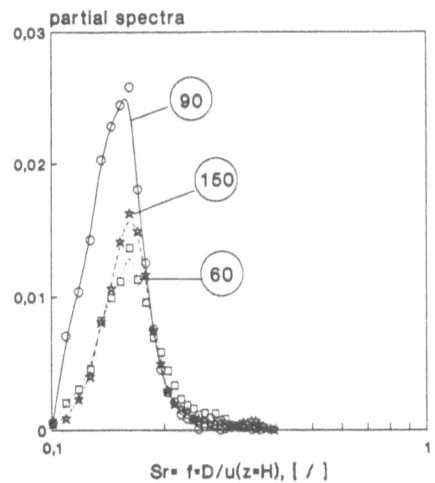
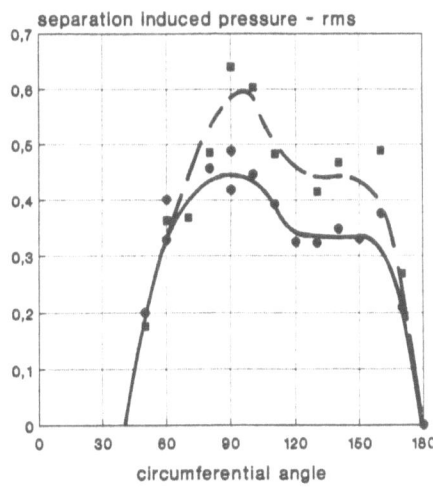

Fig. 5: Separation induced pressure fluctuations,
a) pressure spectra: $S(f)/\sigma^2$, b) pressure-rms:
——— $\sigma^2_{vortex}/\sigma^2_{total}$, peak value ---$S(f_{sr})\cdot f_{sr}/\sigma^2_{total}$

• The pressure fluctuations are not perfectly correlated in
spanwise direction. Due to the distortion of turbulent eddies,
at the stagnation line, pressure correlation exceed the vertical correlation of longitudinal velocity fluctuations considerably. Around the flanges, the turbulent eddies are further
elongated. Therefore, spanwise pressure correlation lengths for
a reference point at mid height have a maximum of $\approx 0.8\cdot H$ close
to the minimum mean pressure. Even in the wake, correlation
lengths are comparatively high. Furthermore, significant
correlations between pressures in the stagnation and wake region are evident. As suggested by [9] this indicates, that
turbulent eddies penetrate into the wake. At first, small scale
fluctuations increase the instability of the separated free
shear layers. Thereby the cavity is accessible to large scale
eddies.

This interpretation of an indirect turbulence influence is
confirmed by frequency dependent correlation lengths. They are
determined on the basis of the modulus of cross-spectral den-
sities. It is inferred, that the correlation pattern is maintained especially by low frequency, respectively large scale
fluctuations. In the frequency range influenced by the vortex
shedding a significant increase of the axial coherence is
evident. An associated coherence length of $\approx 0.85\cdot H$ is evaluated. Quasi-periodic flow separation occurs without any phase in
spanwise direction. However, in the time domain the onset and
the breakdown of vortex shedding does not occur simultaneously
along the cylinder span.

FINAL REMARKS

The investigations of the fluctuating flow so far, have been concentrated on the influence of the oncoming turbulence and on the quasi-periodic vortex separation. It is recommended to extend this study to the wake flow. In addition to surface pressure measurements more direct methods as e.g. the sublayer fence technique [7] should be employed.

ACKNOWLEDGEMENT

The financial support of the German Research Foundation (DFG) is gratefully acknowledged. By means of the research group "Physik abgelöster Strömungen" it provided a unique opportunity for a continuous and intense research. In particular, the cooperation of theoretical and experimental research has been extremely useful.

REFERENCES

1 **Wember G.**, Hitzdrahtanalyse dreidimensionaler turbulenter Strömungen, Ruhr-Universität Bochum, Studienarbeit, 1988
2 **Hölscher N.**, Ein multipler Ansatz für die aerodynamische Übertragungs-funktion der Winddrücke in atmosphärischer Grenzschichtströmung, Thesis, Ruhr-Universität Bochum, 1992, (is to be published)
3 **Kämpf K.-D.**, Laser-Lichtschnittaufnahmen an einem kurzen Zylinderstumpf, Ruhr-Universität Bochum, Studienarbeit Nr. 241, 1992
4 **Beigang L.**, Einfluß atmosphärischer Modellgrenzschichten auf die quasi-stationäre Belastung zylindrischer Bauwerke, Ruhr-Universität Bochum, Studienarbeit Nr. 149, 1988
5 **Hunt J.C.R.**, Kinematical studie of the flows around free or surface mounted obstacles, JFM 1978, Vol. 86, part 1
6 **Hölscher N., Niemann H.-J.**, Some Aspects about the Flow Around a Surface-Mounted Circular Cylinder in Turbulent Shear Flows, Int. Conf. on Turbulent Shear Flows, France, Toulouse, Sept. 1987
7 **Niemann H.-J., Hölscher N.**, Einsatz von Unterschichtzäunen in dreidimen-sionaler, hochturbulenter und abgelöster Strömung, DFG 1990, Ni 217/3-1
8 **Niemann H.-J., Hölscher N.**, A Review of Recent Experiments on the Flow Past Circular Cylinders, JWIA, No. 37, 1988
9 **Basu R.I.**, Aerodynamic Forces on Structures of Circular Cross-Section, Part 2, JWIA, 24(1986) 33-59
10 **Kavamura T., et.al.**, Flow around a Finite Circular Cylinder on a Flat Plate, Bulletin of JSME, Vol. 27, No. 232, 1984
11 **Farivar Dj.**, Turbulent Uniform Flow around Cylinders of Finite Length, AIAA 81-4056, Vo. 19, No. 3, March 1981
12 **Baban F., So R.M.C.**, Aspect ratio effect on flow-induced forces on cir-cular cylinders, Exp. in Fluids, Vol. 10, 1991
13 **Sakamoto H., Arie M.**, Vortex shedding from a rectangular prism and a circular cylinder, J. Fluid Mech., Vol. 126, 1983
14 **Okuda Y., et.al.**, Flow Visualization Around a Three Dimensional Square Prism, JWIA, No. 37, Oct. 1988
15 **Niemann H.-J.**, Dynamic response of cantilevered structures to wind tur-bulence, The European Conf. on Structural Dynamics, Bochum, June 1990

NON-PARAMETRIC IDENTIFICATION OF A
MULTIPLE AERODYNAMIC PRESSURE ADMITTANCE

N. Hölscher, H.-J. Niemann

Ruhr-Universität Bochum, Institut für Konstruktiven
Ingenieurbau
Aerodynamik im Bauwesen
Universitätsstr. 150, D-4630 Bochum, Germany

SUMMARY

Fluctuating pressures on bluff bodies in turbulent flows are
caused by the upstream turbulence and velocity fluctuations,
produced by flow separations and wake turbulence. Although the
gross properties of the flow around a body and in particular
the separated flow are affected by the intensity and the scale
of the incident turbulence, both sources of pressure fluctuations are thought to be statistically independent. On this
basis, a multiple aerodynamic pressure admittance has been
developed to account for turbulence induced pressure fluctuations in the first instance. This concept has been applied for
the flow around a finite circular cylinder in a simulated
atmospheric boundary layer flow.

INTRODUCTION

The aerodynamic admittance specifies the frequency dependent
relation of turbulent velocity fluctuations and the associated
wind load fluctuations. It is defined as the quotient of the
spectral quantities of the considered random processes. The
lattice plate theory estimates the force admittance by means
of the lateral coherence of the longitudinal velocity component, [1]. Concerning pressure fluctuations this concept implies a pressure admittance equal to unity for all frequencies. This simplification is approximately valid for pressures
in the attached flow region up to a cut-off frequency of
$f \cdot D/u \approx 0.1$, depending on both, the intensity and scale of the
approaching turbulence. In particular, the lattice plate
theory assumes a coherence of unity between the unsteady velocity and the pressures. But as a matter of fact, this coherence collapses beyond the cut-off frequency, which makes evident that pressure fluctuations are not only related to a
single stream filament. To overcome this deficiency and
thereby to clarify the cause and effect relation, a multiple
pressure admittance has been developed. It considers the
spatial and temporal influence of the oncoming turbulent flow
by means of its coherence to the observed pressures.

225

THEORETICAL BACKGROUND OF A MULTIPLE PRESSURE ADMITTANCE

A convolution integral renders possible a weighted super-position of local velocity fluctuations with regard to their spatial and temporal influence on the unsteady surface pressures. In the frequency domain, for a constant parameter linear system with arbitrary stationary random inputs, a multiple admittance function may be defined by a complex valued pair of equations, [2]:

$$S_{ip}(f) = (\varrho \cdot \bar{u} \cdot C_p) \; \underline{S_{ik}(f)} \cdot \underline{\chi_{ip}(f)} \tag{1}$$

$$S_{pp}(f) = (\varrho \cdot \bar{u} \cdot C_p)^2 \; \underline{\chi^T_{ip}(f)} \cdot \underline{S_{ik}(f)} \cdot \underline{\chi^*_{kp}(f)} \; + \; S_{ee}(f)$$

$$= S_{p:i}(f) + S_{ee}(f) \tag{2}$$

The notation is as follows:

$\underline{S_{ip}}$: vector of the velocity/pressure cross-spectra, (nx1)

$\underline{\underline{S_{ik}}}$: cross-spectra of the velocities, (nxn)

$\underline{\chi_{ip}}$: vector of the aerodynamic pressure admittance, (nx1)

S_{pp} : factual spectral density of local pressure fluctuations
$S_{p:i}$: conditioned spectral density cause by i velocity inputs
S_{ee} : spectral density of uncorrelated output noise
     (* : complex conjugate, T : transposed)
$\bar{u}$: arbitrary mean reference velocity
C_p : mean pressure coefficient, $C_p = \Delta p / (\tfrac{1}{2}\varrho \cdot u^2)$
ϱ : mass density

This complex equation represents a black box model where the multiple inputs are given by several velocity fluctuations at various upstream positions. With consideration of their spatial and temporal structure, $S_{ik}(f)$, each fluctuation is linearly transferred to a partial output by means of $\chi_{ip}(f)$. Altogether, those contributions sum up to the conditioned pressure spectrum $S_{p:i}(f)$. An additional noise spectrum $S_{ee}(f)$ accounts for all output components not contained in the linear admittance model. Physically, this noise term represents additional pressure fluctuations, introduced by flow separa-tions and turbulent velocity fluctuations in the wake. It is assumed, that both effects are incoherent with the oncoming turbulence. Furthermore, it is important to confine to those inputs only, which are physically meaningful in that they produce a significant effect on the output. The complete problem of identification subdivides into

- identification of significant velocity fluctuations
- estimation of the admittance vector
- prediction of the conditioned and residual pressure spectrum

THE CORRELATION BETWEEN VELOCITY AND
WALL PRESSURE FLUCTUATIONS

Measurements on a surface-mounted short circular cylinder
provide a data-basis to derive a multiple pressure admittance.
Velocity fluctuations of the longitudinal u(t) and lateral
component v(t) are measured across a cross-section in the undisturbed flow, simultaneously with all wall pressures
(height/diameter= H/D= 2.0, boundary layer thickness/height=
$\delta/H \approx 3.0$, $^x L_u /D \approx 2.0$, transcritical flow, atmospheric boundary
layer flow with a profile exponent $\alpha= 0.18$, [3]). On the
basis of these measurements, correlation between the velocity
and pressure fluctuations are estimated in the time and frequency domain. Typical features of the correlations between
the stagnation streamline and pressures at the same level are:

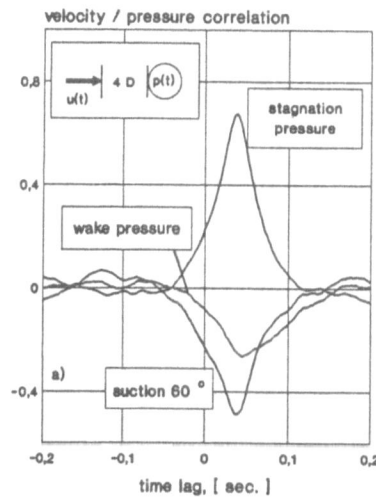
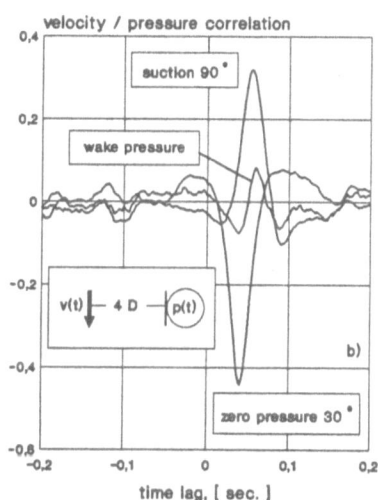

Fig. 1: Cross-Correlation between velocity fluctuations on the
stagnation streamline and unsteady circumferential
pressures: a) longitudinal, b) lateral component, [4]

- Peak values ϱ_p of the velocity/pressure cross-correlation
functions are significantly lower than the quasi-steady predictions [5]. They occur for a positive time lag, fig. 1,
which is determined by the time the mean flow needs to advect
turbulent eddies along a stream filament to the surface. Due
to the displacement effect of the body the advection velocity

is not identical with the mean velocity of the undisturbed flow.

• The cross-correlation functions decrease monotonously and symmetrically with respect to the peak ϱ_p (compare to the observed phase shift). There is no significant periodicity, which would indicate a correlation between turbulence and pressure fluctuations, induced by vortex shedding. This confirms the assumption of uncorrelated residual noise, introduced in eq. (2).

• Varying the upstream position of the reference point for the velocity r_x, in the undisturbed flow region the peak values ϱ_p remain nearly constant. This constancy mainly reflects a frozen turbulence pattern. However, a slight increase to a maximum value ϱ_{max} of peak correlations ϱ_p is observed as the body is approached. Hunt [6] relates the position of this maximum to the onset of the vorticity distortion, which results in a strong decorrelation of velocity and pressure fluctuations further downstream. In small scale turbulence, turbulent eddies are distorted within the mean displacement zone of the body $1.5 \gtrsim r_x/D$, whereas large scale turbulence is blocked at $1.0 \gtrsim r_x/^x L_u$. In the experiments with $^x L_u/D \approx 2.0$, the blockage effect dominates, so that maximum correlations are observed at $r_x \approx ^x L_u$. They are maintained by low-frequency velocity fluctuations with an average frequency of $f \cdot ^x L_u/u \approx 1.0$. For the prediction of the admittance function this correlation maximum ϱ_{max} is used further.

• In frozen turbulence, the upstream position of the reference point has no influence on the velocity/pressure coherence, too. On the other hand, even in the low frequency range, a continuous increase of the phase is observed. Up to a frequency limit of $f \cdot r_x/u \approx 0.5$ this phase shift equals the quasi-steady prediction $\varphi(f) = 2\pi \cdot f \cdot r_x/u$. This limit holds approximately for all circumferential angles and both velocity components. It is concluded, that only in this low frequency range the advection velocity of turbulent eddies is equal to the mean value.

• Velocity / pressure correlations strongly depend on the circumferential angle of the pressure point. Roughly, points with positive (negative) mean surface pressures have positive (negative) correlations with the longitudinal velocity. For the lateral component this change in sign occurs close to the extremes of the mean surface pressures, $(\partial C_p/\partial \varphi = 0)$. Smallest correlations are found for the wake pressures, where the influences of vortex shedding and wake turbulence are dominant.

• According to the quasi-steady theory, the velocity/pressure coherence is determined by a constant factor, which depends on the mean circumferential pressure distribution, and the quotient of the longitudinal and lateral velocity spectra. As this quotient becomes constant for low $f \cdot ^x L_u \lesssim 0.05$ and as well for high frequencies $f \cdot ^x L_u \gtrsim 1$, the theoretical coherence decreases continuously between these constant limits. The experiments confirm the low frequency limit in its global trends,

as to how the coherence depends on the circumferential
position of the pressure point. With respect to the u-component, their absolute values are significantly overestimated
for all frequencies. Typically, the measured coherence
$Ch^2_{vp}(f)$ increases in a certain frequency range. Within this
range, the lateral component is more coherent than predicted.
The maximum is found at $f \cdot {}^x L_u /u \approx 0.4$ and not for the lowest
frequency. The decay of coherence is not continuous. Several
sharp breakdowns indicate the influence of uncorrelated noise.
In the real flow, the coherences completely vanish for
frequencies $f \cdot {}^x L_u /u \gtrsim 2$.

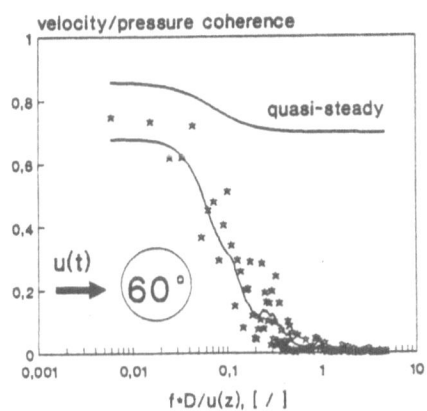 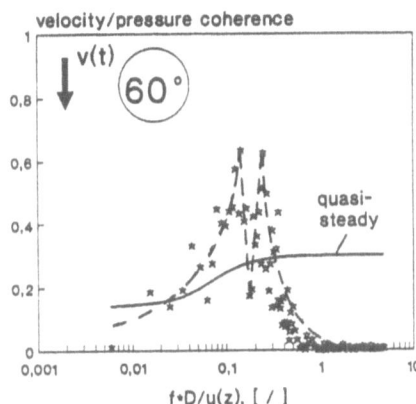

Fig. 2.: Velocity/Pressure coherence function
(pressure point $(z/H)_p = 0.6$, $\varphi = 60°$
velocity point $(z/H)_{u,v} = 0.6$, $r_x/D = 2.0$)

CONDITIONED SPECTRAL ANALYSIS

The classical model of a single input, the u-component,
results in a coherence equal to unity. This was not confirmed
by the tests. Even for the stagnation pressure, the coherence
with the u-component is at most $Ch^2_{up} \approx 0.85$ at low frequencies. The reasons for this lack of coherence are mainly (i)
the superimposed influence of both velocity components, even
at the stagnation point, (ii) the spatial extension of a turbulent eddy and (iii) effects, which are not linearly related
to the oncoming turbulence.

A multiple pressure admittance evades those restraints. It
summarizes pressure fluctuations, conditioned upon all
significant velocities. To evaluate the precision of a multi-
ple model, a multiple coherence $Ch^2_{p:i}$ (correlation $\varrho_{p:i}$) is
used, instead of the ordinary. It is defined by the quotient
of the conditioned and the observed pressure spectrum . There-

fore, the residual spectrum is given by $S_{ee} = (1-Ch^2_{p:1})\cdot S_{pp}$. In order to optimize the admittance model, the multiple coherence has to be optimized $Ch^2_{p:1}(f) \to 1$, by considering further input signals. A Cholesky decomposition of eq. (1) provides an effective approach for the iterative identification of all significant velocities [7].

The quasi-steady theory may be understood as the simplest, i.e. frequency independent, version of a multiple admittance model. It satisfies a multiple correlation (coherence) of unity, which cannot be confirmed experimentally. E.g., considering both, longitudinal and lateral velocity fluctuations, maximum cross-correlations ρ_{max} of the stagnation stream line lead to multiple correlation coefficients of $\approx 55\%$ in the attached flow region and $\approx 35\%$ in the wake. In the frequency domain, the multiple coherence is approximately 15 % higher than the ordinary $Ch^2_{up}(f)$.

Further increase of multiple coherence provides an extended model, considering velocity fluctuations occurring beside the stagnation line. This spatial influence of turbulence is indicated by significant velocity/pressure correlations, observed over the whole cross-section of the undisturbed flow. Surface pressures on the flanges of the cylinder are significantly stronger coherent with longitudinal velocity fluctuations at a particular lateral distance r_y, e.g. for $\varphi = 30°$ peak values are found at $r_y/D \approx 0.8$, fig. 3, [8]. Those asymmetric lateral correlation profiles have been observed for circumferential angles where the mean pressure distribution has strong gradients $\partial C_p/\partial\varphi$.

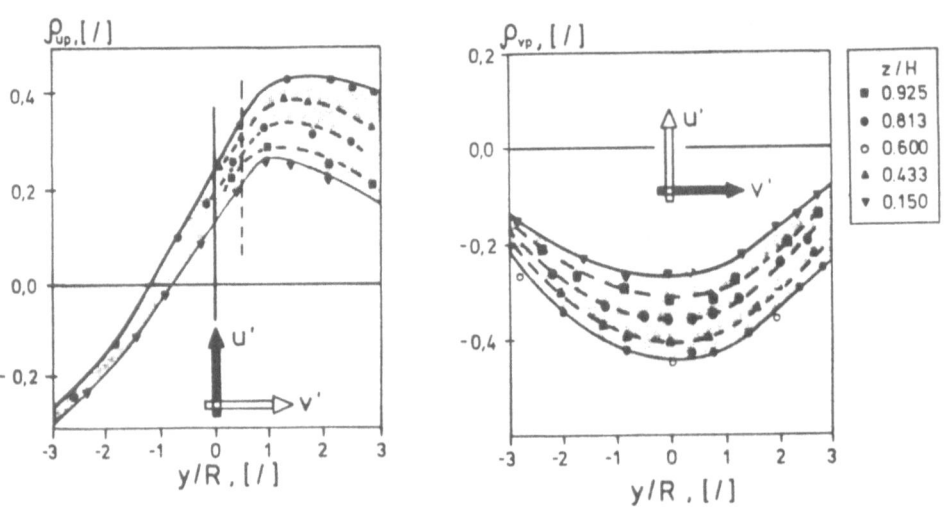

Fig. 3: Spatial influence of correlated velocity fluctuations
($(z/H)_p = 0.6$, $\varphi = 30°$; $(r_x/D)_{u,v} = 2.0$)

If all system inputs are well defined, eq. (1) can be solved. In a first approximation, the longitudinal and lateral component of the stagnation streamline are assumed to be the only decisive inputs. As both velocities are uncorrelated, the gain factors of the admittance vector is given by $|\chi_{ip}(f)|^2 = Ch^2_{ip} \cdot S_{pp}/S_{ii}$, (i= u, v), fig. 4. In comparison to the empirical scalar definition of a pressure admittance $|\chi_{ip}(f)|^2_{scalar} = S_{pp}/S_{ii}$, the correct admittance relates cause and effect. By means of the conditioned pressure spectrum $S_{p:i} = Ch^2_{p:i} \cdot S_{pp}$ and its related velocity spectrum S_{ii} turbulence induced pressure fluctuations are identified. These fluctuations are strictly separated from body induced fluctuations, fig. 5.

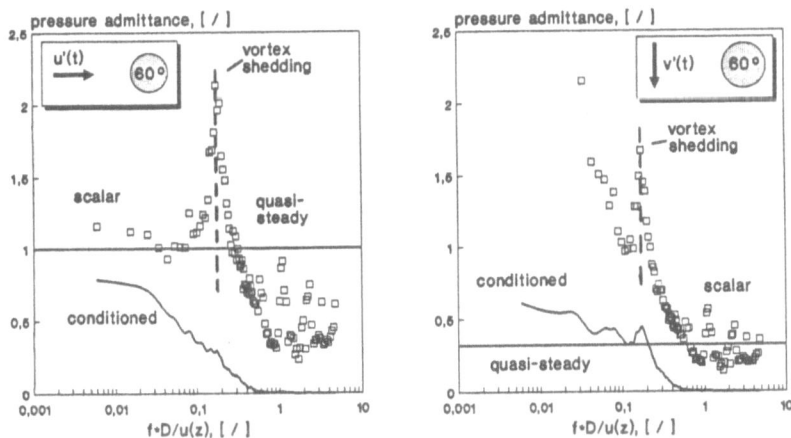

Fig. 4: Admittance functions for uncorrelated velocities

A complete set of admittance functions are presented in [2]. Empirical approximations of these functions enable an engineering application with sufficient accuracy.

FINAL REMARKS

Particularly for a multivariate system with correlated input-signals, further sophistication of the admittance model is requested. The resulting noise spectra should be analysed in comparison to the partial pressure spectra, induced by the vortex shedding, [3]. An attempt is made in [2]. Remaining deviations are associated to non-linear effects and should be the topic of extended research.

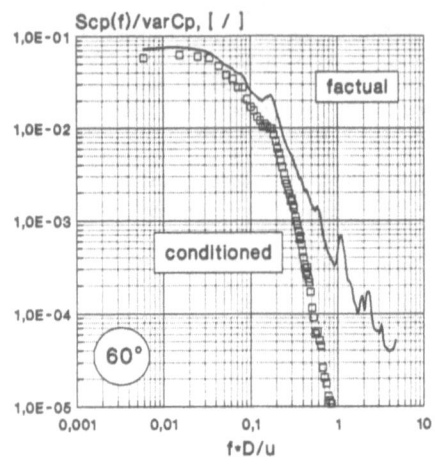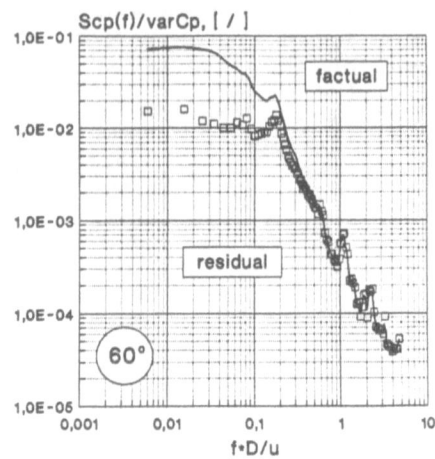

Fig. 5: Conditioned and residual pressure spectrum resulting
from a frequency dependent strip model
(pressure point $(z/H)_p = 0.6$, $\varphi = 60°$,
velocity points $(z/H)_{u,v} = 0.6$, $r_x/D = 2.0$)

ACKNOWLEDGEMENT

This research was supported financially by the German Research
Foundation (DFG). This support is gratefully acknowledged.

REFERENCES

1 Vickery B.J., On the flow behind a coarse grid and its use as a model of
   atmospheric turbulence, NPL-report 1143, 1965
2 Bendat J., Engineering Applications of Correlation and Spectral Analysis, John Wiley&Sons, 1991
3 Hölscher N., Niemann H.-J., On the Flow Around Finite Circular Cylinders
   in Turbulent Shear Flows, Notes on Num. Fluid Mechanics, Vieweg, 1992
4 Hölscher N., Niemann H.-J., Multiple Aerodynamic Pressure Admittance,
   Int. Conf. on Wind Engineering, Canada, 1991
5 KAWAI H., Pressure Fluctuations on Square Prisms, JWIA, 13 (1983)
6 HUNT J.C.R., et.al., A Review of Velocity and Pressure Fluctuations in
   Turbulent Flows around Bluff Bodies, JWIA, 35 (1990), 49-85
7 HÖLSCHER N., Ein multipler Ansatz für die aerodynamische Übertragungs-
   funktion der Winddrücke in atmosphärischer Grenzschichtströmung, Ph.D.
   Thesis, Ruhr-Universität Bochum, 1992, (to be published)
8 Hölscher N., Niemann H.-J., A Multiple Aerodynamic Pressure Admittance
   to Account for Turbulence Induced Pressure Fluctuations, Int. Conf. on
   Bluff Body Aerodynamics and its Applications, Melbourne, 1992,
   (submitted for publication)

232

Experimental Investigation of the Wake Past Bluff Bodies

K. Gersten   A. Becker   T. Demmer

Instiutut für Thermo- und Fluiddynamik
Ruhr-Universität Bochum
D-4630 Bochum, Germany

Abstract

The asymptotic solution of a high-Reynolds-number flow with separation around a bluff body consists of an external potential flow around the body and an additional fictitious body representing the dead air region. The shape of the fictitious body is supposed to be universal when scaled appropriately. Systematic detailed three-component laser-Doppler velocimetry (LDV) measurements in the wake of different axisymmetric and one three-dimensional body have been performed to support this hypothesis. A universal shape would enable the drag computation without a turbulence model. On the other hand, the data may be used to validate turbulence models.

1 Introduction

Investigations of backward-facing step flows reveal that certain flow characteristics are universal, [1]. Scaling the displacement line $\eta(\xi)$ with the reattachment length ξ_R and the effective step height $H_e = H + \delta_{10} - \delta_{1\infty}$ yields a universal wake function $F(\xi/\xi_R) = (\eta - \delta_{1\infty})/H_e$, where ξ and η are the wake coordinates with the origin at the base. It is assumed that this holds not only for a backward-facing step flow, but generally for all flows past bluff bodies, [2].

This would enable a rather simple method for computing the forces on the body. Such a method takes into account that at high Reynolds numbers the flow field is divided into various zones and hence is called the zonal method. The governing equations for each region are solved separately with an appropriate procedure. Finally, the different solutions have to be matched to get an overall solution. In the cases considered here three different zones have to be distinguished (see Figure 1):

- The inviscid external flow,
- The boundary layer on the body,
- The wake.

The knowledge of the relevant contour, i.e. the contour of the fictitious body, is necessary for the computation of the external flow. In the front part the fictitious body consists of the real body enlarged by the displacement thickness of the boundary layer. In the rear part, i.e. in the wake region, the fictitious body contour $\eta(\xi)$ is modelled with the wake function $F(\xi/\xi_R)$.

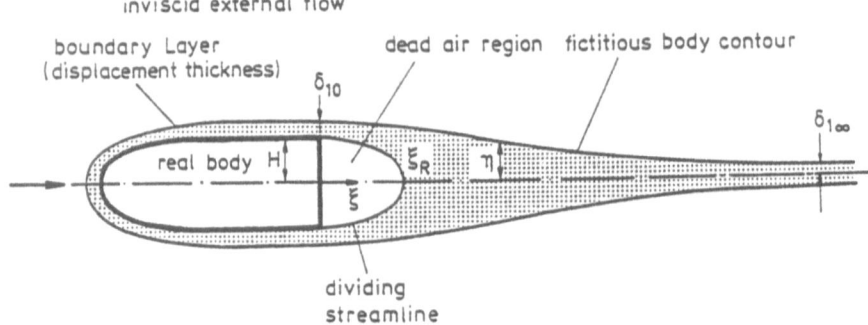

Figure 1: Various zones of the flow field past bluff bodies

A computer program was developed which implements this zonal concept for axisymmetric bluff bodies in axial flow. The potential flow around the fictitious body was calculated by using a vortex ring method, [3]. The laminar boundary layer was handled with the Walz method [4], the turbulent one was determined by Green's method [5]. The laminar-turbulent transition was fixed a priori at the locations of the trip wires. The displacement thickness at the base δ_{10} is a result of the boundary layer calculations. The following relations were used to determine to further unknowns $\delta_{1\infty}$ and ξ_R in the universal wake function $F(\xi/\xi_R)$. The global momentum balance yields

$$c_D = \frac{D}{\frac{\rho}{2}V^2\pi H^2} = 2\left(\frac{\delta_{2\infty}}{H}\right)^2 \approx 2\left(\frac{\delta_{1\infty}}{H}\right)^2,\tag{1}$$

where the total drag D can be derived by integration of the pressure forces as well as of the friction forces on the body.

The turbulent boundary layer is forced to separate at the body base, but will continue as turbulent free shear layer. Assuming that this free shear layer behaves like a turbulent mixing layer where the total pressure is constant along the dividing streamline, a relationship between the pressure coefficient c_{pB} at the base and the pressure coefficient c_{pR} at the reattachment point can be derived, [2, 6].

The main result of the theory is the calculation of the drag coefficient c_D as function of the Reynolds number and of the geometry of the body and to compare theoretical and experimental results for axisymmetric bodies.

The main purpose of this investigation was to measure in detail the velocity field in the wakes of bluff bodies in order to check the generality of the above mentioned wake function.

2 Body Geometry and Measurement Set-Up

Modells of two different shapes were investigated in the wind tunnel, an axisymmetric and a symmetrical three-dimensional body. The axisymmetric body has a half-ellipsoidal front part with an aspect ratio of 1:6 and a length of 336 mm, and a cylindrical section at the rear, giving a total length of $L = 372$ mm. With another cylindrical attached part the total length could be increased to $L = 560$ mm, thus the length/radius ratios were 6.64 and 10, respectively. The model was fixed with a cylindrical sting with a length of 600 mm. Various stings with different diameters were used to estimate their influence. The ratios sting radius/base radius (R_S/R) were 0.286, 0.357, and 0.446, respectively. These values were assumed to be small enough to neglect the sting's influence on the base pressure. The static pressure behavior was analyzed with 1.2 mm pressure taps that were aligned along the contour. Three additional circumferential pressure taps allowed the alignment to a 0°-angle of attack. Trip wires at the front part of the bodies assured that the boundary layer was turbulent at the base.

The three-dimensional body was a half-ellipsoid of 500 mm length. The semi-axes of the ellipse of the base were 93.75 mm and 62.5 mm. An attached cylindrical part increased the total length to $L = 1100$ mm. The tail sting had a diameter of 40 mm and a length of 800 mm. A rotatory bearing allowed measurements over the complete circumference. Like in the axisymmetric case, the static pressure was measured with pressure taps on the body contour. Circumferential pressure taps were used for the alignment of the body. Trip wires were installed at the beginning of the cylindrical part of the body to achieve a turbulent boundary layer at the base.

The measurements were accomplished in the closed-loop wind tunnel of the institute. The cross-sectional area of the open test section has a size of 1.2 m×1.5 m, the overall length is 2.6 m. Preliminary measurements showed that the velocity profile of the flow is sufficiently homogeneous, the deviation of the local mean velocity from the main velocity remains less then 0.1%, the turbulence intensity is Tu ≤ 0.007.

The non-intrusive velocity measurements were performed using a three-component LDV system. It is composed of a four-beam 2D-fiber-flow probe and a two-beam 1D-fiber-flow probe. The Doppler signals were analyzed with *Burst-Spectrum-Analyzers* (BSA). In contrast to the often used counters these facilities are able to handle the rather poor measurement conditions, which were

- the large focal length (1200 mm) of the optical system
- the low particle density, especially inside the wake
- measurements close to walls
- coincidental 3D-measurements.

The two fiber-flow systems were mounted on a traversing mechanism that is capable to move 1050 mm, 1250 mm, and 1150 mm in the x-, y-, and z-direction, respectively.

Measurements and data recording were conducted automatically with the aid of a computer (PC-386).

The particle generator used to produce the necessary tracer particles was filled with DEHS ($C_{26}H_{50}O_4$). The advantage of this fluid is that it gives a better signal quality and data rate than the usual water-glycerine mixture. Furthermore, the necessary consumption of fluid was lower. The seeding was added to the flow in the diffusor *behind* the test section, assuring that the particles are homogeneously distributed after one loop in the wind-tunnel.

A description of the body geometry and the measurement set-up may be found in [6].

3 Results for the Axisymmetric Case — Measurement and Computation

The measurement planes were located at 12 positions in the interval $0 \leq \xi/\xi_R \leq 3.5$.

Table 1 lists the experimental results of ξ_R/R and c_{pB} for different sting diameters R_S. These values were obtained from the LDV measurements. The base pressure is practically independent of the sting diameter.

Table 1: Experimental results for ξ_R/R, $Re_R = 3.73 \times 10^4$

	R_S/R	0.286	0.357	0.446
$L/R = 6.64$	ξ_R/R	2.71	2.57	2.39
	$-c_{pB}$	0.121	0.115	0.114
	R_S/R	0.286	0.357	0.446
$L/R = 10$	ξ_R/R	2.48	2.34	2.20
	$-c_{pB}$	0.105	0.108	0.103

The c_p-distribution for the case $L/R = 10$ along the contour is shown in Figure 2. The difference between the measurements (symbols) with and without trip wires is only significant close to the position of the wires. The figure also includes the results of the computations (solid and dashed lines).

Figure 3 shows the normalized u-, v-, and w-profiles at different positions ξ/ξ_R. The u-component is drawn with symbols, the v-component with a solid line, and the w-component with a dashed line. The rotational symmetry can be seen from the negligible w-component.

From the measurements the shape of the universal wake function could be obtained (see Figure 4). The figure also includes results of a two-dimensional car shape (upper and lower side), which are described in in [6]. The own measurements can be compared with the results of Van Wagenen, [7], who measured an axisymmetrical wake without a rear sting.

The computation was performed with the wake function of Figure 4. The computed pressure distribution on the body was already shown in Figure 2. Both cases, finite and infinite Reynolds number, coincide quite well with each other and with the measured values. The trip wires and the sting are not modelled. The difference at the wires is rather large. Nevertheless, the values at the base differ only slightly, as shown in Table 2.

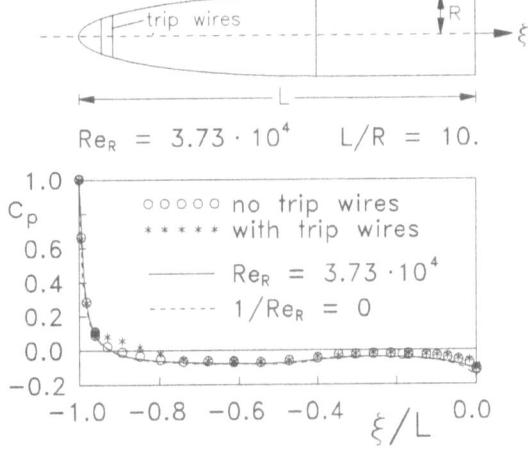

Figure 2: Pressure distribution on the body contour

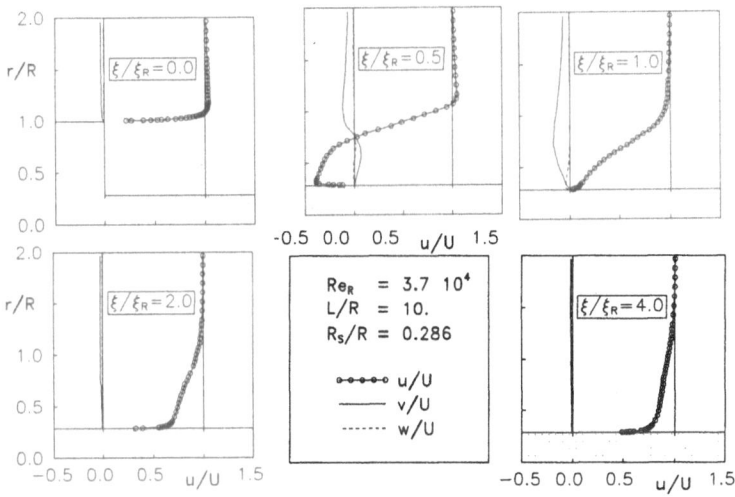

Figure 3: Velocities in the wake

Table 2: Comparison between experiment and theory

	c_{pB} (measured)	c_{pB} (computed)	ξ_R/R (extrapolated)	ξ_R/R (computed)
L/R = 6.64	−0.13	−0.13	3.2	2.9
L/R = 10	−0.12	−0.11	2.9	2.9

The results for the reattachment lengthes differ more. The computations yielded the ratio $\xi_R/R = 2.9$ for both the short and the longer body. They are higher than the measured values, and would coincide with the extrapolated experimental values, see Figure 5.

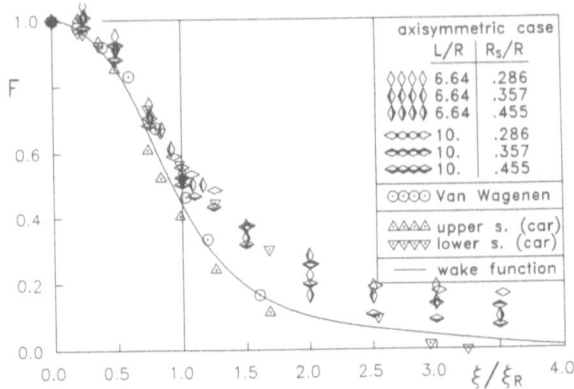

Figure 4: Universal wake function for the axisymmetric case

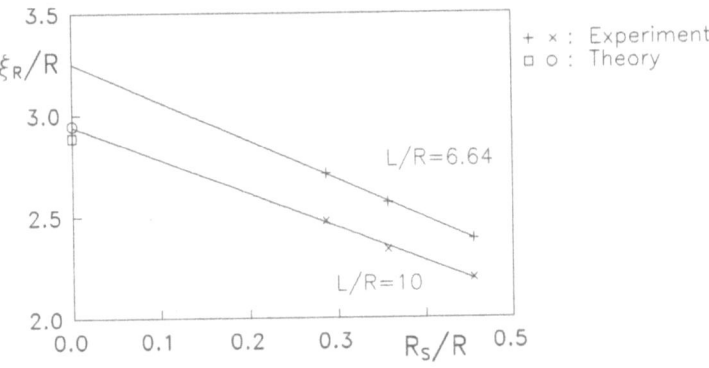

Figure 5: Reattachment lengths

4 Results of the Three-dimensional Case

The velocity measurements were performed in 30 different planes (see Table 3), at four parametric angles, $\varphi = 0°$, $30°$, $60°$, and $90°$, see Figure 6. A detailed description of the measurements can be found in [6].

Table 3 : Velocity measurement planes for the 3D-case

ξ/L	ξ/L	ξ/L	ξ/ξ_R	ξ/ξ_R	ξ/ξ_R	ξ/ξ_R
-0.363	-0.136	-0.005	0.00	0.50	1.05	2.0
-0.318	-0.091	-0.002	0.05	0.75	1.10	2.50
-0.273	-0.045	0.000	0.10	0.90	1.25	
-0.227	-0.023		0.15	0.95	1.50	
-0.182	-0.009		0.25	1.00	1.75	

$$\xi_R/L = 0.16$$

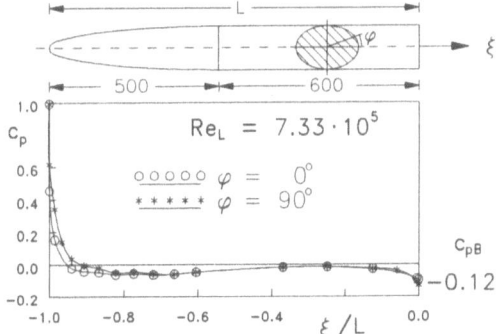

Figure 6: Pressure distribution for the 3D-body

The pressure distribution on the body contour is shown in Figure 6.

Determining the separating streamlines in the 0° and 90° plane and normalizing them with the reattachment length ξ_R and the step height $H(\varphi) = R(\varphi) - R_S$ leads to the results shown in Figure 7. One reason for the differences between the curves at two angles might be that the base cross-section is ellipsoidal, whereas the sting has a circular cross-section. The results for the axi-symmetric measurements are indicated by symbols. They

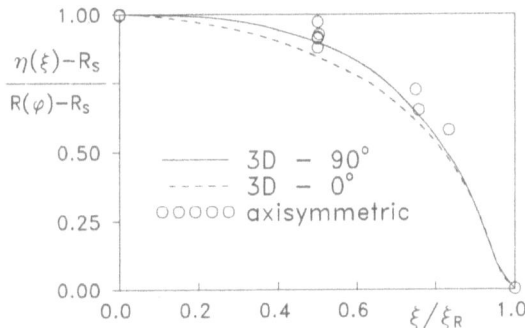

Figure 7: Normalized dividing streamlines in the 0° and 90° planes

show basically the same tendency for the shape of the separating streamline. Although Figure 7 shows only the separating streamlines it is plausible that the behaviour of the displacement lines will be equivalent.

Therefore, the similarity of the separating streamlines is a strong evidence for the assumption of self-similar cross sections of the fictitious body.

5 Conclusions

The assumption of the existence of a universal function describing the wakes of bluff bodies could be confirmed experimentally for the axisymmetric case.

For the three-dimensional case the measurements show strong evidence that the concept of the zonal method can be applied also to this case.

References

[1] Gersten,K., Herwig, H., and Wauschkuhn, P.: Theoretical and experimental investigations of two-dimensional flows with separated regions of finite length. AGARD CP 291, 23-1-23-13, 1980.

[2] Gersten,K., Papenfuß, H.D.: Separated flow behind bluff bodies at low speeds including ground effects. In: Proceedings of the second Caribbean Conference on Fluid Dynamics. The University of West Indies, Trinidad, January 1992, pp. 115–122.

[3] Vandrey, F.: A direct iteration method for the calculation of the velocity distribution of bodies of revolution and symmetrical profiles, ARC R.&M. No. 3374, 1964.

[4] Gersten,K., Herwig, H.: Strömungsmechanik, Grundlagen der Impuls- Wärme und Stoffübertragung aus asymptotischer Sicht, Vieweg Verlag, Wiesbaden, 1992.

[5] Green, J.E. : Application of Head's entrainment method to the prediction of turbulent boundary layer and wakes in compressible flow, RAE Tech. Rep. 72079, 1972.

[6] Becker, A.: Messungen im Nachlauf von Körpern mit stumpfem Heck mittels Laser-Doppler-Anemometrie, Doctoral Thesis, Ruhr-Universität Bochum, 1992.

[7] van Wagenen, R.G.: A study of axially-symmetric subsonic base flow, Ph. D. Thesis, University of Washington, 1968.

Three-Dimensional Separated Flow Around Automobiles with Different Rear Profile: Application of the Zonal Method

Heinz-Dieter Papenfuss and Peter Dilgen

Ruhr-Universität Bochum
Institut für Thermo- und Fluiddynamik
D-4630 Bochum, Germany

Summary

The zonal method for the three-dimensional separated flow around bluff bodies with particular application to automobile aerodynamics is developed. The method takes advantage of the fact that at high Reynolds numbers, the flow field is structured. For each flow zone, an asymptotic version of the full time-averaged Navier-Stokes equations for large Reynolds numbers must be solved. The concept of the fictitious body is used to simulate the displacement of the outer inviscid flow induced by the wake behind the body and by the boundary layers. The fictitious body of the wake zone is assumed to attain a universal, experimentally determined contour with geometry dependent constants. The zonal method makes it possible to determine the pressure and shear stress distributions on the body surface and, thus, the drag and lift coefficients as functions of the Reynolds number.

The method is applied to determine the flow around automobiles with different rear profile (squareback, fastback, notchback) but otherwise with identical geometry. For the calculation of the inviscid outer flow a panel method (doublets and sources) is used, whereas for the calculation of the turbulent boundary layer a three-dimensional integral method is employed. The results for the pressure distribution are in excellent agreement with corresponding experimental results for the plane of symmetry and for planes perpendicular to the main flow direction. An exception is the case of the notchback version; it is characterized by the peculiarity that part of the rear body contour is buried under the fictitious body of the wake zone. Although the pressure distribution in the plane of symmetry is predicted with satisfactory accuracy in this case, further extensions of the zonal method are necessary to determine the pressure distribution in the transverse direction with equal quality.

1. Introduction

The flow field past bluff bodies at high Reynolds numbers consists of inviscid and viscous domains. The largest domain is formed by the outer inviscid flow. The boundary layer at the body contour and the separated flow region behind the body, i.e. the wake, form the viscous domain. The wake interacts strongly with the inviscid outer flow and, consequently, affects the pressure distribution on the body, particularly at the rear end. In the case of a road vehicle, the boundary layer developing along the ground is an additional zone within the viscous domain; see **Fig. 1**. The zonal method takes advantage of this structure of the flow field. It is equivalent to an asymptotic approach to the full time-averaged Navier-Stokes equations for large Reynolds numbers in analogy to the classical boundary-layer theory. The separated region, however, is included in this method. The present work is an application of the zonal method to the separated flow around automobiles where the Reynolds number is typically of the order of magnitude of 10^7. The geometries which will be considered are restricted to cases where the position of the flow separation is fixed due to sharp corners at the rear end.

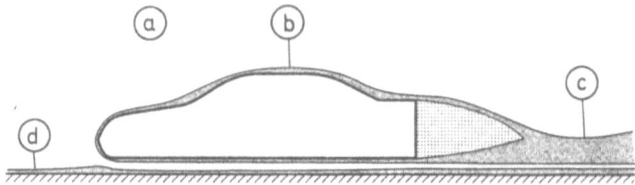

Fig. 1: Zones in the flow past bluff bodies with separation.
a = inviscid outer flow
b = boundary layer attached to the body
c = wake zone including deadwater
d = boundary layer on the moving ground

As a consequence of the asymptotic structure of the flow field, the Navier-Stokes equations can be simplified for each zone. The final solution for the different zones must be matched asymptotically (Kline [2]). The most difficult element of the calculation is related to the separated region in the wake of the body which in principle requires the solution of the full time-averaged Navier-Stokes equations. To avoid the problems inherent in such a solution, the much simpler concept of the "fictitious body" will be applied here to simulate the displacement effect induced by the viscous zones, particularly by the separated region, on the inviscid outer flow.

For a given geometry, the shape of the fictitious body is unknown a priori. For the two-dimensional problem it is known, however, that the fictitious body in the wake zone has a universal shape if a proper scaling is used (Wauschkuhn [6], Bauhaus [1]). A corresponding universal shape of the fictitious body for the wake zone in the three-dimensional case has not been reported. Therefore, experiments were carried out which made it possible to determine the shape of the fictitious body for the three-dimensional case. These results close the system of equations. Calculations are performed using the zonal method for the three-dimensional flow around automobiles with different rear profile: square back, hatch back and notch back. The close proximity of the automobile to the ground is taken into account. The results of the calculations are compared with experiments which were carried out using different techniques to simulate the ground effect: 1) a moving belt installed in a wind tunnel and 2) a moving-model technique in which the model was launched pneumatically along a stationary track (Papenfuss and Kronast [4]).

2. Zonal Method

The purpose of the fictitious body is to represent the displacement effect of the viscosity dominated zones on the inviscid outer flow. The displacement effect of the attached boundary layers is related to the streamwise development of the displacement thickness, which can also be interpreted as an equivalent source distribution. The displacement effect of the wake must be modelled. For two-dimensional problems, an analogy exists to the flow over a backward-facing step where the displacement function F is universal for all step heights when the coordinate in the main flow direction is scaled with the reattachment length ξ_R and the transverse coordinate with the 'effective step height' H_e. Gersten [2] has shown that the contour function, F, is valid not only for backward-facing steps but also for other flow fields with fixed separation.

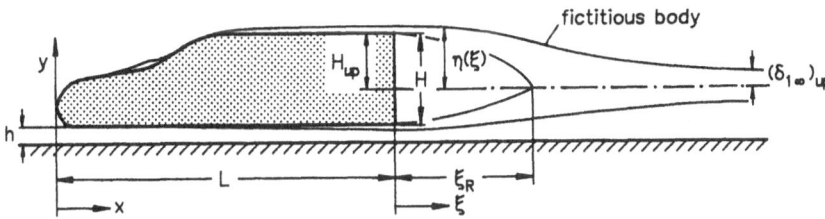

Fig. 2: Fictitious body for the flow past a bluff body

In **Fig. 2** the application of this concept is shown schematically for a bluff body in ground proximity. The global momentum balance requires a finite thickness of the fictitious body far downstream ($\xi \to \infty$) which is equivalent to the drag coefficient of the bluff body. Based on the two-dimensional approach [1], the shape function of the three-dimensional fictitious body can be written as

$$F(\xi/\xi_R) = \frac{\eta(\xi,\varphi) - \delta_{1\infty}(\varphi)}{H_e(\varphi)} = \frac{\eta(\xi,\varphi) - \delta_{1\infty}(\varphi)}{H(\varphi) + \delta_{1B}(\varphi) - \delta_{1\infty}(\varphi)} \qquad (1)$$

$$H_e(\varphi) = H(\varphi) + \delta_{1B}(\varphi) - \delta_{1\infty}(\varphi) \qquad (2)$$

where H_e is the local effective step height determined at an azemuthal angle φ. This angle is defined in a polar coordinate system with its origin at the base of the body (see **Fig.**). A global momentum balance leads to

$$c_D = 2 \cdot \Delta_{1\infty} / S_F \qquad (3)$$

where c_D is the drag coefficient based on the projected front area, S_F, of the body. $\Delta_{1\infty}$ is the displacement area of the fictitious body far downstream which can be calculated by integration of $\delta_{1\infty}$ over the angle φ.

The turbulent boundary layer attached to the body surface is forced to separate at the edges of the base and extends as a free turbulent mixing layer in the wake. In this manner a continuous change of the velocities between the inviscid outer flow and the reversed flow in the inner part of the separated region is established. From the theory of the turbulent mixing layer it follows that the total pressure is constant along the dividing streamline. Using this fact, a linear relationship between the pressure coefficient at the base, c_{pB}, and the pressure coefficient at the reattachment point, c_{pR}, can be derived:

$$\frac{\delta_{2B}}{\xi_R} = 0.009 \, \frac{1 - 3\,c_{pR} + 2\,c_{pB}}{c_{pR} - c_{pB}} \,. \tag{4}$$

δ_{2B} is the momentum thickness at the separation point. The factor 0.009 is a result of the theory for the turbulent mixing layer.

In the presence of the ground or in cases where the body has lift, the condition must be satisfied that the fictitious body in the wake zone has locally no lift. Consequently, the pressure at any position ξ of this part of the fictitious body must be independent of the azemuthal angle φ.

The complete system of equations (Bauhaus [1]) has been converted into a numerical code which solves the problem iteratively. At the beginning of the iteration, the parameters which determine the shape of the fictitious body are estimated (c_D, ξ_R, H_{up}) and the inviscid outer flow is calculated using a panel method (combination of doublets and sources). Then, the boundary layer is calculated. These calculations provide quantities needed to check the initial estimates and the global balances mentioned above. The drag is obtained by integrating the pressure and shear stress distributions along the contour of the original body. For the separation region, it is assumed that the pressure at the base is constant and equal to the value at the base perimeter (= separation line). This assumption is based on theoretical considerations and has been confirmed by experiments [5].

3. Experimental Determination of the Fictitious Body in the Wake Zone Behind Automobiles with Different Rear Profile

It has been mentioned above that the shape of the fictitious body in the wake zone must be modelled. In order to obtain a reasonable idea of the shape of the fictitious body for automobiles with different rear, experimental investigations of the total pressure in the wake have been carried out using a Kiel probe. For each geometrical variant, the outline of the deadwater was determined by systematic search of the border of the recirculation zone (see Fig. 3). Although the recirculation zone is not identical with the deadwater, the difference between the two is relatively small. This is due to the fact that the streamwise velocities in the outer part of the deadwater have much higher values and, therefore, a smaller displacement effect than the velocities in the recirculation zone (where the velocities are almost zero). In order to obtain the final shape of the fictitious body, the displacement thickness of the turbulent mixing layer must be added locally to the boundary of the deadwater. This becomes possible by combining the measured profiles of the total pressure with calculated profiles of the static pressure which, consequently, enables the determination of the velocity profiles.

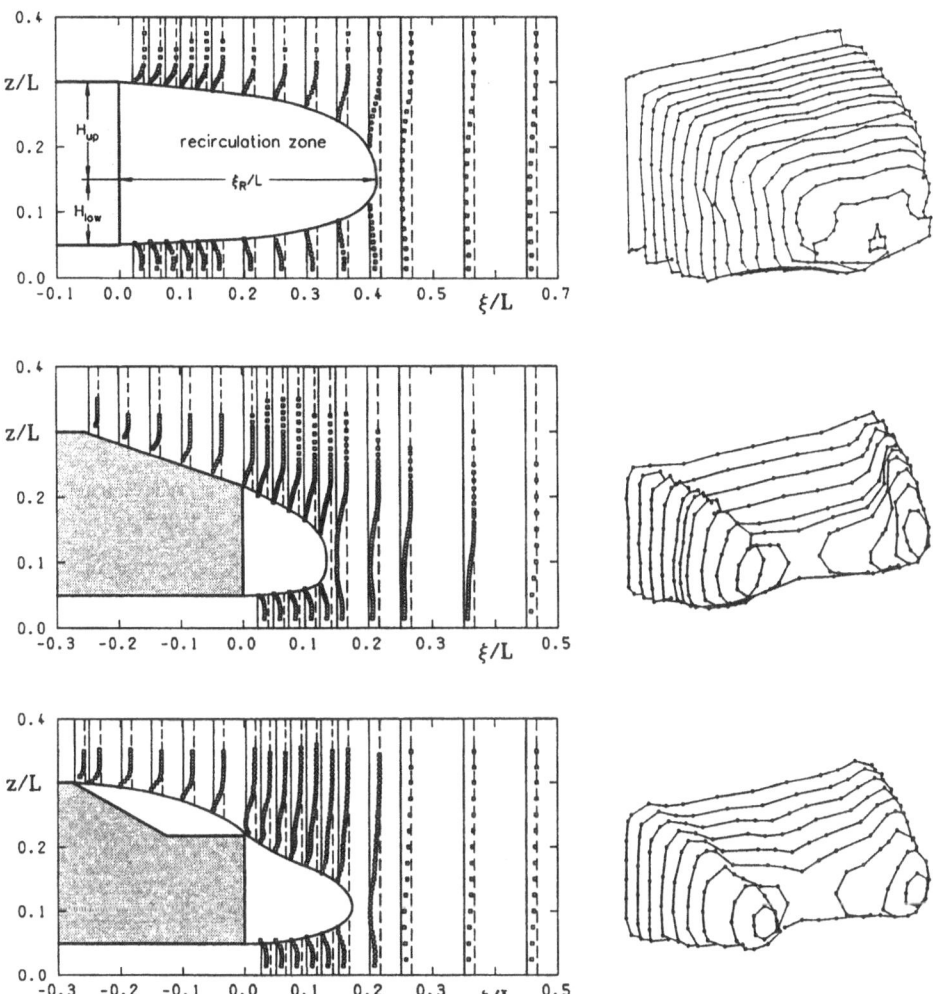

Fig. 3: Distribution of the total pressure in the wake behind automobiles with different rear profile. Three-dimensional representation of the recirculation zone

For the automobile with a squareback, it was found that the cross-sections of the deadwater were similar to the base area of the body. Therefore, the three-dimensional shape was non-dimensionalized for a local angle φ by the effective step height $H_e(\varphi)$ following Eq. (1). As a result of a regression analysis, the contour function $F(\xi, \xi_R)$ was obtained.

4. Results of the Zonal Method and Comparison with Experiments

Experimental results for the three-dimensional pressure distribution have been obtained using car models (scale 1:10) with different rear profile. The moving-model technique (Papenfuss and Kronast [4]) was employed to simulate the ground effect (ground clearance $h/L = 0.05$). The Reynolds number based on the model length was $2.5 \cdot 10^5$. In **Figs. 4** to **6**, the fictitious body and the results for the pressure distribution (predicted and measured) are shown for the geometries studied.

Fig. 4 refers to an automobile of the **squareback** type. In this case, the flow separates at the trailing edge of the roof. The pressure distributions in the center-section and in a plane perpendicular to the main flow direction are in good agreement with the experiments. The base pressure is also predicted well. An accurate prediction of the base pressure is necessary for a correct determination of the drag coefficient. The drag coefficient was calculated as c_D = 0.226. The lift coefficient c_L is negative which implies a force directed toward the ground. Calculations of the flow for different Reynolds numbers showed that the pressure distribution is not affected by this parameter. Changes in the drag coefficient due to changes in the Reynolds number are, therefore, a consequence of the friction drag only.

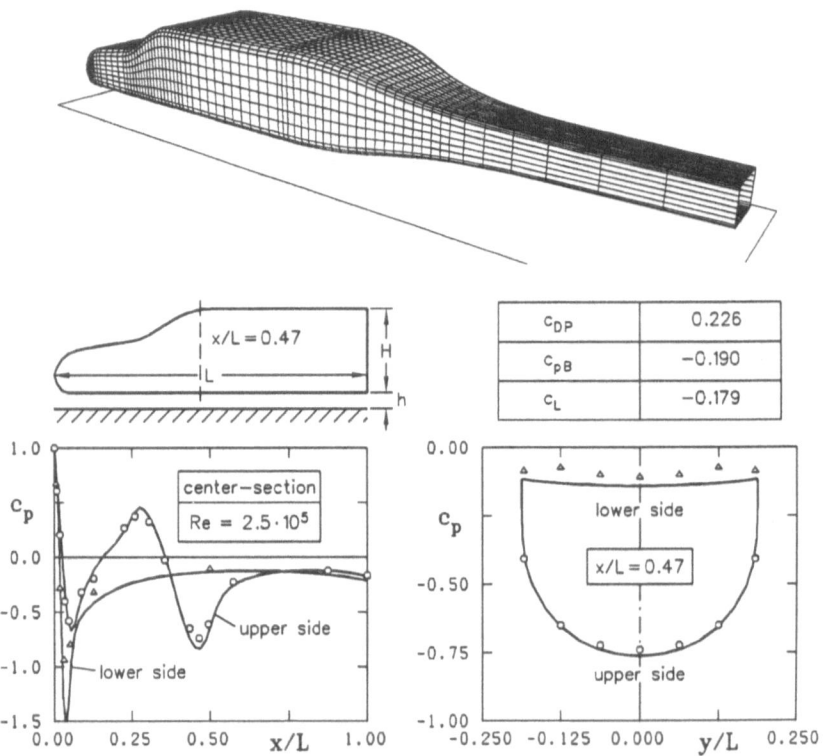

Fig. 4: Pressure distribution on an automobile of the squareback type

Fig. 5 shows the results for a car of the **fastback** type with a slant angle of 19°. The flow separates at the trailing edge of the automobile. Pressure distributions are shown for the center-section and for two transverse cross-sections near the roof trailing edge. The measured pressure profile in the center-section is reproduced well by the theory. It is interesting to note from the right side of **Fig. 5** that the theory also predicts the drastic change in the pressure characteristics occuring on the slanted surface (back-light). This change is induced by the flow around the relatively sharp rear side edges.

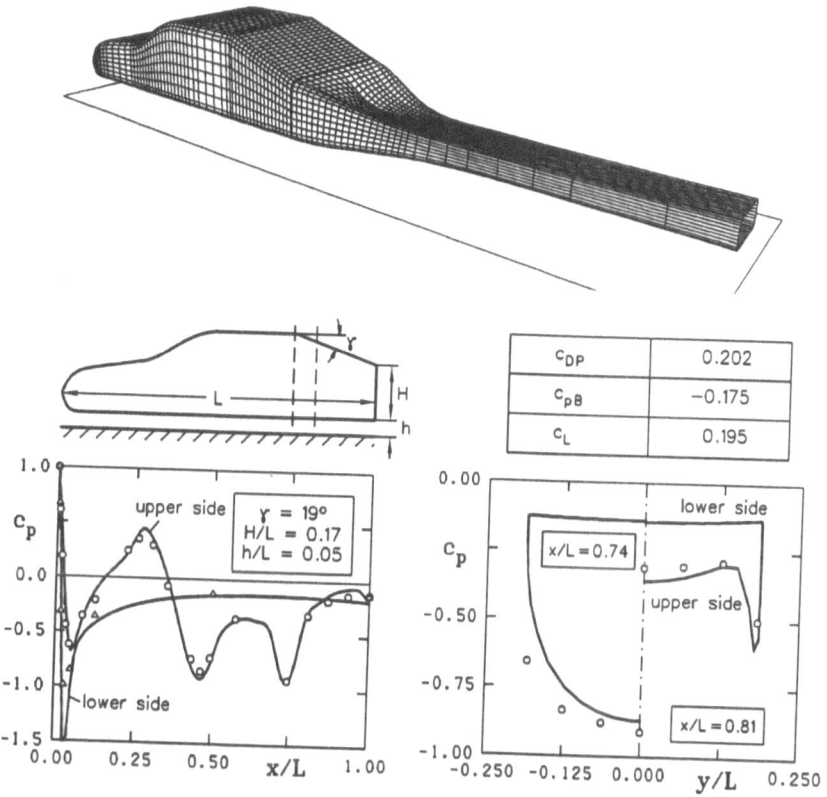

Fig. 5: Pressure distribution on an automobile of the fastback type

The most complicated flow situation occurs when the car of the **notchback** type is considered (**Fig. 6**). In this case, the flow separates straight off from the roof edge, and transverse vortices trapped in the concave corner and strong longitudinal edge vortices appear. The zonal method fails to reproduce the complicated pressure distribution in the separated flow region. This is a consequence of the absence of the longitudinal edge vortices in the present model and the peculiar shape of the fictitious body. The fictitious body of the wake zone is not a simple continuation of the blunt rear end, as in the case of the squareback and the fastback, but envelopes completely the rear part of the automobile including the back-light and the trunk. Further extensions of the zonal method are necessary to solve these problems. The predicted pressure distribution in the region of the attached flow agrees well with the experiments.

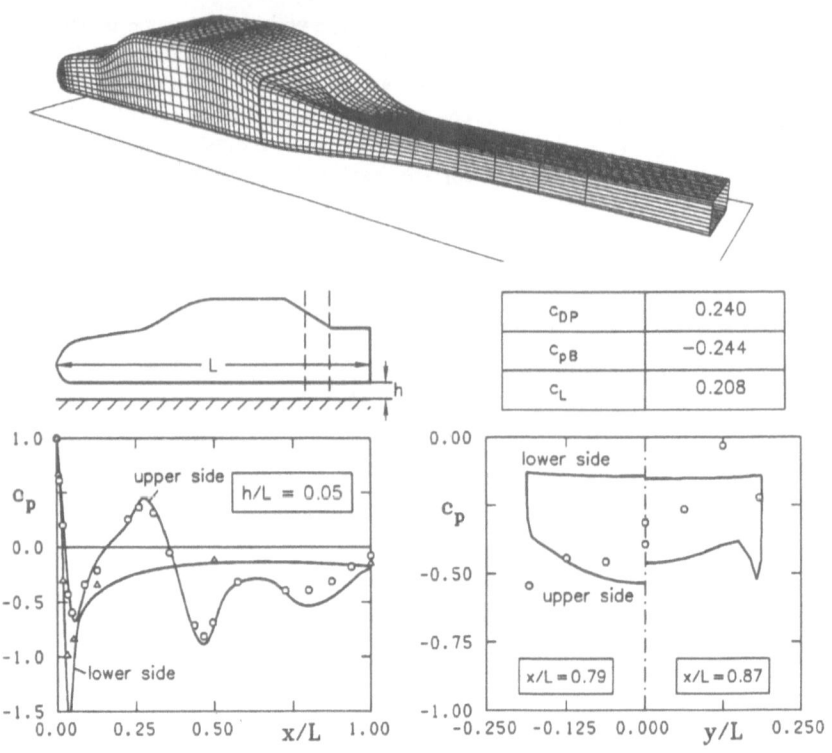

c_{DP}	0.240
c_{pB}	−0.244
c_L	0.208

References

[1] Bauhaus, F.-J.: Theoretische Untersuchung von Strömungen um Kraftfahrzeuge mit Hilfe der Zonenmethode. Ph.D. Thesis, Ruhr-Universität Bochum, 1991.

[2] Gersten, K., Herwig, H. and Wauschkuhn, P.: Theoretical and experimental investigations of two-dimensional flows with separated regions of finite length. AGARD CP 291, 23-1 to 23-13, 1980.

[3] Kline, S.J.: Universal or zonal modeling - the road ahead. In: S.J. Kline, B.J. Cantwell, G.M. Lilley (Eds.): Proceedings of the 1980-81 AFOSR/HTTM Stanford Conference on Complex Turbulent Flows. Stanford University, Stanford, Cal., Vol. II: Comparison of Computation and Experiment, pp. 991-1015, 1982.

[4] Papenfuss, H.D. and Kronast, M.: Moving-model technique used in automobile aerodynamics for measurement of ground effects. Experiments in Fluids 11, 161-166, 1991.

[5] Ramm, G.: Experimentelle und theoretische Untersuchungen der Strömung um Fahrzeuge. Ph.D. Thesis, Technische Universität Braunschweig, 1991.

[6] Wauschkuhn, P.: Ein Beitrag zu den ebenen turbulenten Strömungen mit begrenztem Ablösungsgebiet. Ph.D. Thesis, Ruhr-Universität Bochum, 1982.

248

Analysis of the Changing Topological Structures of Three-Dimensional Separated Flows

Uwe Dallmann, Wilhelm Kordulla
Heinrich Vollmers and Burkhard Schulte-Werning

DLR - Institute of Theoretical Fluid Mechanics
D-3400 Göttingen, Germany

Summary

Theoretical work has revealed elementary topological vortex and vorticity structures to be used for an analysis of three-dimensional separated flow simulations.
It is pointed out that topological flow analysis provides a meaningful concept for
CFD-validation. Numerical simulations using different Navier-Stokes codes, different approximations and grids exhibit that the transonic, laminar separated flow
around a hemisphere-cylinder configuration at $Re = 212500$, $Ma = 0.9$ is unsteady
and three-dimensional even at zero angle of attack. The calculated separation
lines of this hemisphere-cylinder flow, of the numerically simulated flow around a
sphere at low Reynolds numbers and of the experimental flow around a cylinder
at high Reynolds numbers remain almost stationary despite of unsteadiness of the
wake flows. Topological flow analysis exhibits that a "granulation" of wall-flow
patterns can be caused by improper resolution or incorrect boundary conditions
for the wall-near pressure field. The usefulness of a Galileian invariant vortex
definition, based on complex eigenvalues of the velocity gradient tensor, is
stressed. Since the assumption of laminar flow in high Reynolds number simulations is questionable it is important to find that the origin of large-scale, spanwise non-uniformity of a flow around a cylinder is coincident with the onset of
laminar-turbulent boundary-layer transition at $Re \approx 3.5 \times 10^5$, the critical value for
the drag crisis. Direct numerical simulations of the laminar flows around a sphere
up to $Re = 2000$ have revealed unsteady, three-dimensional, separated vortex
chains in good agreement with experiments.

1. Introduction

In 1984 the physical understanding of three-dimensional flow separation and the
subsequent formation of vortices was still at its infancy despite of numerous
experimental investigations. The main body of knowledge was based on wall-flow
information like oil-flow visualizations and wall-pressure measurements. Most of
the field information consisted of cross-sectional data and smoke or dye visualizations. Hence, direct numerical simulations were thought to be able to provide
the missing informations necessary for physical modelling of three-dimensional,
separating vortex flow structures.

Topological concepts had been adopted earlier [19] in order to characterize
three-dimensional flows by their "footprints", namely the skin-friction patterns.
Qualitative changes in surface flows were recognized to occur due to changes in
parameters like Reynolds number, Mach number, angle of attack, body shape, etc.
A description of surface (wall) flow topology turned out to be a usefull means to
reduce the qualitative information provided by the surface (wall) flow data (or
visualizations) and to give a precise description of wall-flow changes in terms of

249

topological changes. This can be of practical use for experimental control of three-dimensional flow separations and interactions [14] [15] [16]. However, surface-flow trajectories, especially wall shear stress informations are not sufficient to conjecture qualitative features within the outer (mid-air) flow field. The so-called open separation is one example where the wall shear stress topology does not change when separated vortices are formed.

The first author introduced the concept of elementary flow structures and their topological changes to characterize the formation of structures and "vortices" in real space [11] [4].[1]) Due to the local character of every creation and anihilation of a flow structure its topological description necessarily involves well-defined ensembles of critical points (rather than single, generic critical points considered by other authors). Such ensembles define a finite set of elementary flow structures which allow to dissect every complex flow field into such generic topological elements. Topological flow analysis is based on identifying such elements and their spatio-temporal connections. However, it requires much more than instantaneous streamline plotting.

A variation of parameters like Reynolds number, Mach number, angle of attack, etc., as well as physical or numerical disturbances leads to discrete changes in the topologies of various fields like velocity, vorticity, pressure gradient, etc. Hence, *a flow topology consists of the topological properties of all the fields of dependent variables and herewith defines the flow's structure*. It follows that the topological equivalence or non-equivalence of these fields is a necessary ingredient of every flow characterization. In order to get better insight into the complicated structure of three-dimensional separated flows we have started this investigation, supported by DFG.

The second author utilized a numerical code designed to solve the complete Navier-Stokes equations for three-dimensional transonic flows around blunt bodies [1] [2]. Therefore our first attempt was to analyze numerical simulations of the flow around a hemisphere-cylinder configuration for answering the following questions which we thought to be important also for flows around other configurations:

- How (and where) does the separating three-dimensional vorticity field form the separated vortices?
- How are flow-separation lines connected to the steady or unsteady separated vortical flow structures?
- Under which conditions does axisymmetric or three-dimensional, steady or unsteady flow separation take place?
- What kind of simplifications can be made in the equations of motion in order to achieve equivalent results by a minimum of CPU-time?

The numerical simulation of the flow around a hemisphere-cylinder configuration gave rise to additional theoretical, experimental and other numerical investigations in order to answer the additional questions:

- Is the observed spatio-temporal flow change of physical nature such that a solution bifurcation occurs due to changes of physical parameters like Reynolds number, Mach number, etc.?

---

[1]) Topological flow structures can form and change either locally at single degenerate critical points or globally by break-ups of singular surfaces connecting critical points. We recall that a critical point in a velocity (or any other vector) field is a point where the velocity (or any other vector) is zero and the local flow direction is non-unique. The topological structure of a velocity field is then defined by all the streamlines connecting critical points and the singular streamsurfaces which are solely defined by those streamlines which leave or reach these critical points.

- Is the calculated flow structure, i.e. the topology of all of the dependent variables of the flow unique and stable or will it change by altering the numerical method, the grid and the time stepping or by changing the initial and boundary conditions?

Instabilities and bifurcations of separated flows are little understood. The sensitivity to perturbations which leads from a two-dimensional/axisymmetric to a three-dimensional flow structure could result either from initial conditions, from laminar-turbulent transition in the boundary layer and the wake or from disturbances which enter through the boundary conditions. Three-dimensional and/or unsteady perturbations, either of non-physical, i.e. numerical nature or real physical perturbations present in every environment, will be amplified above critical sets of parameters like Reynolds number, Mach number, angle of attack, etc. Flow bifurcations result and changes of flow topologies can be identified in various quantities. Hence, topological flow analysis allows to identify the effect of any physical or non-physical perturbation. The necessity of appropriate use of computer graphics for simulation data analysis is obvious [3].

Answering questions like those above requires a comparison between at least two sets of data fields or of qualitative informations such as flow visualizations. Hence, a philosophy for such a comparison and, therefore, also for the validation procedure of computational fluid dynamics' results is required. In his contribution to the AGARD Symposium on CFD-Validation [20] Bradley defines: "CFD-Code Validation := Detailed surface-and-flow-field comparisons with experimental data to verify the code's ability to accurately model the critical physics of the flow. Validation can occur only when the accuracy and limitations of the experimental data are known and thoroughly understood and when the accuracy and limitations of the code's numerical algorithms, grid-density effects, and physical basis are equally known and understood over a range of specified parameters." There is one more statement that we would like to add, namely: Any validation approach requires a well-defined technique which allows to reduce the information contained in the huge data sets provided by CFD.

Topological flow analysis provides a meaningful way to achieve a physical reduction of information. It allows a precise comparison of qualitative features of one data field with another and provides precise definitions of flow structures, vortices, etc.. Hence, we suggest that the validation of CFD-results should be based on a comparison of the flow topologies and its changes in physical and numerical parameters' spaces. The topological validation procedure provides a clear concept to first consider qualitative agreement, namely topologcal equivalence between data in the flow field and on the wall, before the code is finally validated by quantitative comparisons.

The present paper summarizes only the most important results obtained in this DFG-project. Further details have been published in [1] to [10] and in additional papers cited there.

2. Numerical Simulation of Flow Separation on a Hemisphere-Cylinder Configuration

The numerical simulation of the transonic flow past a hemisphere-cylinder configuration at high Reynolds numbers, based on the integration of either the complete or the thin-layer approximated Navier-Stokes equations, has attracted the interest of a number of researchers since more than a decade (see references in [2]). In most cases the prediction of pressure distributions was the prime purpose. It was expected that the numerical solutions would converge to a steady state in agreement with measured pressure data. However, our investigations [1] revealed that unsteady flows persist in the laminar flow calculations although the pressure data on the windward side had converged to a steady value close to the measured one and although the residuum had dropped by several orders of magnitude. Unfortunately it was not possible to detect a periodic flow behaviour within reasonable computation times, but it is likely that either periodic bubble unsteadiness or even periodic vortex shedding sets in at a critical set of Reynolds and Mach numbers.

It was also detected that the solution for axisymmetric boundary conditions, with the body at zero angle of attack, did not converge to an axisymmetric flow with a steady, closed separation bubble. The discussion in [2] comparing results obtained by two, completely different numerical methods came to the conclusion that both codes failed to calculate an axisymmetric separation bubble at high Reynolds numbers. It seems reasonable to believe that a physical flow bifurcation from an axisymmetric separation bubble to a three-dimensional (azimuthally periodic) separated flow occurs for a critical set of Reynolds and Mach number. Any experimental evidence for this axisymmetric → non-axisymmetric flow bifurcation would be highly appreciated. It is of interest to recognize that the instantaneous lines of primary flow separation remain almost stationary although the separated flow is unsteady.

The hemisphere-cylinder computations performed at an angle of attack of $\alpha = 19°$ [1] [2] revealed the following: With increasing mesh resolution the skin friction pattern in the nose region became more and more "granular", which means that additional critical points appeared, where the wall-shear stress dropped to zero. Oil-flow patterns showed rather simple structures. This "granulation" problem deserved a thorough analysis which will be summarized in chapter 3.

The wall-pressure distribution in the leeward symmetry plane exhibited pronounced time-dependent spatial variations while the windward side pressure remained fairly steady in good agreement with experiments. The concept of local time steps did not enforce an expected asymptotically steady flow behaviour. The chosen globally constant time step, on the other hand, did not resolve this insuperable, physical flow unsteadiness.

Of course, the assumption of a laminar flow throughout the entire flow field including the separated flow regions had to be questioned. Nevertheless, we had to leave this problem unsolved until transitional flow calculations can be done. But even then the calculation of a laminar basic state will be necessary for instability and transition prediction.

The above mentioned findings led us to concentrate further efforts on the following five research topics:
1. The wall-flow "granulation"-problem in numerical simulations.
2. Disturbance influence on the time-averaged flow structure at laminar-turbulent transition Reynolds numbers.
3. Simultaneous formation of unsteady and three-dimensional flow separation around axisymmetric bodies like spheres.
4. Identification of "vortices" and vortex flow structures in vorticity fields.
5. Quasi-stationarity of separation lines despite unsteady separated flow and vortex shedding.

From an applicational point of view the choice of simulating the flow around the hemisphere-cylinder was justified. However, it would have been advantageous to have studied lower Reynolds-number flows first, especially when the separated flow field is physically unsteady.

3. Topological Flow Analysis.
Three-dimensional Vortex Structures and Vorticity Topology.

Since topological properties change only at discrete sets of bifurcation parameters, i.e. at critical values of Reynolds number, Mach number, angle of attack, aspect ratio, etc., we may restrict our investigation to the vicinity of bifurcating flow states and to the vicinity of points in the flow where this flow undergoes structural, i.e. topological changes. If we then look for possible topological changes (bifurcations) we have to distinguish between local and global bifurcations. Local bifurcations change the number of critical points where the velocity and/or vorticity, etc., drops to zero and from whereon the flow cannot be continued uniquely. Global bifurcations change the connections of these critical points. Both bifurcations lead to structurally stable flow structures whose topology is completely defined by the set of so-called singular streamsurfaces (one of which is the surface of the body around which we analyze the flow). Three-dimensional flow topologies cannot be identified by sectional flow patterns but they can be located by identifying these singular surfaces within the velocity-, vorticity-field, etc.. Computer aided topological flow analysis requires the computer graphical representation of such and only such singular surfaces on graphic workstations. This, however, is a non-trivial task due to the definition of the singular surfaces [1]).

Let us briefly mention a topological flow analysis to identify the origins of the "granular" flow structure. It is known that the three invariants of the velocity gradient tensor define the local properties of the velocity field in the vicinity of generic, hyperbolic critical points. "Granulation" of the skin friction pattern requires local incipient, marginal flow separations. Only such local events can produce sets of additional critical points on the wall. Hence, the "granular" structures bifurcate from local degeneracies (non-hyperbolic points) of the veloctiy field at the wall. This, in turn, requires the search for spatial-temporal changes of the invariants of the velocity gradient tensor $grad\vec{v}$ at the wall. On the wall this requires the investigation of the shear stress gradient tensor $\partial(\tau, \sigma)/\partial(x, y)$. We look for regions of varying sign of the invariants of this tensor, namely the wall-Jacobian determinant $J_{Wall} = \mu^{-2}(\tau_x \sigma_y - \tau_y \sigma_x)$ and of the wall-normal pressure gradient $p_z = -\tau_x - \sigma_y$. The locus $J_{Wall} = 0$ then defines locations where critical points ($\tau = 0, \sigma = 0$) can be generated either by saddle-point/nodal-point bifurcations ($p_z \neq 0$) or by saddle point-center or higher-order bifurcations ($p_z = 0$). This Jacobian determinant is given by Dallmann [10]

$$J_{Wall} \cdot 4\mu^2 = \mu^2 v \Delta\Delta W_z + \mu \partial_t(p_{xx} + p_{yy}) + (p_z)^2 - \tau p_{zx} - \sigma p_{zy} =$$
$$= \mu v \Delta\Delta p + 4(p_z)^2 + 2\tau p_{zx} + 2\sigma p_{zy} .$$

It follows immediately that "granulation" of the wall flow pattern associated with $J_{wall} = 0$ can occur due to inadequate numerical treatment of the near-wall velocity field (W = wall-normal velocity), the temporal variation of the wall-pressure distribution and the pressure gradient normal to the wall causing spatial variations of the biharmonic pressure-function $\Delta\Delta p$ on S. Due to this topological consideration of the "granulation" problem we suggest a thorough examination of the near-wall pressure field as well as a study of the correct wall pressure boundary conditions for Navier-Stokes equations.

The invariants of the velocity gradient tensor provide important means to define and identify "vortices". Vortices form on a wall at foci. Such foci are critical points with complex eigenvalues of the wall shear stress gradient tensor $\partial(\tau, \sigma)/\partial(x, y)$. But how can we define the location of "vortices" or "vortex cores" when no critical point defines the origin of the vortex and so-called open flow separation takes place? There is no topological change of the skin-friction and wall-pressure patterns associated with the formation of - for instance - leeside vortices on prolate spheroids or round edged delta wings (Dallmann et al. [6], Herberg et al. [13]). Nevertheless, topological analysis helps to locate even those vortices as follows:

An instantaneous, "swirling motion", which we call a "vortex" within a certain Galileian frame of reference, will appear only in regions where the local velocity gradient tensor $grad\dot{v}$ exhibits complex eigenvalues around those critical points which appear within that frame of reference. Hence, a necessary condition for local vortex formation within an arbitrary frame of reference is the change of sign of the local Jacobian determinant of $grad\dot{v}$, and a sufficient criterion is fulfilled if complex eigenvalues of the velocity gradient tensor exist in some regions in physical space (Dallmann [11], Vollmers et al. [12], Dallmann et al. [6], Herberg et al. [13]). Regions of complex eigenvalues exhibit only one real eigenvector of $grad\dot{v}$ which defines the local axis of rotation of the fluid seen by an observer moving with the flow. The analytically given condition for the existence of such a vortical flow region also shows that vorticity is a necessary ingredient of a vortex.

"Vortices" within an instantaneous velocity field do not (in general) coincide with "vortices" identifyable in the iso-vorticity field or iso-pressure surfaces, etc. In addition, vortex lines can exhibit structures quite different from iso-vorticity surfaces, even in steady flows. For a meaningful characterization of the different separated flow structures, i.e. the flow topologies it is important that only a finite set of elementary topological flow structures can form locally on a body where the no-slip boundary condition holds. Certain steady velocity and vorticity field structures of incompressible fluids have been obtained by Dallmann [4] [5] as solutions of nonlinear dynamical systems which are locally equivalent to the equations of motion. Such elementary structures can be used as building blocks for larger-scale flow structures [14] [15] [16].

4. On the Changes of the Separating Flow Structure at Laminar-turbulent Transition Reynolds Numbers

The observation of the formation of a non-axisymmetric, separated, high Reynolds number flow despite of axisymmetric boundary conditions for the hemisphere-cylinder at zero angle of attack led us to the following question:

• Is it possible that laminar-turbulent transition causes or triggers somehow large-scale non-uniformities in the wall-flow patterns?

To answer this question the first author in collaboration with G. Schewe [7] analyzed oil-flow visualizations of the flow around a circular cylinder between subcritical and supercritical Reynolds numbers. The most surprising result was as follows: At the subcritical value $Re = 1.85 \times 10^5$ perfectly two-dimensional flow separation can be achieved, while at the supercritical value $Re = 4.1 \times 10^5$ a spanwise cellular structure appeared at the line of flow separation. This indicated the creation of three-dimensional, streamwise vortices which form along the transitional separation line. This coexistence of laminar-turbulent transition, drag crisis and onset of a large scale, periodic, spanwise coherence of the separated flow deserves further investigations and, to the first author's knowledge, the investigation [7] has shown this for the very first time.

If a similar phenomenon would occur in the nose region of our hemisphere-cylinder at a critical Reynolds number remains open. The experimental observations of Bippes [17] have, indeed, shown a sensitive dependence of the structures of separated vortex flows around a hemisphere-cylinder on laminar, free or forced, transitional or turbulent boundary layer development.

5. Stationary Separation and Unsteady Vortex Shedding.
Unsteady and Three-Dimensional Wake Formation Behind Spheres.

The observation of stationarity of separation lines, mentioned in chapter 2 raised the question: How stationary can separation and attachment points be in unsteady separation bubbles or during von Kármán-vortex shedding? It is often stated that during the vortex shedding process the instantaneous separation and attachment lines move on the body and annihilate each other at the very instant when a vortex is shed. However, this picture does not fit to those observations in surface flow visualizations where sharp, almost stationary separation lines, but oscillating or blurred attachment lines are seen. We have, therefore, analyzed numerical simulations of von Kármán vortex shedding by Sun and earlier work by Perry et al. in [9]. We recognized that the vortices seen in the instantaneous velocity field are formed within the near wake and not at the wall. The shedding process leaves separation lines almost stationary.

As mentioned in chapter 2, flow unsteadyness <u>and</u> three-dimensionality has been identified even under axisymmetric boundary conditions, namely in the calculated separated flow around the hemispere-cylinder at zero angle of attack. An equivalent phenomenon was to be expected in the near-wake behind the geometrically simpler spheres. Hence, we studied the formation of these flow structures [8] [9] [10] [18]. Here we only give a summary of first results derived within this DFG-supported project:

The formation of three-dimensional vortex structures in laminar, slightly compressible flow fields around a sphere has been numerically simulated. The implicit numerical algorithms used were of the Beam-Warming type and solved the complete, unsteady, three-dimensional, compressible Navier-Stokes equations. Experimental investigations of the sphere flow show that three-dimensional flows develop around spheres at rather low Reynolds numbers $Re = 0(200)$. In contrast to the flow around cylinders the unsteady vortex shedding from spheres (vortex chains) is preceded by the formation of a three-dimensional vorticity field. For $Re \approx 200$ the numerical simulations revealed a first bifurcation to a three-dimensional, topologically stable flow. With increasing Re a spatial left-right symmetry appeared, a nearly constant wall pressure was established in the separated flow region and the closed near-wake bubble with recirculating fluid broke up and time periodicity was observed. Sofar the simulations performed up to $Re = 2000$ indicate, that the so-called primary separation line remains an almost stationary and axisymmetric circle around the sphere although the separated flow is highly three-dimensional, and unsteady vortex chains form.

These investigations have been partly supported by *Deutsche Forschungsgemeinschaft (DFG)* under *Da 183/1-6*.

7. References

[1] Kordulla, W., Vollmers, H., Dallmann, U.: *Simulation of three-dimensional transonic flow with separation past a hemisphere-cylinder configuration.* Conference Proceedings No. 412 of 58th Meeting of the AGARD Fluid Dynamics Panel Symposium, April 10, 1986, Aix-en-Provence, France.

[2] Kordulla, W., Müller, B., Vollmers, H.: *Comparison of Two Different Navier-Stokes Methods for the Simulation of 3-D Transonic Flows with Separation.* AIAA-89-0559, 1989.

[3] Vollmers, H.: *The recovering of flow features from large numerical data bases.* Proc. VKI Lectures Series on "Computer Graphics and Flow Visualization in Computational Fluid Dynamics", Brüssel, Belgien, ISSN0377-8312, 1991.

[4] Dallmann, U.: *On the Formation of Three-Dimensional Vortex Flow Structures.* DFVLR-IB 221-85 A13, 1985.

[5] Dallmann, U.: *Three-Dimensional Vortex Structures and Vorticity Topology.* In: Proc. of the IUTAM Symposium on Fundamental Aspects of Vortex Motion, Tokyo, Japan, 1987, H. Hasimonto & T. Kambe (eds.), North Holland - Amsterdam (1988), 183-189.

[6] Dallmann, U., Hilgenstock, A., Riedelbauch, S., Schulte-Werning, B., Vollmers, H.,: *On the footprints of three-dimensional separated vortex flows around blunt bodies. Attempts of defining and analyzing complex flow structures.* Proc. of the 67th Meeting of the Fluid Dynamics Panel Symposium on Vortex Flow Aerodynamics, Scheveningen, The Netherlands, Oct.1-4, 1990. AGARD CP 494, 1991, pp. 9.1 - 9.13.

[7] Dallmann, U., Schewe, G.: *On Topological Changes of Separating Flow Structures at Transitions Reynolds Numbers.* AIAA-87-1266, 1987.

[8] Dallmann, U., Schulte-Werning, B.: *Topological changes of axisymmetric and non-axisymmetric vortex flows.* In: Topological Fluid Mechanics, (Moffat, H.K. & Tsinober, A., eds.), Cambridge Univ. Press, Cambridge, 1990, 372-383.

[9] Dallmann, U., Schulte-Werning, B.: *On the three-dimensionality and unsteadiness of separated flows.* Proc. Third International Congress of Fluid Mechanics, Cairo, Egypt, January 1990, Vol.I, 117-132, 1990.

[10] Dallmann, U.: *Analysis of simulations of topologically changing three-dimensional separated flows.* In: Separated Flows and Jets (V.V. Kozlov, A.V. Dovgal, eds.) Proc. IUTAM Symp. Novosibirsk, USSR, July 1990, Springer (1991), pp.167-171.

[11] Dallmann, U.: *Topological structures of three-dimensional flow separations.* DFVLR-IB 221-82 A 07 (1982).

[12] Vollmers, H., Kreplin,H.-P., Meier, H.U.: *Separation and vortical-type flow around a prolate spheroid - Evaluation of relevant parameters.* AGARD-Symp. Aerodynamics of Vortical Type Flows in Three Dimensions, Rotterdam, AGARD-CP-342, 1983.

[13] Herberg, T., Dallmann, U., *Wirbelbildung in einem dreidimensionalen Wirbelstärkefeld am Beispiel des Deltaflügels.* DLR-IB 221-92 A11 (1992).

[14] Dallmann, U., Doerffer, P.: *Three-dimensional flow separation caused by normal shock-wave / turbulent boundary-layer interaction.* Proc. IUTAM Symposium Transsonicum III (J. Zierep, H. Oertel, Eds.), Göttingen, Springer-Verlag, 1989, 429-438

[15] Doerffer, P., Dallmann, U.: *Reynolds number effect on separation structures at normal shock wave / turbulent boundary layer interaction.* AIAA Journal. Vol. 27, No. 9 (1989), 1206-1212.

[16] Doerffer, P., Dallmann, U.: *Spatial and Temporal Features of a separated flow field at a convex wall induced by normal shock wave turbulent boundary layer interaction.* AIAA 21st Fluid Dynamics, Plasma Dynamics and Lasers Conference, Seattle, WA, June 18-20, 1990. AIAA Paper 90-1457, 1990.

[17] Bippes, H., Turk, M.: *Oil flow patterns of separated flow on a hemisphere cylinder at incidence.* DFVLR-FB 84-20, 1984.

[18] Schulte-Werning, B.: *Numerische Simulation und topologische Analyse der abgelösten Strömung an einer Kugel.* DLR-FB 90-43, 1990 (Dissertation Univ. München, 1990)

[19] Peake, D.J., Tobak, M.: *Three-dimensional interactions and vortical flows with emphasis on high speeds.* AGARD-AG-252 (1980).

[20] Bradley, R.G.: *CFD validation philisophy.* AGARD-CP-437, Vol. I, Symp. Validation of Computational Fluid Dynamics, Lisbon, Portugal, May 1988.

FLOW STRUCTURE AND HEAT TRANSFER IN CHANNELS WITH WING-TYPE VORTEX GENERATORS

M. Fiebig and N.K. Mitra
Institut für Thermo- und Fluiddynamik
Ruhr-Universität Bochum

SUMMARY

A computational scheme has been adapted and modified for the solution of the three dimensional Navier-Stokes and energy equations in a rectangular channel with punched or mounted wing-type longitudinal vortex generators. Exemplary results of the flow structure in the channel with a pair of punched delta winglets are reported here. Results show that even at large angle of attack ($65°$) stable vortices are generated. The spiraling motion induced by these vortices can enhance the Nusselt number by 52 % on an area which is 15 times larger than the vortex generator area. Various parameter studies have been performed and reported elsewhere [4,5,7-12]. Some of these results have been summarized here.

INTRODUCTION

Wing-type vortex generators (VG) such as delta or rectangular wings or winglets (see fig. 1) can be mounted on or punched out of the channel wall in order to introduce longitudinal vortices in channel flows. The secondary (azimuthal) velocity induced by a longitudinal vortex causes a continuous mixing of the fluid in the core and the fluid near the wall of the channel and thereby improves the convective heat or mass transfer. Because of the favorable pressure gradient in the channel flow, the longitudinal vortices are stable and can influence a large area compared to the VG area.

In fin-plate or plate-tube heat exchangers the enhancement of the heat transfer coefficient due to the swirling motion of the vortices reduces the thermal resistance. The reduction of the gas side thermal resistance of a gas-liquid heat exchanger is of prime economical interest since this is the dominant part of the total thermal resistance of the heat exchanger. However, a price in form of the increased flow losses has to be paid for the reduction of the thermal resistance with VGs. The investigations of the potential of VGs regarding the heat transfer enhancement and the associated flow losses in compact heat exchangers provide the background of our project.

Typical spacing of the fins and the fin-thickness in compact heat exchangers are of the order of 5mm and 0.5mm respectively. The mean gas velocity is of the order of 2-3m/s. Hence the gas flow is laminar. Wing-type VGs can be punched out of the fin, so that while remaining attached to the fin at the base it will stick out with an angle of attack to the main flow direction. Hence the flow between fins can be modeled as a three dimensional flow in a rectangular channel with vortex generators attached on the wall. The main purpose of our project is to investigate the complex three dimensional interaction of the counterrotating longitudinal vortices generated by wings or winglets with laminar flows in rectangular channels in order to shed some light on how this interaction affects the temperature field and enhances the heat transfer coefficient. Some experimental investigations on the effect of vortex generators on the heat transfer have been presented in refs. [1-3]. The present investigations are carried out numerically.

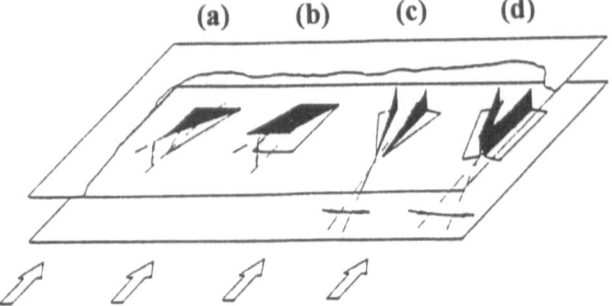

Fig. 1 : Schematic of the vortex generators on a channel wall; (a) delta wing, (b) rectangular wing, (c) delta winglets, (d) rectangular winglets.

BASIC EQUATIONS, METHOD OF SOLUTION

Because of the three dimensionality and the possible upstream effect due to the flow separation and dead water zone at the juncture of the VG and the channel wall complete three dimensional Navier-Stokes and energy equations have been used to simulate the flow. The flow medium can be a gas with constant or temperature dependent density and transport properties (viscosity, heat conductivity). The pressure dependence of the density can be neglected since the Mach number in typical applications in heat exchangers is extremely small (≈ 0.01). This justifies also the neglect of the dissipation terms in the energy equation. The basic equations have been presented in refs. [4,5]. Figure 2 shows the computational domain consisting of a rectangular channel with a pair of punched delta winglets.

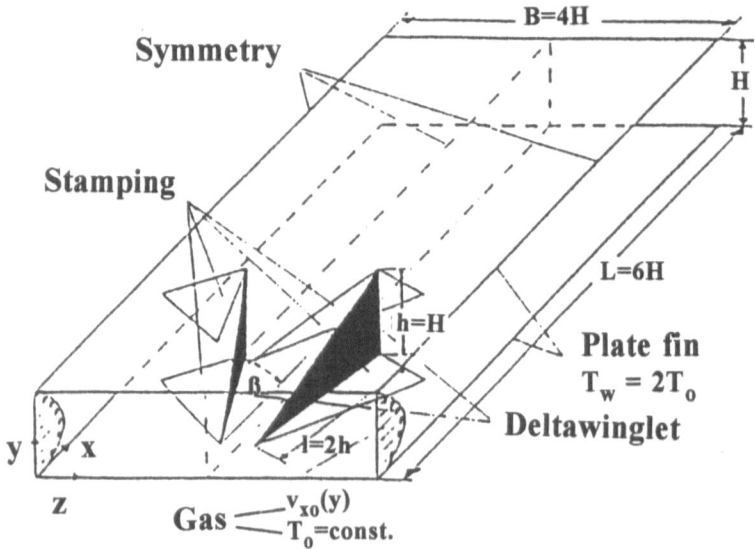

Fig. 2 : Computational domain comprising of a rectangular channel with a pair of delta winglets.

In practical applications rows of VGs will be used. Hence the investigation of an element of the channel with symmetry conditions on the side boundaries will shed significant light on the influence of VG on flow structure and heat transfer.

Computations need to be performed only for the half width of the channel because the symmetry on the vertical midplane can be imposed. No-slip conditions with constant wall temperature T_w are used on the channel walls and on the VGs. At the inlet a homogeneous or fully developed 2D velocity profile with a constant temperature T_∞ ($T_\infty \neq T_w$) is used. At the exit the second derivatives of the dependent variables in the flow direction are set equal to zero and the periodicity condition is imposed at the stampings.

The basic equations are solved by a modified version of the Marker and Cell technique [6]. The details of the computational scheme can be found in ref. [4].

RESULTS AND DISCUSSION

A number of computations have been performed in order to study the effect of the VG-form, Reynolds number, angle of attack and temperature dependent properties on the flow field and the heat transfer.

Here we present some exemplary computational results mainly with a pair of delta winglets in the channel in order to show the structure of the flow. The channel and the winglet geometries are shown in fig. 2. The flow at the inlet is hydrodynamically fully developed. Figure 3 elucidates the formation and the transport of the longitudinal vortices by showing the secondary velocity vectors at different channel cross sections for the Reynolds number Re (based on the channel height and the average velocity) of 500 and the angle of attack ß of 25°. The vortex cores are flattened near the exit. The azimuthal velocity near the vortex core can be as high as 35% of the average axial velocity in the channel. Furthermore we notice the formation of secondary vortices near the bottom wall. The secondary vortices appear due to the entrainment effect of primary vortices and the walls.

Figure 4 shows the contours of the cross flow velocities at different cross sections. Without VG the cross flow velocity will be absent since the flow at the inlet is fully developed.

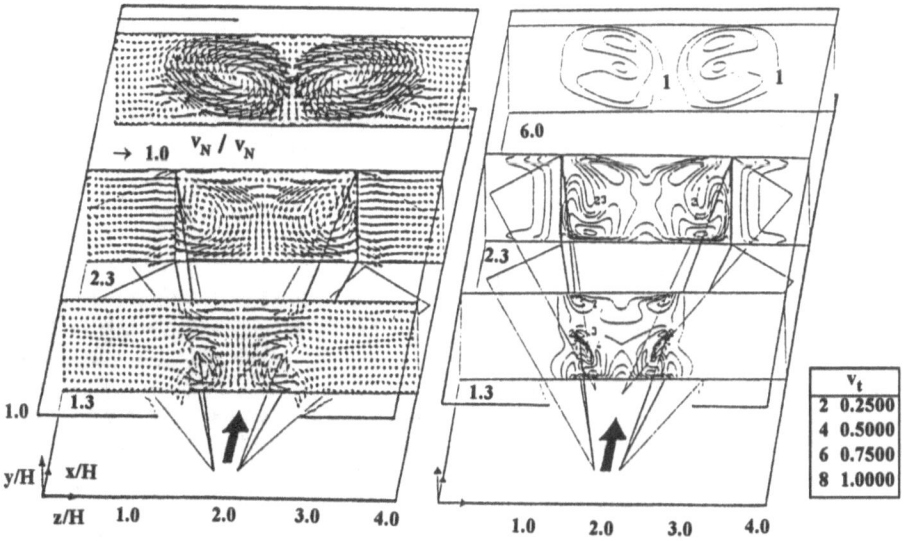

Fig. 3 : Vectorplot of cross flow velocity at different cross-sections, Re=500, ß=25°.

Fig. 4 : Contours of cross flow velocity, Re=500, ß=25°.

The contours of constant cross flow velocities give a measure of the mixing due to the VG. Figure 5 shows contours of constant axial velocity at different cross sections. The lines on the right side of the figure show the corresponding contours in the channel without VG. The axial velocity shows a velocity defect in the vortex core. The wake-like character of the axial velocity distribution is in contrast to the jet-like axial velocity in the core of the longitudinal vortex in an infinite medium.

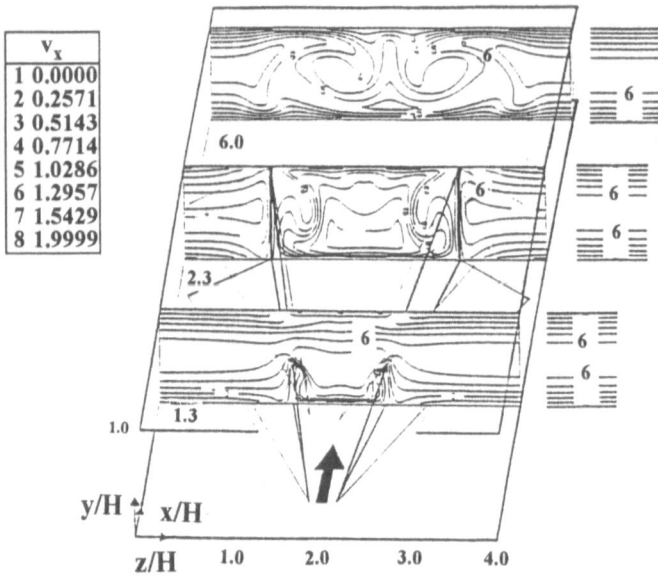

v_x	
1	0.0000
2	0.2571
3	0.5143
4	0.7714
5	1.0286
6	1.2957
7	1.5429
8	1.9999

Fig. 5 : Contours of axial velocity at different cross sections, Re=500, ß=25°

Figure 6 shows isotherms in the channel at different cross sections for T_w/T_∞=2 where T_w is the wall temperature and T_∞ is the gas temperature at the inlet. The isotherms for a thermally developing flow in the channel without VG are shown on the right. Structurally the isotherms in the channel with VG have some similarity to the corresponding contours of the axial velocity. Comparison of the isotherms for the cases with and without VG shows again the extent of mixing due the longitudinal vortices. Without VG the ^ore remains thermally undisturbed. With VG the area surrounded by the isotherm 2 or less moves towards sides of the channel, see the isotherms at x/H=6. In the middle of the channel the smallest isotherm for x/H=6 is 3, T/T_o=1.400.

Figure 7 compares the cross-averaged Nusselt numbers Nu_x for a thermally developing channel flow with and without VGs. As VGs a pair of delta winglets (DWP) and rectangular winglets (RWP) have been used. The Nusselt number is defined with the difference of the wall and the bulk temperatures. The Nusselt number distribution directly downstream of the inlet shows the typical fast decrease of a thermally developing flow. At the forward edge of the VG the Nu_x increases again to a second peak immediately behind the rear edge of the VG. The peak value for DWP is 12 and that for RWP is 11. At x/H=4 the Nu_x for the RWP and the DWP becomes the same and roughly equal to 7.4 compared to the corresponding Nu_x of 5.5 for a channel without VG. For x/H>4 Nu_x decreases slowly for all three configurations. Surprisingly even at x/H=10, the difference between the values of Nu_x for flows with and without VG remains nearly equal to 2.

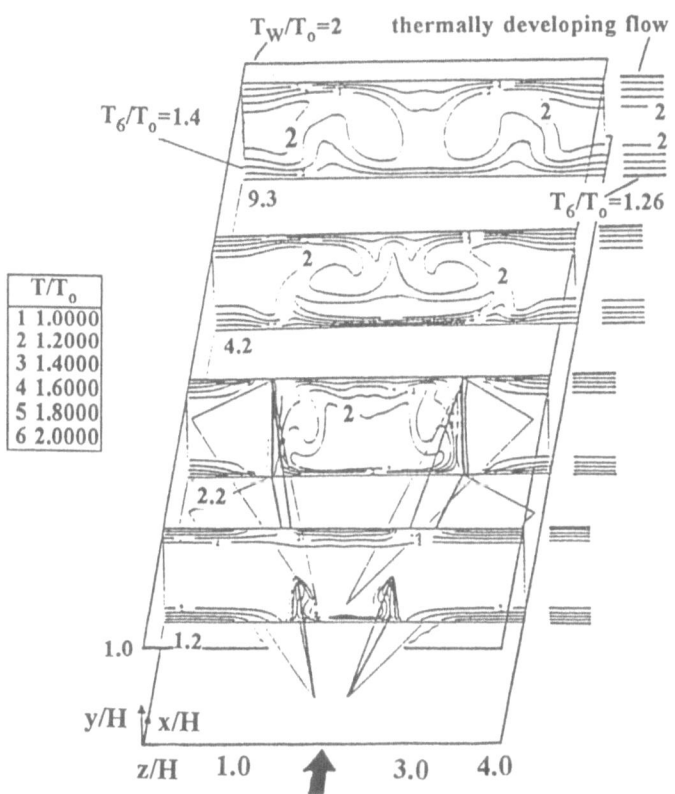

Fig. 6 : Isotherms at different cross sections, Re=500, ß=25°

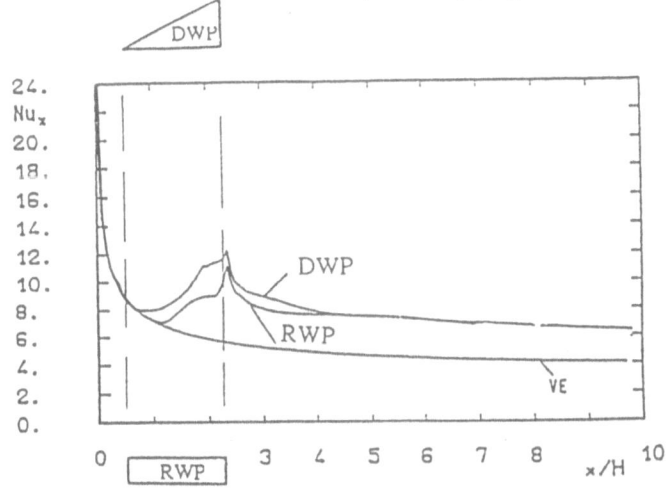

Fig. 7 : Cross averaged Nusselt number, Re=500, ß=25°

Table 1 compares the global results of the heat transfer and flow losses in a channel without and with a pair of punched delta winglets. The Reynolds number is 500 and the channel length is 10H. The heat transfer is characterized by the Nusselt number averaged over both walls. The flow losses are characterized by the apparent friction factor f_{app}.

Table 1 : Comparison of the bulk temperatures, Nusselt numbers and f_{app} for Re=500 in a channel without and with a pair of delta winglets (DWP)

	Bulk temperature at x/H=10	Mean Nusselt numbers	$f_{app} \cdot Re$
without VG	1.258	5.27	12
with DWP	1.393	8.01	33

An increase of 52% of the Nusselt number causes an increase of 175% of f_{app}.
As already mentioned, a limited number of parametric studies have been carried out. Refs. [4,5,7] present details for delta wings and delta winglets in a channel and investigate also the influence of temperature dependent properties. Ref. 8 presents the parametric studies with regard to the angle of attack for delta and rectangular winglets. Some of the above comparisons have also been presented in ref. [9-12]. Here we present the influence of the angle of attack ß on the bulk temperature T_B at the exit (x=6H). Figure 8 compares the bulk temperature distributions for a pair of deltawinglets (DWP) and a pair of rectangular winglets (RWP) for different ß at Re=500. Without VG, the bulk temperature for the thermally developing flow at the channel exit is 1.18. Figure 8 shows that at 45°≤ß≤55° both forms of VGs give maximum heat transfer enhancement.

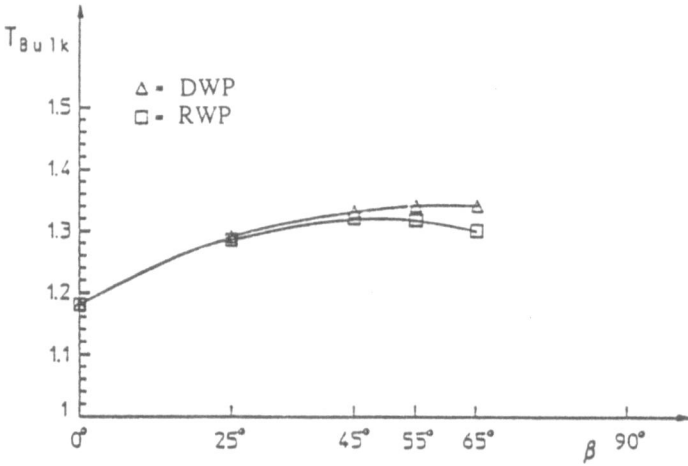

Fig. 8 : Bulk temperature at the exit, Re=500, T_{Bulk}=1.18 for channel without VG

Table 2 compares the increases in f_{app} and Nu over the corresponding values of a channel of length 6H without VG for RWP and DWP for different ß.
The maximum enhancement in Nu is obtained at ß=55° for DWP and at ß=45° for RWP. With further increase in ß, Nu decreases but f_{app} still increases.

Table 2 : Comparison of the average Nu, f_{app} at different β; Re=500
Nu$_o$=5.27; $f_{app,o} \cdot$Re=12

Increase %		β=25°	β=45°	β=55°	β=65°
Nu/Nu$_o$	DWP	24.6	40.2	42.3	40
	RWP	23.3	39.5	37.6	26.6
$f_{app}/f_{app,o}$	DWP	191.7	554.2	800	1000
	RWP	149.2	443.3	608.3	800

SUMMARY AND CONCLUSION

Velocity and temperature fields of laminar channel flows with a row of built-in longitudinal vortex generators in the form of wings and winglets have been computed. For this purpose an existing computer code has been adapted and modified. The VGs generate a pair of counterrotating stable vortices even at an angle of 65°.

The influence of the parameters (Re,β), temperature-dependent properties and the VG-forms on the flow and heat transfer have been presented in refs [4,5,7-12]. The gross structure of the flow is similar for all four types of VGs; delta wing, rectangular wing, delta winglet pair and rectangular winglet pair. Drawing on the above references we notice that the generated longitudinal vortices in a channel have the following characteristics that make them different from the corresponding vortices generated by VGs in an infinite medium :

i)   An axial velocity defect relative to the basic channel flow exists. This defect occurs in the vortex core whereas in free vortex cores a velocity excess may exist.

ii)   Severe structural changes of the vortex or breakdown as is typical of free vortices at large angles of attack (β>25°) have not been observed even at an angle of attack of 65°. The favorable pressure gradient in a channel flow stabilizes the vortices. Already a longitudinal vortex generated at small angles of attack β<25° extends over the whole cross section.

iii)   The axes of the vortex cores and the form of their cross sections are strongly dependent on the vortex generator form and the ratio of the channel height to the wing span. Behind the delta wings the vortex cross sections become flattened. Behind the winglets, this flattening is less pronounced.

iv)   Because of the entrainment between the main vortices and the channel walls, secondary vortices are generated.

The heat transfer enhancement induced by vortex generators to the basic channel flow increases with Reynolds number and with angle of attack β up to an optimum β~50°. The average friction factor on the other hand increases monotonously with β.

Variable properties have strong local influences on the velocity field due to the density variation [4,5]. A 30% density variation can reduce the heat transfer by 10%.

REFERENCES

[1]   RUSSEL,C.M.B., JONES,T.V., LEE,G. : "Heat Transfer Enhancement Using Vortex Generators", Proc. 7th IHTC, vol. 3, 1982, pp. 283-288.

[2]   TURK,A.J., JUNKHAN,G.H. : "Heat Transfer Enhancement Downstream of Vortex Generators on a Flat Plate", Proc. 8th IHTC, vol. 6, 1986, pp. 2903-2908.

[3]   FIEBIG,M., KALLWEIT,P., MITRA,N.K., TIGGELBECK,S. : "Heat Transfer Enhancement and Drag by Longitudinal Vortex Generators in Channel Flow", ETF Science, vol. 4, 1991, pp. 103-114.

[4]   BROCKMEIER,U. : "Numerisches Verfahren zur Berechnung dreidimensionaler Strömungs- und Temperaturfelder in Kanälen mit Längswirbelerzeugern und Untersuchung von Wärmeübergang und Strömungsverlust", Dissertation, Ruhr-Universität Bochum, Germany, 1987.

[5]   FIEBIG,M., BROCKMEIER,U., MITRA,N.K., GÜNTERMANN,T. : "Structure of Velocity and Temperature Fields in Laminar Channel Flows with Longitudinal Vortex Generators", Num. Heat Transfer, A, vol. 15, 1989, pp. 281-302.

[6]   HIRT,C.W., NICHOLS,B.D., ROMERO,N.C. : "SOLA - A Numerical Solution Algorithm for Transient Fluid Flows", Los Alamos Scientific Laboratory Report, LA - 5652, 1975.

[7]   BROCKMEIER,U., FIEBIG,M., GÜNTERMANN,T., MITRA,N.K. : "Heat Transfer Enhancement In Fin-Plate Heat Exchangers by Wing-Type Vortex Generators", Chem. Eng. Technol. 12, pp. 288-294, 1989.

[8]   BRENNER,A. : "Entwurf modularer Programmbausteine zur numerischen Untersuchung des optimalen Anstellwinkels ausgestanzter Deltawinglet und Rechteckwinglet in laminarer Kanalströmung", Interner Bericht 90/8, 1990 Lehrstuhl für Wärme- und Stoffübertragung ITF, Ruhr-Universität, Bochum.

[9]   FIEBIG,M., GÜNTERMANN,T. : "Heat Transfer Enhancement by Longitudinal Vortex Generators", Rio de Janeiro, Proc. 10th Brazil. Cong. Mech. Eng., pp. 445-448, 1989.

[10]   GÜNTERMANN,T., FIEBIG,M. : "Wechselwirkung zwischen Längswirbeln und laminarer Kanalströmung", Z. angew. Math. Mech. 70 (1990) 5, T 466-T 468.

[11]   GÜNTERMANN,T., FIEBIG,M., MITRA,N.K. : "Calculation of Heat Transfer Augmentation and Flow Losses in Laminar Channel Flows Influenced by Longitudinal Vortex Generators", Bochum, Proc. Eurotherm Seminar No. 9 : "Heat Transfer in Single Phase Flows", July 10.-11. 1989, ENTROPIE, Créteil, France, pp. 17-20.

[12]   GÜNTERMANN,T., FIEBIG,M., MITRA,N.K. : "Heat Transfer Enhancement in Heat Exchangers by Longitudinal Vortex Generators", Fluid Machinery Components, Fed-Vol 101, ed. D.L. Rhode, J. Tuzson, ASME WAM, pp. 83-90, 1990.

Experimental Investigation of the Three-Dimensional Separated Flow in an Annular Compressor Cascade

H.E. Gallus, M.D. Schulz, C.A. Poensgen

Institut für Strahlantriebe und Turbomaschinen
RWTH Aachen
Templergraben 55, D-5100 Aachen

Introduction

One of the major objectives in aircraft engine design is to achieve a high power to weight ratio. Since the weight of the compressor in modern aircraft engines amounts 2/3 of the entire engine, its weight reduction is of high economical interest to aircraft industries.

This trend leads to a reduced number of stages in the compressor and hence to higher stage loading and closer stage spacing. Increased loading can be achieved by either higher rotor tip speed or higher blade row turning. Higher tip speeds result in transonic or supersonic flow subsequently in shock losses, shock boundary layer interactions and the well known limitation of stage operation range. On the other hand, blade turning is limited by the blade row diffusion ([1] and [2]) and the onset of flow separation. Flow separation though reduces the effective area of the flow channel (Fig 1.), generates high losses and consequently leads to low engine efficiency. Losses and flow deviation caused by flow separation are also very hard to predict, and they are difficult to account for in the design process.

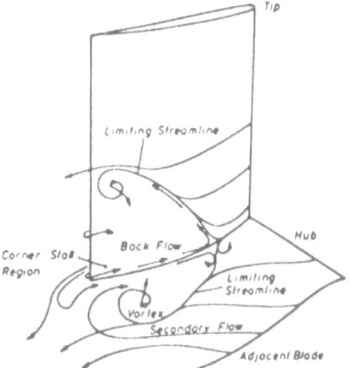

Fig. 1: Flow Separation at Hub Corner

Therefore, there is a need to a better insight into the very complex flow in highly loaded compressor blade rows and the physical phenomena associated with flow separation. In particular, only very little is known about the effects, which rotor-stator interaction, the fluctuation of the free stream velocity, the periodically changing loading of the blade profiles and the enhanced turbulence caused by the wakes shed off an upstream blade row, have on the development and the extent of separated regions.

Experimental Facility

The measurements were carried out in the annular compressor cascade described in detail by [3] and [4]. The cascade consists of 24 untwisted blades mounted on the hub, with a hub to tip ratio of 0.75 and a tip diameter of 428 mm. The aspect ratio is 0.86 and the solidity at midspan is 0.78. Blade metal angles at inlet and exit are 44° and 15°, respectively (measured from the axial); other geometrical parameters are shown in Fig. 2. The blade profiles are radially stacked at their center of gravity, and the zero incidence flow angle is 42°.

Forty-eight variable inlet guide vanes far upstream of the cascade provide an inlet swirl.

Rotor-stator viscous wake interaction was generated by a rotor with rotating cylinders (simulating blades) located at 57 percent chord upstream of the cascade. The rotor incorporates 24 rods with a diameter of 5.3 mm. For the investigations reported here, the rotor speed was kept constant at 3000 RPM.

For the investigations with steady, uniform inflow the rotor was removed and replaced by a non rotating hub. For both test series, measurements at different blade loadings, denoted in the following by the measured midspan inlet flow angle, have been performed.

Measuring the unsteady, three-dimensional flow inside a compressor cascade is an enormous task, which - as is inherent in separated flows might give ambiguous data. It is therefore necessary to apply various techniques to the same flow in order to

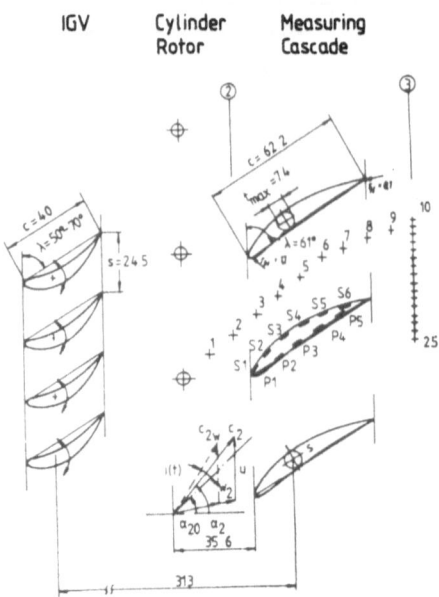

Fig. 2: Cross Section of the Annular Cascade Test Rig

corroborate the measurements and to avoid conclusions from erroneous data with respect to the limited length of this paper. The used measurement techniques are not described here. This can be found in [4], [5] and [6].

Experimental Results

The inlet flow angle was varied from 39° up to 54°. Only some representative results will be discussed here. For steady, homogeneous stator inlet flow (without rotor), Fig. 3 shows the flow along the hub at an inlet flow angle of 44.2°, visualized by dye injection. The most striking feature is the region of separated flow that originates near the hub endwall and the blade suction side at all flow angles. This region extends in size with increasing angle of attack, and in case of high incidence angle it starts at the 20 percent axial chord location and extends radially outward to 75 percent span at the trailing edge. There appears to be a large amount of radial outflow on the vane surface within and outside of the separation zone. A region of back flow, very similar to that seen by [12] and [13], is also visible within the separated flow region on the suction surface and on the hub. Some of the back flow originates in the pressure surface boundary layer of the blade, turns around the trailing edge, and moves upstream towards the leading edge.

The flow visualization shows a significant overturning of the hub boundary layer flow due to the cross passage pressure gradient, even more pronounced at high angles of attack. Close to the suction surface the cross-passage flow moves upstream into the separation region, joining the back flow and generating a vortex with a clockwise orientation.

This complex flow structure is very well represented by the numerical calculation. More details about the structure of the three-dimensional flow inside of the stator passage can be found in [4],[7] and [10]. The following discussion shows the influence of the unsteady flow - due to the presence of a rotor - on the downstream stator flow field.

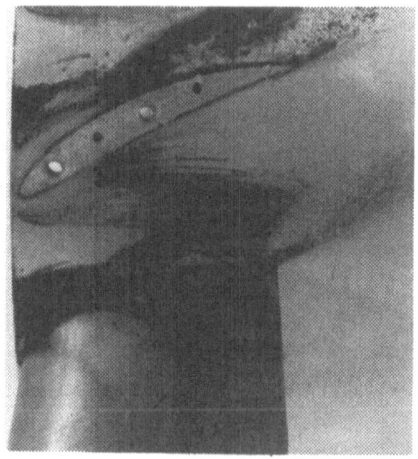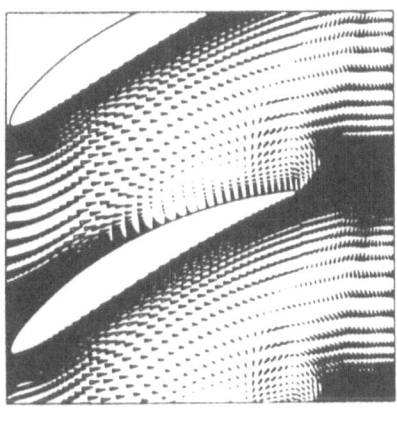

Fig. 3: Comparison of the Surface Flow at the Hub;
Flow Visualization <-> Navier Stokes Calculation [11]

The major objective of these investigations was to determine the impact of the wake produced unsteadiness on the development of the stator losses. In a first step the rotor with the cylindrical spokes was removed. The total pressure loss distribution at the stator exit plane was measured using five-hole probes. Detailed boundary layer investigations based on hot-wire and hot-film anemometry gave insight into the development of the blade boundary layers. In a second step the rotor with cylindrical spokes was mounted upstream of the stator blade row and the tests were repeated.

In order to simulate the periodic wake flow, shed from abladed rotor, a rotor with cylindrical rods was chosen. The goal was to obtain wake parameters (e.g. dimensionless width, depth and turbulence) which are typical for those observed in real machines. Thus, the influence of the wake disturbed inlet flow on the profile boundary layer can be simulated very well.

The results of the blade boundary layer measurements are focussed in Fig. 4. Here, the development of the boundary layer with chord is plotted against the inlet flow angle. The upper figure represents the results obtained at midspan of the blading and the lower one shows the effects close to the hub, where a large corner flow separation occurs. The effect of the rotor wakes on the transition process has been discussed in detail by [8] and [9] and will be summarized here.

Comparing the results with steady inlet flow and wake disturbed inlet flow, three major changes can be observed. First, the three-dimensional laminar separation bubble (LSB), found in the case without the rotor, disappears at all incidence angles at the suction side with an upstream rotor. Secondly, the regions of turbulent separated flow in the vicinity of the endwalls decrease in the rotor case, especially at smaller inlet angles (region TS in Fig. 4). Thirdly, the midspan boundary layer seems to separate earlier at the trailing edge in the rotor case (region TS in Fig. 4).

Fig. 5 displays the development of the total pressure loss contours at inlet flow angles of 44.2° (2° incidence) and 49.2° (7° incidence) for both test cases. For the rotor test case the contour lines indicate a higher profile loss near midspan and lower losses in the wake of the

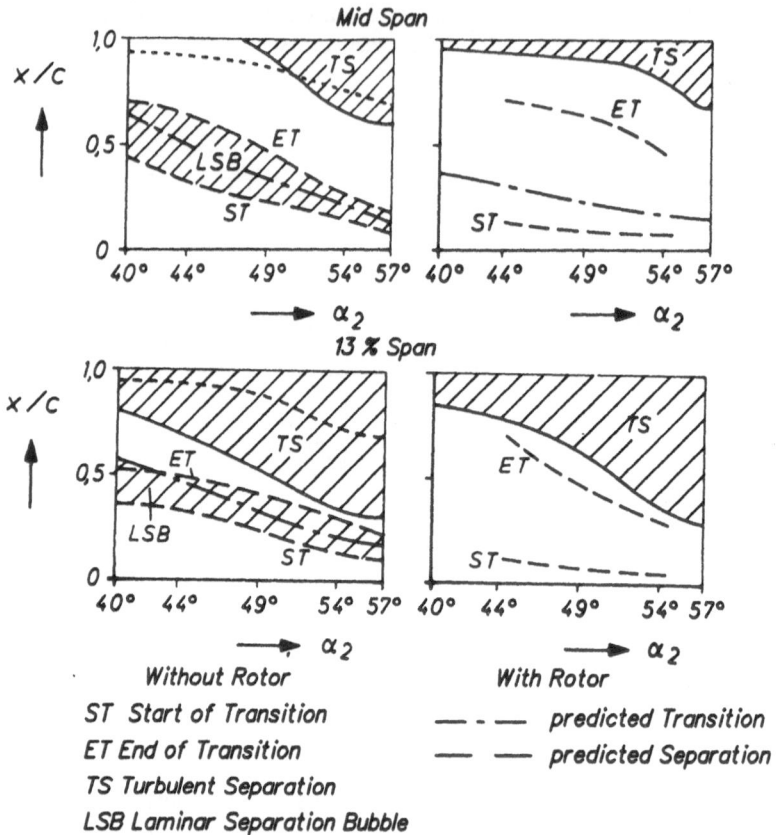

Mid Span

13 % Span

Without Rotor

With Rotor

ST Start of Transition
ET End of Transition
TS Turbulent Separation
LSB Laminar Separation Bubble

— · — predicted Transition
— — predicted Separation

Fig. 4: Summary of the Blade Boundary Layer Investigations on the Blade Suction Side

hub corner stall. Similar to the distribution of the pressure losses the results shown in Fig. 4 indicate a decrease of the hub corner stall in the rotor case.

The variation with incidence of the mass-averaged loss coefficient at midspan ω and the overall loss coefficient ω is shown in Fig. 6 for both test cases. As it was already discussed, with rotor the earlier onset of transition and the subsequent development of the turbulent boundary layer, as well as the enhanced channel diffusion produce higher mid span losses. Despite this increase in profile losses, the overall losses decrease as much as 40 percent due to the decreased hub corner stall, except at negative incidence.

The unsteady flow field downstream of the stator was examined using multi-sensor hot-wire anemometry. Thus, it was possible to resolve the flow downstream of the stator also in the time-domain. Fig. 7. displays the influence of the rotor wakes on the velocity field downstream of the stator for two different rotor/stator positions. The different phase angles represent the time dependent location of the cylindrical rotor rods. As the rotor turns at 3000 rpm and as there are 24 rotor rods, one cycle lasts 0.83 milliseconds. The influence of the tip leakage flow can be seen clearly. In the lower part of the blade passage, a periodic counterrotating vortex system is visible. At a phase angle of 270° the intensity of the vortex system reaches its maximum. After the rotor has been moved half its rod spacing, this vortex system is getting smaller. Corresponding to this vortex motion, a strong periodic fluc-

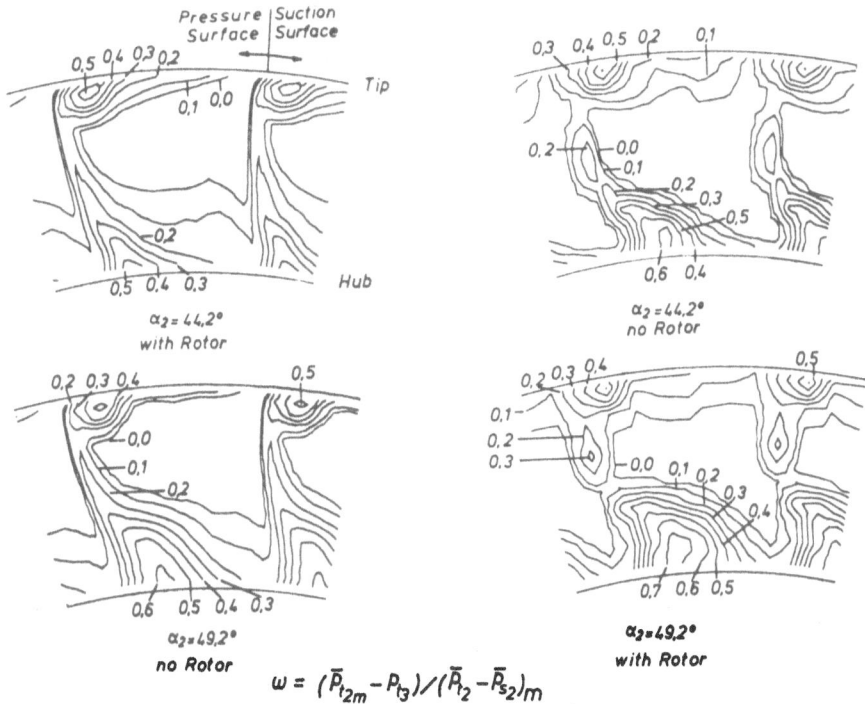

$$\omega = (\bar{P}_{t2m} - P_{t3})/(\bar{P}_{t2} - \bar{P}_{s2})_m$$

Fig. 5: Total Pressure Contours Downstream of the Stator

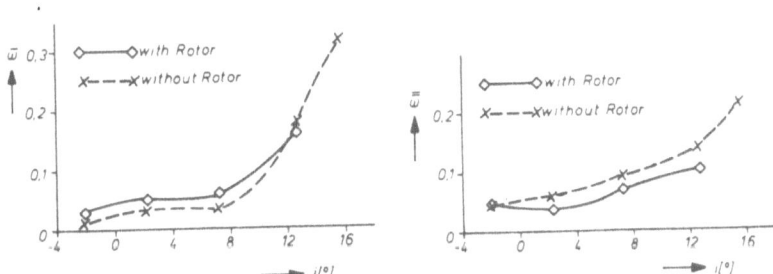

Fig. 6: Mass Averaged Cascade Losses with and without Rotor
Losses at Midspan (left), Overall Losses (right)

tuation of the velocity was found at the boundaries of the stator wake downstream of the hub corner separation. The isocontour diagrams of the velocity (Fig. 7) show that the size of the hub corner separation bubble changes periodically with time. At a phase angle of 90°, the hub corner separation reaches its spatial minimum. The inverse is true at a phase angle of 270°. More details about the influences of the fluctuating hub-corner stall on the aerodynamic loading and aerodynamic blockage are given by [10].

In order to assess the exciting energy of the rotor wake, which reaches the region of the hub corner stall, the wake decay was recorded along a S2M surface through the stator

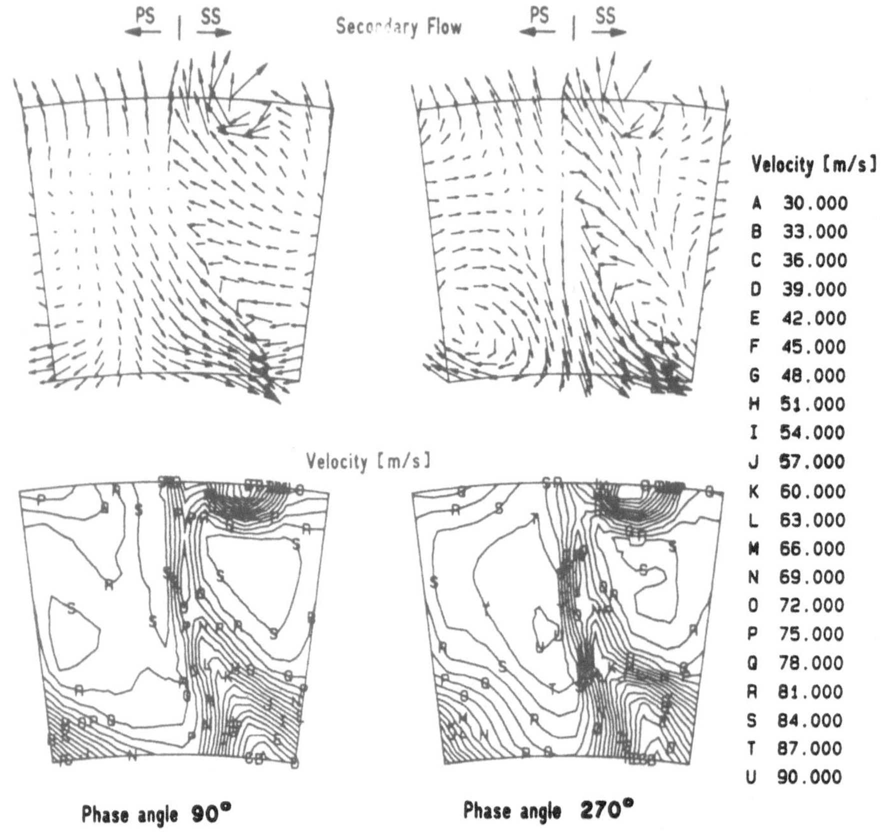

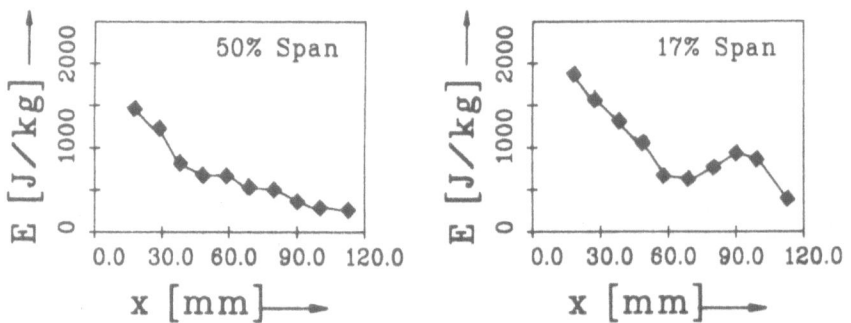

PS
SS
Secondary Flow
PS
SS
Velocity [m/s]
A 30.000
B 33.000
C 36.000
D 39.000
E 42.000
F 45.000
G 48.000
H 51.000
I 54.000
J 57.000
K 60.000
L 63.000
M 66.000
N 69.000
O 72.000
P 75.000
Q 78.000
R 81.000
S 84.000
T 87.000
U 90.000
Velocity [m/s]
Phase angle 90°
Phase angle 270°

Fig. 7: Unsteady Flow Field Downstream of the Compressor Stator Blade Row; Seconadary Flow (Upper Row) and Velocity Field (Lower Row)

50% Span
E [J/kg]
0
1000
2000
0.0
30.0
60.0
90.0
120.0
x [mm]
17% Span
E [J/kg]
0
1000
2000
0.0
30.0
60.0
90.0
120.0
x [mm]

Fig. 8: Development of the Defect of the Kinetic Energy During the Wake Decay Process at Midspan and 17 Percent Span (α_2 = 49.2°)

passage. The rotor wake causes a temporal defect of kinetic energy. Fig. 8 displays the development of the wake decay with axial length and with respect to the defect of the kinetic energy inside the rotor wake. At midspan, the wake decays nearly in an exponential manner.

Close to the hub at 17% span an amplification of the rotor wake is determined (Fig. 8). Here the hot-wire sensors were close to the hub corner stall region. The periodic motion of the hub corner stall causes locally large periodic velocity fluctuations. Thus the rotor wakes, which decay while they are transported through the stator passage to about half their size compared to the stator inlet plane, are amplified. This effect represents a nonlinear amplification of the rotor wake and will convect rotor wake information further downstream.

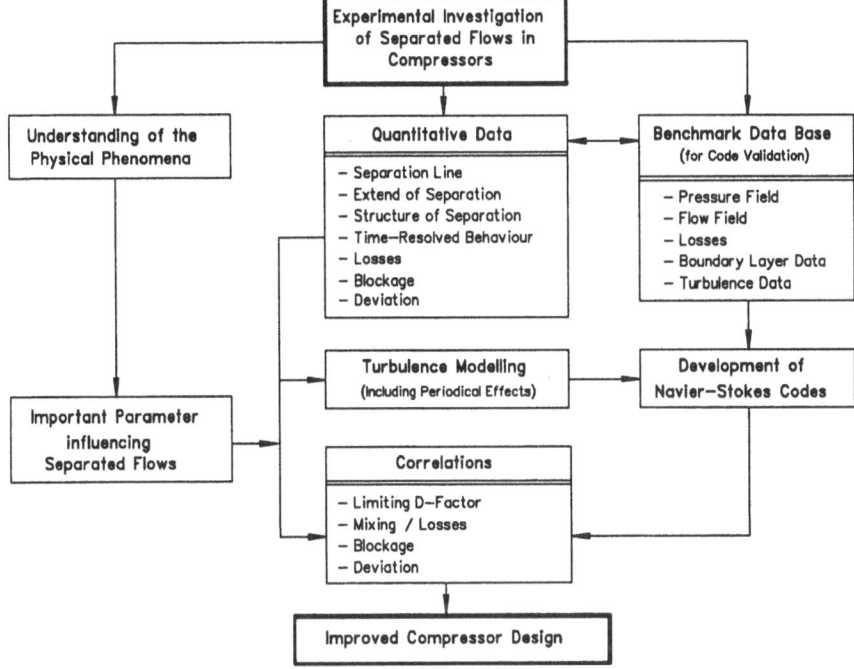

Fig. 9: The Importance and the Benefit of Experimental Investigations of Separated Flows in Compressors

Concluding Remarks

The physical phenomena associated with separated flows are very complex, and only very little is known about its structure and the parameters which describe the extent and the appearance of separated flows. As Fig. 9 summarizes, experimental investigations of separated flows in compressors lead to an improved understanding of the physical phenomena and, based on this, allow to define appropriate parameters which have a major influence on flow separation. They also provide quantitative data about the onset, the extent and the structure of separated flows. The data can be used for either correlations for the onset of separation in a three-dimensional flow and for the associated losses. They are very important for compressor design procedures, too. Those are still mostly based on the assumption of axisymetric flow, and their applicability to the flow in compressors therefore largely depends on the quality of implemented correlations for losses, flow blockage, mixing and flow deviations.

Detailed information about the structure of separated flows taken from the experimental data can be used to improve turbulence models, which are necessary for computational codes based on the Reynolds-averaged Navier-Stokes-equations. These turbulence models should account for the periodical unsteadiness of the flow in compressor stages. Results from numerical calculations then can be compared against the acquired benchmark data base as it was demonstrated by [11]. The experimental investigation of separated flows is therefore a necessary and rewarding task in turbomachine research, and it should be continued in the future.

References

[1]   De Haller P. (1955) "Das Verhalten von Tragflügelgittern in Axialverdichtern und im Windkanal" VDI-Berichte, Bd. 3

[2]   Lieblein S. (1953) "Diffusion Factor for Estimating Loses and Limiting Blade Loadings in Axial Flow Compressor Blade Elements" NACA RM E53D01

[3]   Grollius H.W. (1981) "Experimentelle Untersuchung von Rotor Nachlaufdellen und deren Auswirkungen auf die dynamische Belastung axialer Verdichter- und Turbinengitter", Dissertation RWTH Aachen,

[4]   Schulz H.D. (1989) "Experimentelle Untersuchung der dreidimensionalen abgelösten Strömung in einem Axialverdichterringgitter", Dissertation RWTH Aachen, 1989

[5]   Poensgen C.A (1989) "Ein Verfahren zur Vermessung der instationären dreidimensionalen Strömungsvektoren in Turbomaschinen", Institut für Strahlantriebe und Turboarbeitsmaschinen RWTH Aachen, Mitteilung Nr. 89-01, 1989

[6]   Poensgen C.A. (1991) "Experimentelle Untersuchung der Strömung in einer Unterschall-Axialverdichterstufe bei hoher Drosselung und im Rotating Stall", Dissertation RWTH Aachen, 1991

[7]   Schulz H.D., Gallus H.E. (1988) "Experimental Investigations of the Three-Dimensional Flow in an Annular Compressor Cascade", Journal of Turbomachinery, Vol. 110

[8]   Schulz H.D., Gallus H.E., Lakshminarayana B. (1989) "Three-Dimensional Separated Flow Field in the Endwall Region of an Annular Compressor Cascade in the Presence of Rotor-Stator Interaction"
Part I: Quasi-Steady Flow Field and Comparison with Steady-State Data
Part II: Unsteady Flow and Pressure Field
Journal of Turbomachinery, Vol. 112

[9]   Schulz H.D., Gallus H.E., (1989) "Experimental Investigation of the Influence of Rotor Wakes on the Development of the Profile Boundary Layer and the Performance of an Annular Compressor Cascade" AGARD CPP-468/469, Conference on Unsteady Aerodynamic Phenomena in Turbomachines, Luxembourg 1989

[10]   Poensgen C.A., Gallus H.E. (1990) "Three-Dimensional Wake Decay Inside of a Compressor Cascade and its Influence on the Downstream Unsteady Flow Field"
Part I: Wake Decay Characteristics in the Flow Passage
Part II: Unsteady Flow Field Downstream of the Stator
Journal of Turbomachinery, Vol 113, April 1991, pp 180 ff

[11]   Gallus H.E., Hah C., Schulz H.D. (1990) "Experimental and Numerical Investigation of Three-Dimensional Viscous Flows and Vortex Motion inside an Annular Compressor Blade Row" Journal of Turbomchinery, Vol. 113, April 1991

[12]   Joslyn H.D., Dring R.P., (1985) "Axial Compressor Stator Aerodynamics", Journal of Engineering for Gas Turbines and Power, Vol 107

[13]   Dong Y., Gallimore S.J., Hodson H.P., (1987) "Three-Dimensional Flows and Loss Reduction in Axial Compressors" Journal of Turbomachinery, Vol. 109

LASER MEASUREMENTS OF ROTATIONAL FLOWS FOR THE VALIDATION OF EULER–CODES

U. Ganzer, R. Kelm

Institut für Luft– und Raumfahrt der Technischen Universität Berlin,
Marchstr.14, D–1000 Berlin 10

ABSTRACT

The goal of this study was to determine the lee–side flow of a delta wing experimentally and to compare this data with results of currently available theoretical models. Velocity measurements using a Laser–2–focus velocimeter had to be carried out in addition to flow visualization using Schlieren, Oil–film and Laser–light–sheet techniques. Three–dimensional flow–velocity vectors had to be determined using two angels at a perticular position.

The experiments were carried out in the high–speed–wind–tunnel at the Institut für Luft– und Raumfahrt in Berlin for Mach Numbers M = 0.85 and 1.2. Laser measurements in the immediate region of the vortex core proved to be extraordinary difficult. The region of the vortex center could not even be located. The reason for this was that the tracer particles necessary for the measurements exhibited great lag with respect to the high acceleration of the flow. Therefore, all measurements of the region near the core of the vortex have not been successful.

The slip effect was quantified through a theoretical simulation of the particle behaviour. Previous efforts had given incorrect results because of false assumptions. Extensions of previous assumptions using the Knudsen Number in the description of the particle drag were implemented.

Reduction of the lag between particles and flow was attempted through the development of particles of low density and better drag characteristic. The use of SiO_2

particles made it possible to achieve laser measurements in the core of transonic vortices for the first time. The experimental results on the delta wing demonstrated a new flow: a shock–induced secondary separation. The laser measurements were augmented by flow visualization techniques. Quantitative comparisons of the experimental results with numerical solutions of the Euler equation showed good agreement in areas where the viscous effects are not important.

REFERENCE

[1]   *Kelm, R.*: Laser–Messungen in transsonischen Wirbelströmungen. VDI–Fort schrittbericht Nr. 182, Reihe 7: Strömungstechnik, 1990, VDI–Verlag, Düsseldorf

LDA-INVESTIGATIONS OF THE SEPARATED FLOW OVER SLENDER WINGS

S. Kommallein, D. Hummel

TU Braunschweig, Institut für Strömungsmechanik
Bienroder Weg 3,D - 3300 Braunschweig, Germany

Summary

The flowfield over inclined slender delta wings has been investigated using a 3D-Laser-Doppler Anemometer. Preliminary tests led to the result that a sufficient distribution of scattering particles in the whole vortex flowfield can be achieved by a proper combination of particle diameter and free stream Reynoldsnumber. The available 3D-LDA system with three counters leads to reliable results for all three components of the mean velocity but for two components of the velocity fluctuations only. It is demonstrated that the reliability of the third component of the velocity fluctuations can also be achieved by the application of burst spectrum analyzers (BSA) as signal processors.

Vortex breakdown has been analyzed on an A = 2.31 delta wing at α = 16.5° by LDA measurements of the flowfield in two crossflow planes upstream and downstream of the vortex breakdown point. The results confirm former probe measurements to some extend, but concerning the shape of the rolled-up vortex sheet an enlargement of the area within this vortex sheet downstream of the breakdown point has been found. In this region the turbulent fluctuations reach extremely high values and this demonstrates the unsteadiness of the flow which appears in the course of vortex breakdown.

Introduction

The well known vortex formation over slender wings at angles of attack has so far been investigated by means of probe measurements [1], [2]. In this case the local velocity vector as well as the total pressure distribution in the flowfield are determined. Concerning the vortex breakdown phenomenon at high angles of attack the disturbance produced by the probe is likely to effect the structure of the flowfield itself. This can be avoided by the application of the Laser Doppler Anemometry (LDA). Using a 3D-LDA system informations about pressures in the flowfield are no longer available, but the velocity vector as well as the velocity fluctuations can be determined. After the installation of a 3D-LDA system at the 1.3 m windtunnel of Institut für Strömungsmechanik at TU Braunschweig a research program has been initiated in order to investigate the flowfield over delta wings at very high angles of attack without and with vortex breakdown. In the course of this program preliminary investigations were necessary concerning experimental techniques such as proper seeding of the vortex flow with particles and the determination of the flow turbulence. Some results have been reported in Refs. [3] and [4] and the program is still in progress.

Selection of suitable scattering particles

At the beginning of the investigations scattering particles consisting of a mixture of glycerin (10 %) and distilled water (90 %) were used. The water-glycerin mixture was dispersed into 0.6 to 6 µm small particles by means of an ultra-sound atomizer. In experiments on wings with different sweepback at various angles of attack certain combinations of these two parameters turned out for which very strong vortices with high circulation occurred, and in the centres of these vortices no scattering particles were present.

$d = 4\mu m$, $Re = 1.9 \cdot 10^6$ $d = 4\mu m$, $Re = 1.9 \cdot 10^6$

$d = 4\mu m$, $Re = 0.9 \cdot 10^6$ $d = 2\mu m$, $Re = 0.9 \cdot 10^6$

Fig. 1: Particle distribution in a crossflow plane at the trailing-edge of an A = 1 delta wing at α = 20.5° depending on particle diameter d and free stream Reynoldsnumber Re.

For a more detailed investigation of this problem laser light-sheet investigations were carried out by means of another particle generator which produced particles with an adjustable diameter. Fig. 1 shows photos of laser lightsheet visualizations for different flow velocities (circulations) and different particle diameters (particle masses). It turned out that with decreasing particle mass and therefore with a decrease of the particle's centrifugal force the region without particles around the vortex axis is reduced. The same effect can be observed for a decrease of the flow velocity corresponding to a reduction in Reynoldsnumber. In this case the circulation of the vortices and therefore the centrifugal forces on the particles decrease also. This clearly indicates that the problem of the distribution of the scattering particles in vortex flows strongly depends on the equilibrium between the centrifugal force on the one hand and the drag force on the particles on the other hand. The point where the scattering particles are introduced into the flow plays a minor role.

As an outcome of these preliminary investigations a new generator for the production of 1 μm particles was chosen and a continuous distribution of these particles in the vortex flow fields has been achieved by a proper adjustment of the free stream velocity.

Determination of the turbulent fluctuation velocities

In unsteady flow mean values for the velocity components can be determined from the measuring values and the corresponding standard deviation is a quantitative measure for the turbulent fluctuations. The problem concerning the determination of the turbulence level arises from the fact that it is not possible to distinguish between the real turbulence of the flow and an artificial turbulence generated by the uncertainty of the measuring device.

For this reason in the following the accuracy of LDA measurements will be discussed in detail. The available three-component-LDA system determines three non-orthogonal velocity components c_1, c_2, c_3. During the evaluation they are converted into the cartesian velocity components u, v and w of the wing-fixed coordinate system using a coordinate transformation. The measuring uncertainties Δc_1, Δc_2 and Δc_3 lead to corresponding uncertainties Δu, Δv and Δw in the cartesian velocity components which are described by Gauß's law of error propagation as

$$\Delta u = \sqrt{a_{11}^2 \Delta c_1^2 + a_{12}^2 \Delta c_2^2 + a_{13}^2 \Delta c_3^2}$$

$$\Delta v = \sqrt{a_{21}^2 \Delta c_1^2 + a_{22}^2 \Delta c_2^2 + a_{23}^2 \Delta c_3^2} \qquad (1)$$

$$\Delta w = \sqrt{a_{31}^2 \Delta c_1^2 + a_{32}^2 \Delta c_2^2 + a_{33}^2 \Delta c_3^2} \ .$$

The quantities a_{ij} are the coefficients of the transformation matrix. Since identical counters were used in the measurements on all three channels, the three measuring uncertainties can be assumed as $\Delta c_1 = \Delta c_2 = \Delta c_3 = \Delta c$. If the coefficients of the transformation matrix of the existing LDA system are introduced into equ. (1) yields

$$\Delta u = 0{,}7 \cdot \Delta c$$

$$\Delta v = 3{,}4 \cdot \Delta c \qquad\qquad (2)$$

$$\Delta w = 1{,}0 \cdot \Delta c \; .$$

These considerations show, that for the present experimental setup the measuring uncertainty of the v-component is larger then those of the other two velocity components. This means that the artificial increase of turbulence due to measuring uncertainties is unequally distributed on the three velocity components and in the case of a low turbulence level the v-component is the first one to become inaccurate. For the present 3D-LDA system the transformation matrix is given by the optical setup, which cannot be altered. In this situation the only possibility to increase the accuracy of the v-component measurement is to reduce the measuring uncertainty Δc by the application of a better signal processing system.

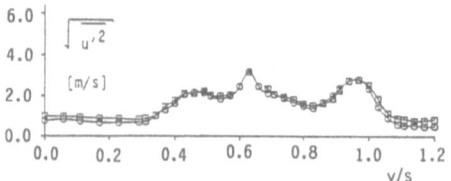

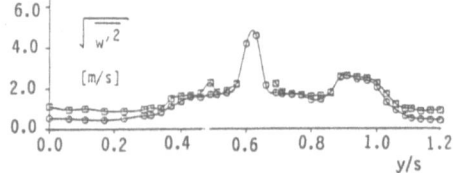

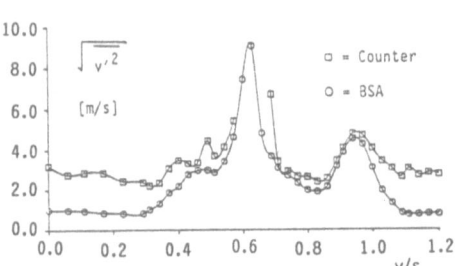

Fig. 2: Velocity fluctuations in the vortex field of an A = 1 delta wing at $\alpha = 20.5°$ in a horizontal section through the vortex centre at the trailing-edge. Comparison of results obtained with different signal processing systems (Counters and Burst Spectrum Analyzers, BSA).

In order to reduce the uncertainty of the measurements the three Counters of the present 3D-LDA system were replaced by Burst Spectrum Analyzers (BSA). Through this measure the uncertainty Δc has been decreased by a factor of 10. The flowfield of an A = 1 delta wing at an angle of attack of $\alpha = 20.5°$ has been analyzed in a horizontal section through the centre of the vortices at the trailing-edge of the wing using both configurations of the signal processing system. Concerning the u and w components of the mean velocity very good agreement turned out and for the v component some minor differences were found. This result indicates that according to the discussions mentioned above the uncertainties of the measurements for the present LDA system are lowest for the u and w components. In general the comparison between the two signal processing systems leads to good agreement concerning the mean velocities. The results for the turbulent fluctuations are shown in Fig. 2. For the standard deviations of the u and w components again good agreement is obtained for both signal processing systems. For the standard deviations of the v component, however, large differences turn out especially in those parts of the flowfield, in which the turbulence level is low, corresponding to that of the free stream flow with isotropic turbulence. In these regions the v' fluctuations evaluated by means of the BSAs are much lower than those determined by the Counter technique. In addition the BSA technique leads to turbulent fluctuations which are in the same order of magnitude for all three cartesian velocity components. These results indicate that with the present LDA system and using the BSA technique all turbulent fluctuations can be measured with good reliability, whereas with the Counter configuration the v component fluctuations have to be omitted. For this reason the following analysis of turbulent fluctuations is based on the u and w components only.

Structure of vortex breakdown

The vortex breakdown phenomenon has been studied by means of 3D-LDA measurements in the flowfield of an A = 2.31 delta wing at an angle of attack α = 16.5°. In this case vortex breakdown occurred within the vortices at $(x/l_i)_{VB}$ = 0.6. The flowfield has been measured in two crossflow planes perpendicular to the free stream located at x/l_i = 0.50 upstream and at x/l_i = 0.75 downstream of the vortex breakdown position. The results are shown and compared with each other in Figs. 3 and 4.

The velocity component within the crossflow plane is shown for both measurements in the uppermost diagram. Each velocity vector corresponds to a measuring point. The completeness of these vector plots indicates that in all positions of the flowfield the available number of scattering particles was sufficient in order to carry out the measurements. Upstream of the breakdown point the velocity components in the crossflow plane increase towards the primary vortex centre. The counter-rotating secondary vortex is clearly indicated in Fig. 3. Downstream of the breakdown point the farfield is virtually the same, but in the region of the vortex centre the crossflow velocities are considerably reduced. According to Fig. 4 the secondary vortex is weakened. These details are already known from probe measurements [2].

The velocity distributions in the two crossflow planes are shown in the central diagrams of Figs. 3 and 4. Plotted are lines of constant mean velocity V_{abs}/U_∞ = const. Upstream of the breakdown point (Fig. 3) an ordinary vortex structure is present. The velocity increases condiderably towards the primary vortex centre. In the region of the secondary vortex a relative minimum of the flow velocity is found which is due to the viscous effects which are present in this vortex. Another relative minimum of the flow velocity can be identified in the outer region of the flowfield. It starts at the leading-edge of the wing and it indicates the position of the rolling-up vortex sheet. Downstream of the breakdown point (Fig. 4) the farfield is again virtually the same as upstream of the breakdown point.

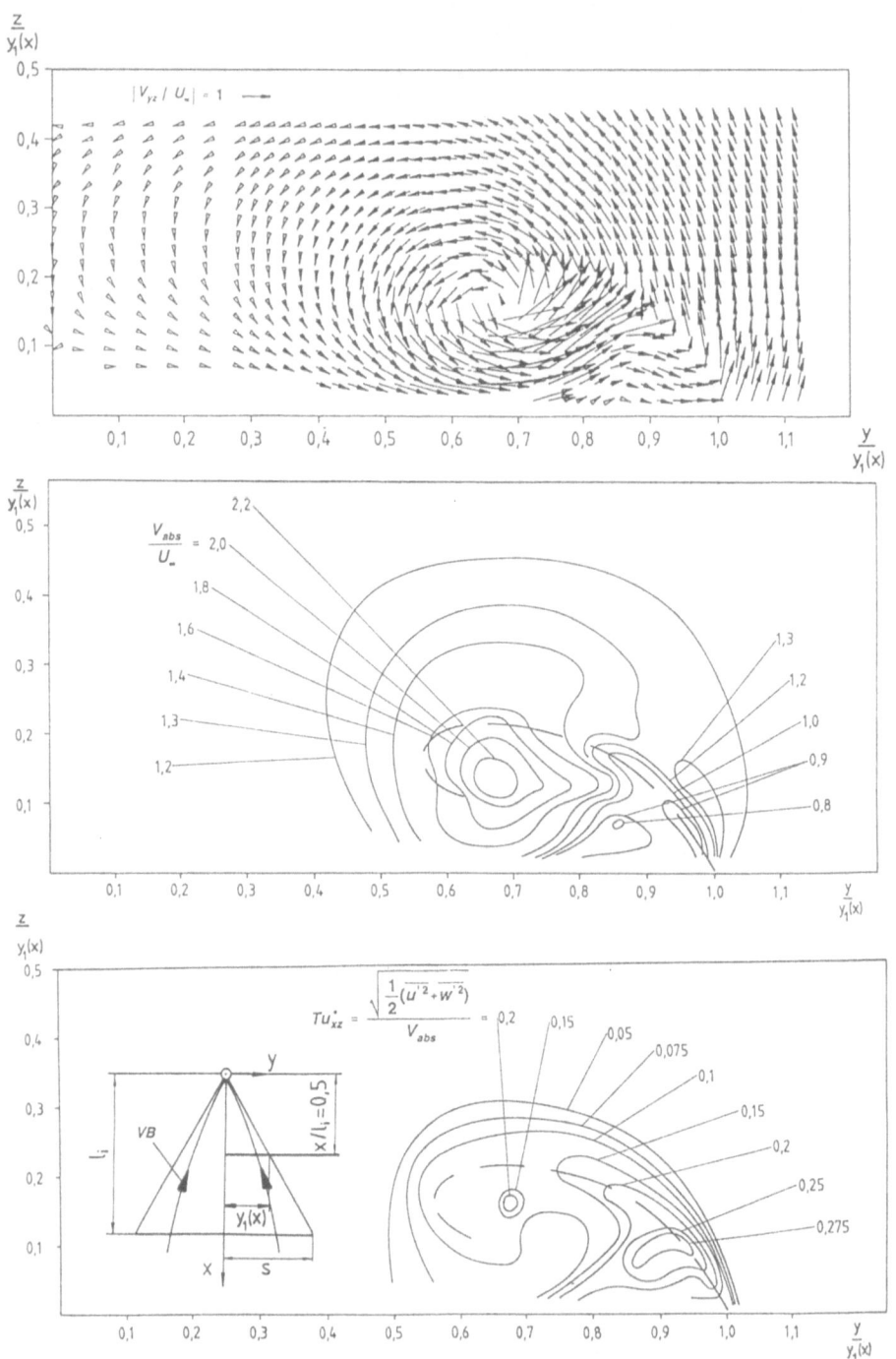

Fig. 3: Flowfield on the upper surface of a delta wing A = 2.31 at α = 16.5° in a crossflow plane at x/l_i = 0.50 upstream of the vortex breakdown point (VB Vortex breakdown).

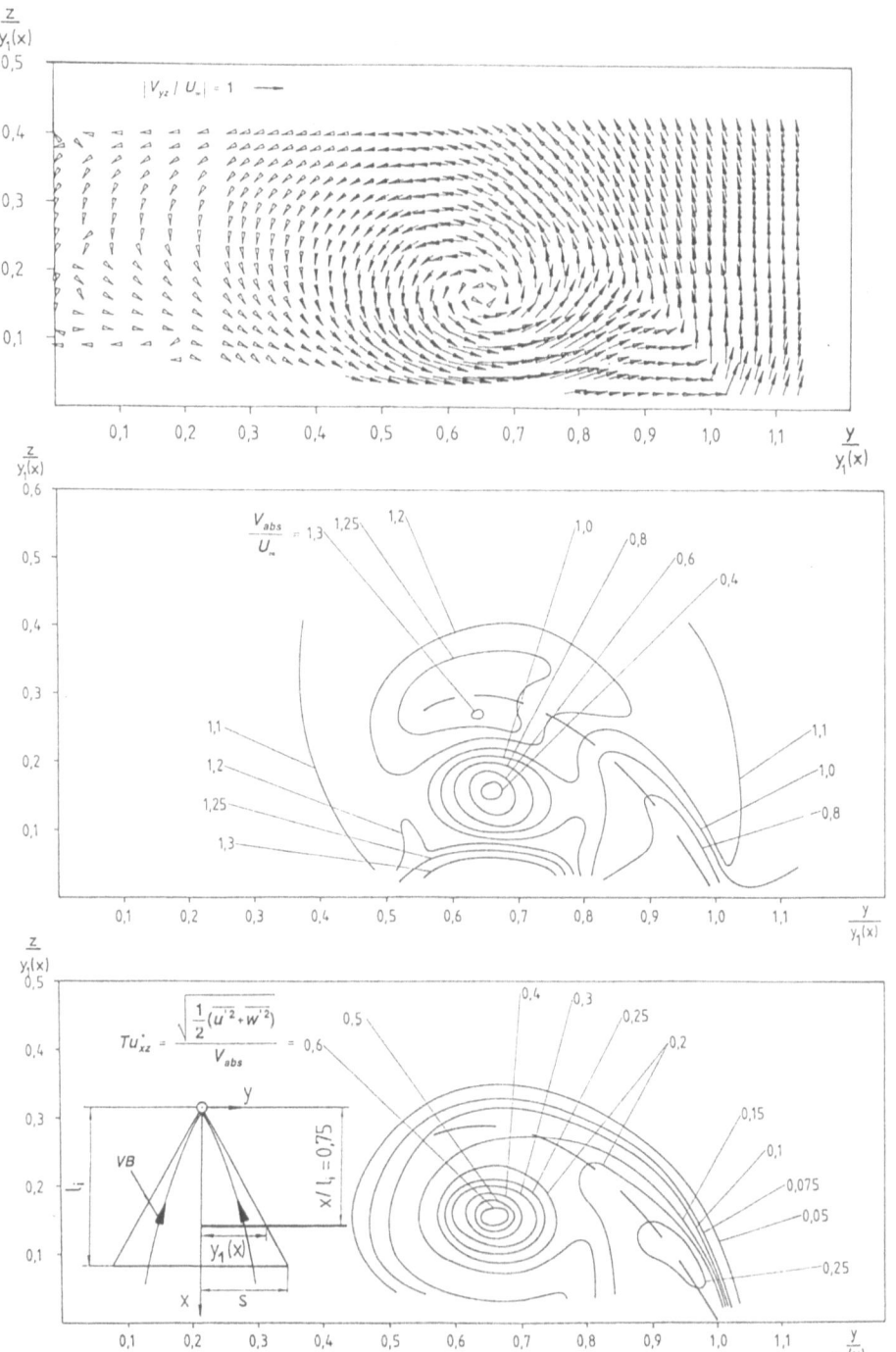

Fig. 4: Flowfield on the upper surface of a delta wing A = 2.31
at α = 16.5° in a crossflow plane at x/l_i = 0.75
downstream of the vortex breakdown point (VB Vortex
breakdown).

In the vicinity of the vortex centre a ring-shaped relative velocity maximum is reached and from this region the velocity decreases towards the vortex centre. The secondary vortex is not distinctly marked, but the relative velocity minimum related to the vortex sheet can again be identified. It turns out that downstream of the breakdown point the area within the rolled-up vortex sheet is enlarged and this result has not yet been documented by probe measurements.

The distributions of the velocity fluctuations in the two crossflow planes are given in the bottom diagrams of Figs. 3 and 4. The turbulence level of the two reliable components u and w is based on the local mean velocity V_{abs} rather than on the free stream velocity U_∞ in order to demonstrate the unsteadiness of the flow. Far away from the wing the turbulence level decreases to the value of the windtunnel flow which is about 1.3 %. Within the primary vortex the fluctuations increase towards the vortex centre as well as towards the vortex sheet and also in the region of the secondary vortex high fluctuations are found. The turbulence level is highest in the vortex sheet close to the secondary vortex and maximum values of 27.5 % are obtained. This high value is not unusual since this level is also reached in ordinary turbulent boundary layers. Downstream of the breakdown point the situation in the vortex sheet and the secondary vortex remains the same. In the region of the vortex centre, however, the local turbulence level has increased very much and on the vortex axis an absolute maximum of 60 % has been found. This means that in this region the velocity fluctuations are nearly in the order of magnitude of the local mean velocity and this demonstrates the high degree of unsteadiness of this part of the flowfield.

References

[1] D. Hummel: On the vortex formation over a slender wing at large angles of incidence. AGARD-CP-247 (1978), 15-1 to 15-17.

[2] D. Hummel: Documentation of separated flows for computational fluid dynamics validation. AGARD-CP-437 (1988), Vol. 2, P 15-1 to P 15-24.

[3] S. Kommallein: Strömungsfeldmessungen an einem Deltaflügel mit einem 3-Komponenten-LDA. DFG SPP "Physik abgelöster Strömungen", AG "Dreidimensionale Strömungen mit Ablösung", 4. Workshop Berlin 1989.

[4] S. Kommallein: LDA-Untersuchungen der abgelösten Strömung an schlanken Flügeln. 5. STAB-Workshop Göttingen 1991.

AERODYNAMIC INVESTIGATIONS ON CLOSE-COUPLED CANARD CONFIGURATIONS

H.-Chr. Oelker, D. Hummel
TU Braunschweig, Institut für Strömungsmechanik
Bienroder Weg 3, D - 3300 Braunschweig, Germany

Summary

Low-speed windtunnel experiments have been carried out on canard-wing-body configurations with two fuselages (flat and round), two delta wings (A_W = 2.31 and 1.38) and two canards (A_C = 2.31 and 1.65). The canard's position relative to the wing-body combination has been varied systematically. The interference of the vortex systems generated by canard, wing and fuselage has been studied by means of six-component-, pressure-distribution- and flowfield-measurements as well as by flow visualizations.

The canard configurations reach higher maximum lift coefficients than the wing-body combinations. At the wing vortex formation is suppressed in the region of the apex and amplified in the tip region. The trailing vortex sheet of the canard merges with the suction side boundary layer of the wing. The interference between canard and wing depends strongly on the height of the canard's trailing-edge above the wing. With increasing setting angle of the canard the tendency towards merging of the vortex systems from canard and wing is increased. In unsymmetrical flow sudden changes of the force and moment coefficients occur due to different vortex interference on both sides and due to a collapse of vortical flow on the windward side of the configuration.

Introduction

It is known since H. Behrbohm [1] that close-coupled canard configurations with canard and wing of small aspect ratio 1 < A < 3 have substantial advantages. The maximum lift coefficient c_{Lmax} and the corresponding angle of attack $\alpha(c_{Lmax})$ can be increased considerably by adding a canard to a slender wing and this is due to favourable interference between the vortex systems of canard and wing. Since the knowledge on the vortex formation over canard-wing configurations was very poor, an experimental research program has been started in 1984 at TU Braunschweig in order to study the interference of the vortex systems from canard and wing and to provide data on aerodynamic coefficients, pressure distributions and flowfield structure for the comparison with numerical solutions of the Euler- and the Navier-Stokes-equations. For this purpose a configuration with a delta wing and a delta canard was chosen. A thin flat fuselage

was used for the investigations in symmetrical flow and a body
of revolution for those in unsymmetrical flow. In both cases a
certain variation of the longitudinal and the vertical position
as well as of the setting angle of the canard relative to the
wing was possible.

At about the same time the "International Vortex Flow Experiment" (VFE) was initiated by G. Drougge [2]. This program was
aimed at an understanding of the physics of the vortex formation
over a canard configuration in compressible flow and a proper
calculation of this flow by means of Euler- and Navier-Stokes-
codes. These investigations were mainly concerned with symmetrical flow around a short-coupled canard configuration described
in Refs. [2] and [3]. Much effort was devoted to the canard-off
configuration but some investigations were also concerned with
the canard-on configuration. In this situation the research
program at TU Braunschweig has been modified. The thin flat

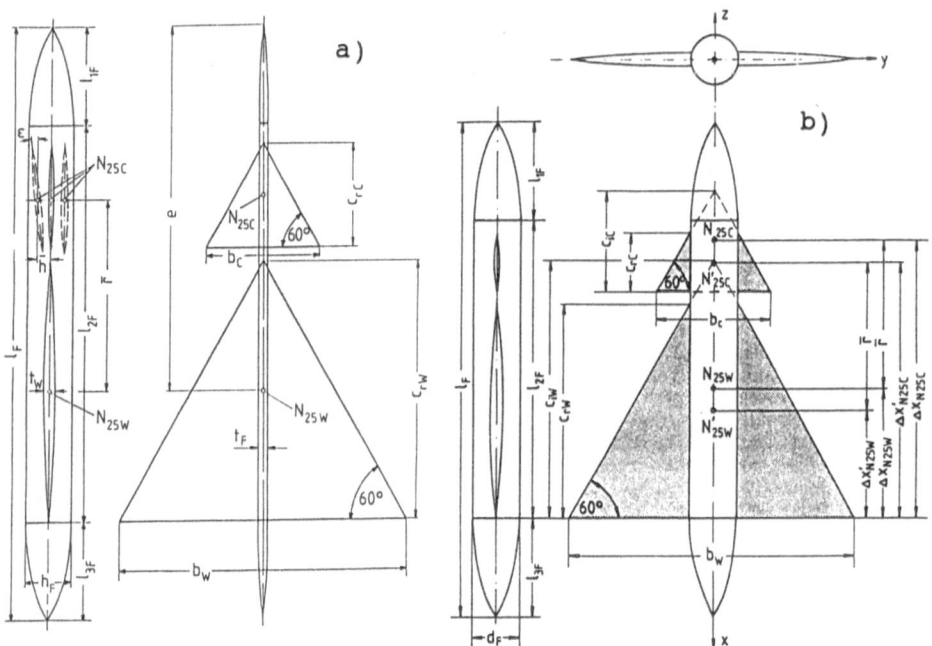

<u>Fig. 1</u>: Delta-canard-configuration with a) thin flat fuselage
[9] and b) body of revolution [11].

fuselage was equipped with the original canard and the original
wing of the VFE configuration and this model has been measured
in incompressible flow. With these investigations as well as
with its closely related experiments on the delta-canard-
configuration TU Braunschweig took part in the "International
Vortex Flow Experiment" and delivered the limiting case $M_\infty \rightarrow 0$.

The results of this research program have been published in
Refs. [4] to [11]. The present paper summarizes some results
obtained for the delta canard configuration. The corresponding
results for the VFE configuration may be taken from Refs. [8]
and [10].

Experimental set-up and test program

The experimental investigations have ben carried out in the 1.3 m low-speed windtunnel of Institut für Strömungsmechanik at TU Braunschweig. The wing-fuselage-canard configuration is shown in Fig. 1. Details of the geometric data may be taken from Refs. [9] and [11]. Wing and canard have delta planforms of aspect ratio $A_C = A_W = 2.31$ and a corresponding leading-edge sweep of $\varphi_C = \varphi_W = 60°$. In both cases symmetric parabolic arc contours have been chosen in chordwise and spanwise directions and the leading-edges were sharp. Both wing and canard are equipped with a tube system underneath the surface and with pressure holes in order to measure the surface pressure distribution. For the investigations in symmetrical flow a very flat fuselage according to Fig. 1a has been used which consists of a cylindrical portion of length $l_{2F} = 8\ h_F$ and attached are front and rear parts of length $l_{1F} = l_{3F} = 2\ h_F$. For the investigations in unsymmetrical flow according to Fig. 1b the fuselage was chosen as a body of revolution with a cylindrical part of length $l_{2F} = 6.05\ d_F$ and attached front and rear parts of length $l_{1F} = l_{3F} = 2\ d_F$. This fuselage was also equipped with a tube system underneath the surface and pressure holes in order to measure the surface pressure distribution. For both fuselages some variations in vertical and longitudinal position as well as in setting angle of the canard were possible. Details may be taken from Refs. [9] and [11].

The windtunnel investigations have been carried out in the Reynoldsnumber range $1.1 \cdot 10^6 \leq Re \leq 1.4 \cdot 10^6$ [6]. In symmetrical flow three-component measurements have been performed for angles of attack $- 5° \leq \alpha \leq 40°$ [4], [5], [9], [10]. The wing alone, the canard-off and the canard-on configuration have been investigated. Additional pressure distribution measurements have been carried out for selected freestream conditions [4], [5], [7], [10], [11]. The flow on the upper surface of the configurations has been studied by means of oilflow patterns [4], [5], [7], [10] and flowfield visualizations according to the laser lightsheet method [10]. Flowfield measurements have been carried out for $\alpha = 8.7°$ in planes perpendicular to the free stream using a five-hole probe, [4] to [7], [10]. A large variety of canard positions relative to the wing has been investigated. For various longitudinal and vertical distances and for different setting angles of the canard, three-component and pressure distribution measurements as well as flow visualizations have been carried out [9]. In unsymmetrical flow six-component measurements have been performed for angles of attack $- 5° \leq \alpha \leq 40°$ and the angle of sideslip has been varied in the range $- 10° \leq \beta \leq 26°$ [11]. In addition comprehensive pressure distribution measurements as well as laser lightsheet flow visualizations have been carried out in order to understand the behaviour of the flow [11].

Results

The three-component measurements for the canard-off and the canard-on configuration are shown in Fig. 2. The considerable increase of the maximum lift coefficient for the canard-on configuration is clearly indicated and the increase of nose-up pitching moments is due to a certain shift of lift from the wing towards the canard. The pressure distribution for $\alpha = 8.7°$

according to **Fig. 3** on the canard is very similar to that for the wing alone and the flow is fairly conical there. On the wing, however, the maximum suction occurs much closer to the leading-edge and the flow is distinctly non-conical. Due to the downwash distribution generated by the canard flow separation at the wing is suppressed in the front part and increased in the rear part of the wing. Due to this mechanism a delayed formation of the wing leading-edge vortex downstream of the wing apex takes place.

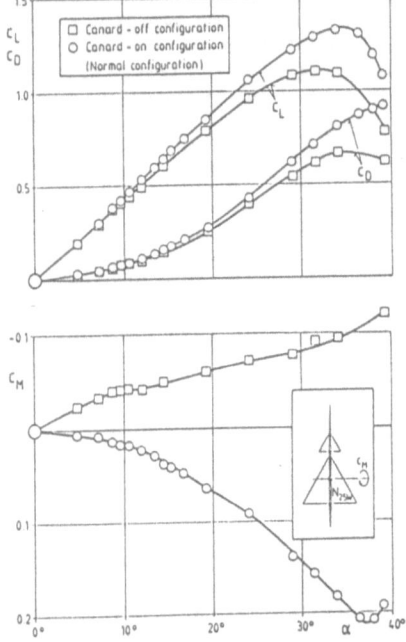

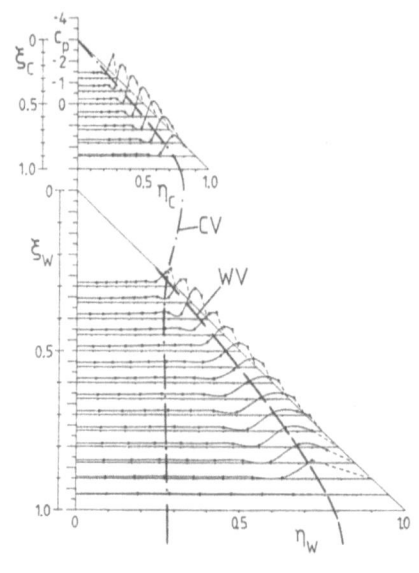

Fig. 2: Results of the three-component measurements for canard-off and canard-on configuration [4].

Fig. 3: Upper surface pressure distribution at $\alpha = 8.7°$ and Re = $1.4 \cdot 10^6$ on the canard-on configuration (CV Canard vortex, WV Wing vortex) [4].

The results of flowfield measurements over the canard-on configuration at $\alpha = 8.7°$ are shown in **Fig. 4**. The vortex systems from canard and wing consist of a primary vortex which is accompanied by a counter-rotating trailing-edge vortex, which is well known from delta wings [12]. Secondary separations are also present in both vortex systems. For the present configuration the two vortex systems remain separate. The canard's trailing vortex sheet merges with the upper surface boundary layer of the wing forming a mixed layer in the inner portion of the wing. The formation of the wing vortex system is delayed up to 30 % of the wing root chord. The canard vortices approach the wing very high above the plane of the wing and under the influence of the developing wing vortices the canard vortices move inboard and downwards. **Fig. 5** shows a schemativ view of the results of these flow studies, taken from [7]. The same results have been obtained also for the configuration of the International Vortex Flow Experiment. Details may be taken from [8].

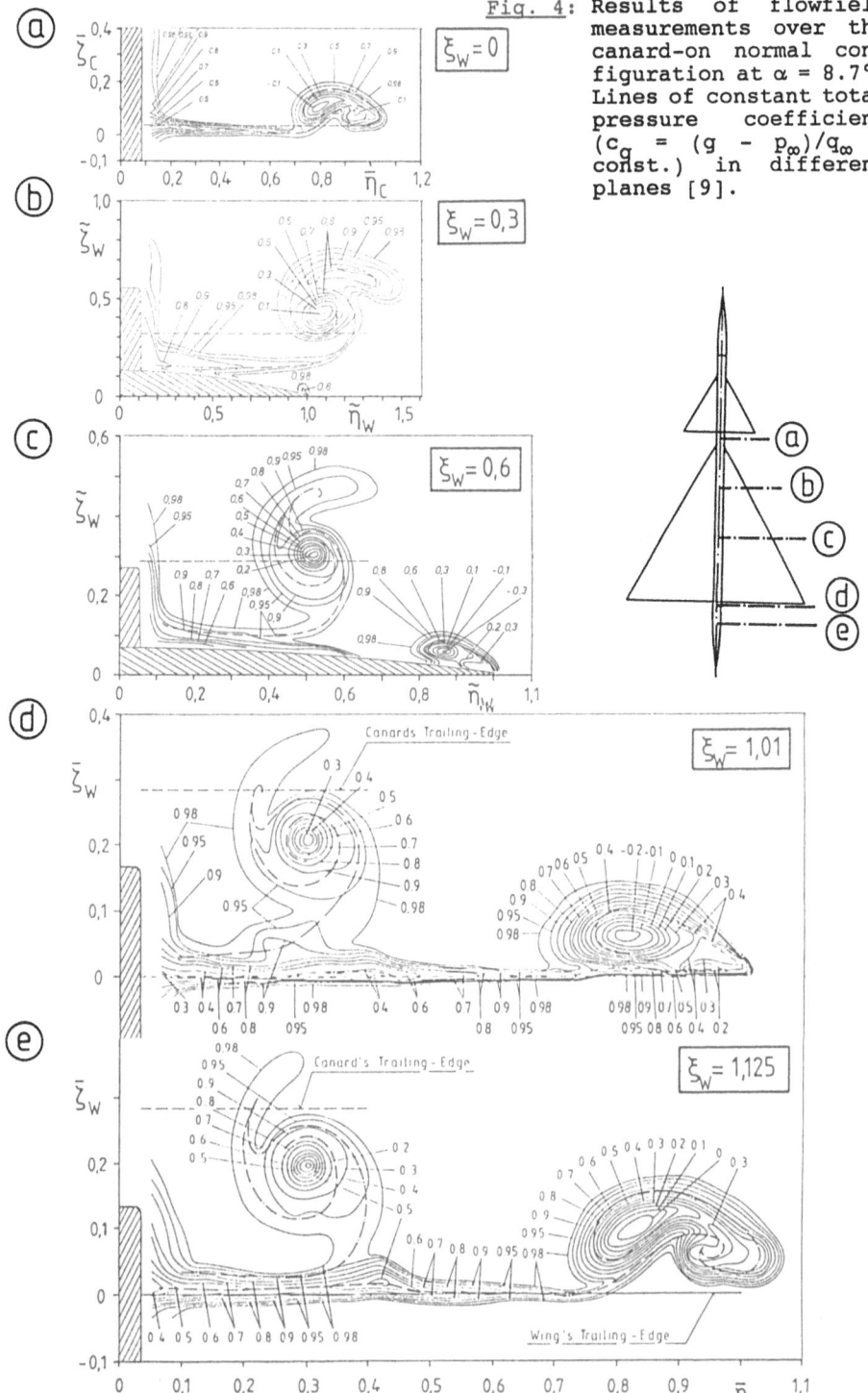

Fig. 4: Results of flowfield measurements over the canard-on normal configuration at $\alpha = 8.7°$. Lines of constant total pressure coefficient ($c_g = (g - p_\infty)/q_\infty =$ const.) in different planes [9].

287

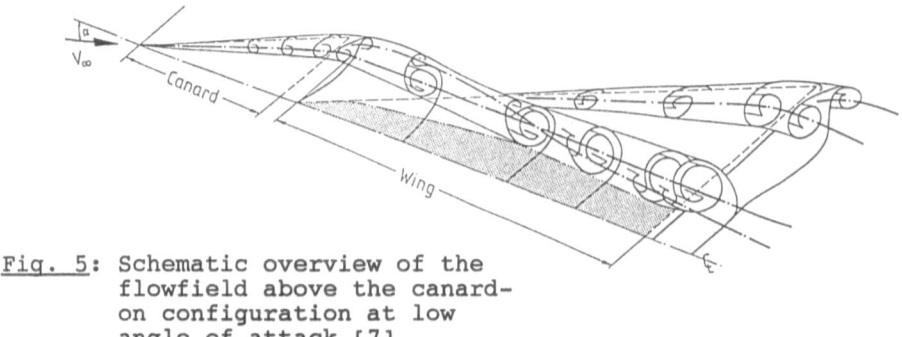

Fig. 5: Schematic overview of the flowfield above the canard-on configuration at low angle of attack [7].

The same interference between the vortex systems of canard and wing has been found for different longitudinal and vertical distances between canard and wing [9]. Large interference effects turned out for high setting angles of the canard. In this case the high loading of the canard leads to a very strong downwash at the wing. At low overall lift this downwash produces negative effective angles of attack in the front part of the wing and this means that flow separations take place in this

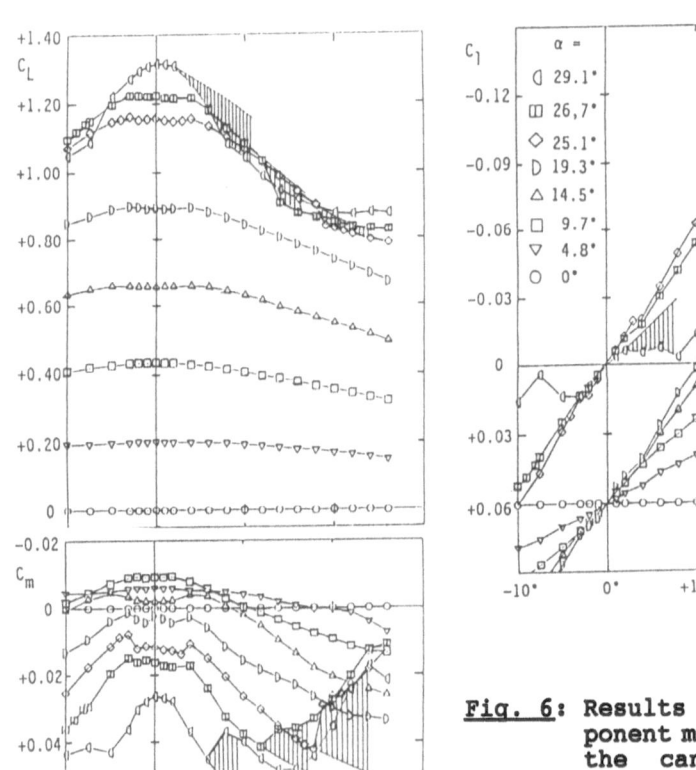

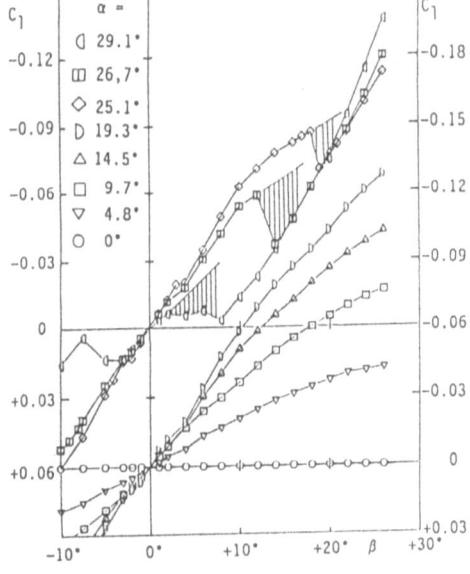

Fig. 6: Results of six-component measurements for the canard-on configuration according to Fig. 1b [11].

region on the lower surface of the wing. At larger angles of attack wing vortices are formed on the upper surface of the configuration only. In this case there exists a tendency towards mixing of the vortex systems from canard and wing. The most important parameter for this effect is the height of the canard's trailing-edge above the wing which decreases with increasing setting angle of the canard. The corresponding flow behaviour is discussed in Ref. [9].

Results of the six-component measurements on the configuration according to Fig. 1b are shown in Fig. 6. Sudden changes of the aerodynamic coefficients occur at certain angles of sideslip β the values of which decrease with increasing angle of attack. The explanation for this behaviour may be taken from Fig. 7. At α = 24.2° in symmetrical flow, β = 0°, there is a tendency towards merging of the two vortex systems over the wing. With increasing angle of sideslip, β = 10°, this tendency reduces on the windward side and increases on the leeward side of the configuration. Vortex breakdown is present on the windward side in both vortex systems but vortex breakdown within the canard vortices is kept close to the canard's trailing-edge due to a favourable upstream influence from the not yet destroyed wing vortex flow.

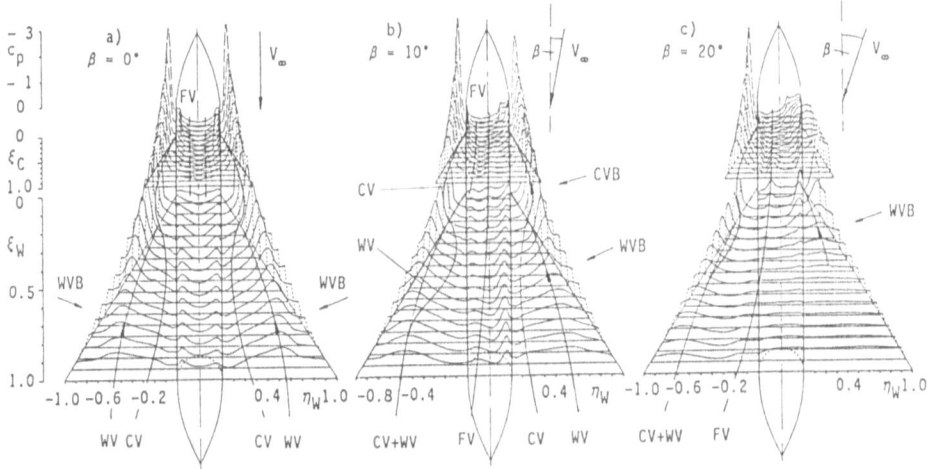

Fig. 7: Suction side pressure distributions for the canard-on configuration at angle of attack α = 24.2° and different angles of sideslip β (WV Wing vortex, FV Fuselage vortex, CV Canard vortex, B Breakdown) [11].

If the angle of sideslip is again increased, β = 20°, this favourable interference effect is no longer able to avoid the upstream movement of vortex breakdwon within the windward canard vortex. Vortex breakdown suddenly occurs over the canard on the windward side and finally a deadwater-type flow is established there. This abrupt change of the flow conditions on the windward side causes the jumps in the aerodynamic coefficients.

References

[1] H. Behrbohm: Basic low speed aerodynamics of the short-coupled canard configuration of small aspect ratio. SAAB TN 60 (1965).

[2] G. Drougge: The international vortex flow experiment for computer code validation. ICAS-Proceedings 1988, Vol. 1, XXXV - XLI.

[3] A. Elsenaar, L. Hjelmberg, K. Bütefisch, W. J. Bannink: The international vortex flow experiment. AGARD-CP-437 (1988), Vol. 1, 9-1 to 9-23.

[4] D. Hummel, H.-Chr. Oelker: Vortex interference effects on close-coupled canard configurations in incompressible flow. Proceedings Symposium "International Vortex Flow Experiment on Euler Code Validation", Stockholm 1986, 47 - 61.

[5] H.-Chr. Oelker, D. Hummel: Experimentelle Untersuchungen an Entenkonfigurationen. DGLR-Bericht 86-03, Bonn 1986, 171 - 191.

[6] D. Hummel: Documentation of separated flows for computational fluid dynamic validation. AGARD-CP-437 (1988), Vol. 2, P15-1 to P15-24.

[7] H.-Chr. Oelker, D. Hummel: Investigations on the vorticity sheets of a close-coupled delta-canard configuration. ICAS Proceedings 1988, Vol. 1, 649 - 662. See also J. Aircraft 26 (1989), 657 - 666.

[8] H.-Chr. Oelker: Windkanaluntersuchungen an der Konfiguration des Internationalen Vortex Flow Experiment. DGLR-Bericht 88-05, Bonn 1988, 141 - 157.

[9] D. Hummel, H.-Chr. Oelker: Effects of canard position on the aerodynamic characteristics of a close-coupled canard configuration at low speed. AGARD-CP-465 (1989), 7-1 to 7-18. See also Z. Flugwiss. Weltraumforsch. 15 (1991), 74 - 88.

[10] H.-Chr. Oelker: Aerodynamische Untersuchungen an kurzgekoppelten Entenkonfigurationen bei symmetrischer Anströmung. Dissertation TU Braunschweig 1990. Zentrum für Luft- u. Raumfahrttechnik d. TU Braunschweig, Forschungsbericht 90-01 (1990).

[11] A. Bergmann, D. Hummel, H.-Chr. Oelker: Vortex formation over a close-coupled canard-wing-body configuration in unsymmetrical flow. AGARD-CP-494 (1991), 14-1 to 14-14.

[12] D. Hummel: On the vortex formation over a slender wing at large angles of incidence. AGARD-CP-247 (1978), 15-1 to 15-17.

SUBJECT INDEX

Notes on Numerical Fluid Mechanics (NNFM)   Volume 40

Addresses of the Editors of the Series "Notes on Numerical Fluid Mechanics"

Brief Instruction for Authors

Manuscripts should have well over 100 pages. As they will be reproduced photomechanically they should be typed with utmost care on special stationary which will be supplied on request.
In print, the size will be reduced linearly to approximately 75 per cent. Figures and diagrams should be lettered accordingly so as to produce letters not smaller than 2 mm in print. The same is valid for handwritten formulae. Manuscripts (in English) or proposals should be sent to the general editor, Prof. Dr. E. H. Hirschel, Herzog-Heinrich-Weg 6, D-8011 Zorneding.